A Dictionary of
PHYSICS

A Dictionary of PHYSICS

A.M. Goldstein

CBS

CBS Publishers & Distributors Pvt. Ltd.

New Delhi • Bengaluru • Chennai • Kochi • Kolkata • Mumbai
Hyderabad • Nagpur • Patna • Pune • Vijayawada

ISBN: 81-239-0882-2

First Indian Edition: 1986
Reprint: 1991, 1995, 1999, 2000, 2002, 2003, 2004

This edition has been published in India by arrangement with
Academic Science, Inc., UK

Published by **Satish Kumar Jain** and produced by **Varun Jain** for
CBS Publishers & Distributors Pvt. Ltd.,
4819/XI Prahlad Street, 24 Ansari Road, Daryaganj, New Delhi - 110002
delhi@cbspd.com, cbspubs@airtelmail.in • www.cbspd.com
Ph.: 23289259, 23266861, 23266867 • Fax: 011-23243014

Corporate Office: 204 FIE, Industrial Area, Patparganj, Delhi - 110 092
Ph: 49344934 • Fax: 011-49344935
E-mail: publishing@cbspd.com • publicity@cbspd.com

Branches:
- ***Bengaluru:*** 2975, 17th Cross, K.R. Road, Bansankari 2nd Stage, Bengaluru - 70 • Ph: +91-80-26771678/79 • Fax: +91-80-26771680 E-mail: cbsbng@gmail.com, bangalore@cbspd.com
- ***Chennai:*** No. 7, Subbaraya Street, Shenoy Nagar, Chennai - 600030 Ph: +91-44-26681266, 26680620 • Fax: +91-44-42032115 E-mail: chennai@cbspd.com
- ***Kochi:*** Ashana House, 39/1904, A.M. Thomas Road, Valanjambalam, Ernakulum, Kochi • Ph: +91-484-4059061-65 Fax: +91-484-4059065 • E-mail: cochin@cbspd.com
- ***Kolkata:*** 6-B, Ground Floor, Rameshwar Shaw Road, Kolkata - 700014 Ph: +91-33-22891126/7/8 • E-mail: kolkata@cbspd.com
- ***Mumbai:*** 83-C, Dr. E. Moses Road, Worli, Mumbai - 400018 Ph: +91-9833017933, 022-24902340/41 • E-mail: mumbai@cbspd.com

Representatives:
- Hyderabad: 0-9885175004
- Patna: 0-9334159340
- Vijayawada: 0-9000660880
- Nagpur: 0-9021734563
- Pune: 0-9623451994

Printed at:
India Binding House, Noida, UP (India)

Preface

This dictionary contains accurate definitions of the working vocabulary of A Dictionary of Physics while compiling the dictionary, the author has concentrated on terms which seem to him to be in the widest use. A very useful system of cross references assists the reader to find his way from one related topic to another. It also allows the greatest account of information to be given in a small space. Thus many of the entries contain words which are themselves defined elsewhere. SI units are used throughout this dictionary.

In compiling a book of this kind it is necessary to draw upon the work of many authorities and seek the advice of colleagues, to all of whom the author is deeply indebted. All comments from users on omissions or shortcomings will be most welcome.

A

ab- A prefix attached to the name of a practical electrical unit to provide a name for a unit in the *electromagnetic system of units, e.g. abampere, abcoulomb, abvolt. The prefix is an abbreviation of the word 'absolute' as this system is also known as the *absolute system*. *Compare* stat-. In modern practice both absolute and electrostatic units have been replaced by *SI units.

aberration 1. (in optics) A defect in the image formed by a lens or curved mirror. In *chromatic aberration* the image formed by a lens (but not a mirror) has coloured fringes as a result of the different extent to which light of different colours is refracted by glass. It is corrected by using an *achromatic lens. In *spherical aberration*, the rays from the object come to a focus in slightly different positions as a result of the curvature of the lens or mirror. For a mirror receiving light strictly parallel with its axis, this can be corrected by using a parabolic surface rather than a spherical surface. Spherical aberration in lenses is minimized by making both surfaces contribute equally to the ray deviations, and can (though with reduced image brightness) be reduced by the use of diaphragms to let light pass only through the centre part of the lens. *See also* astigmatism; coma. **2.** (in astronomy) The apparent displacement in the position of a star as a result of the earth's motion round the sun. Light appears to come from a point that is slightly displaced in the direction of the earth's motion. The angular displacement $\alpha = v/c$, where v is the earth's orbital velocity and c is the speed of light.

abscissa *See* Cartesian coordinates.

absolute 1. Not dependent on or relative to anything else, e.g. *absolute zero. **2.** Denoting a temperature measured on an *absolute scale*, a scale of temperature based on absolute zero. The usual absolute scale now is that of thermodynamic *temperature; its unit, the kelvin, was formerly called the degree absolute (°A) and is the same size as the degree Celsius. In British engineering practice an absolute scale with Fahrenheit-size degrees has been used: this is the Rankine scale.

absolute expansivity *See* expansivity.

absolute humidity *See* humidity.

absolute permittivity *See* permittivity.

absolute pitch (perfect pitch) The ability of a person to identify and reproduce a note without reference to a tuned musical instrument.

absolute temperature *See* absolute; temperature.

absolute value (modulus) The square root of the sum of the squares of the real numbers in a *complex number, i.e. in the complex number $x + iy$, the absolute value is $\sqrt{(x^2 + y^2)}$.

absolute zero Zero of thermodynamic *temperature (0 kelvin) and the lowest temperature theoretically attainable. It is the temperature at which the kinetic energy of atoms and molecules is minimal. It is equivalent to −273.15°C or −459.67°F. *See also* zero-point energy; cryogenics.

absorptance Symbol α. The ratio of the radiant or luminous flux absorbed by a body to the flux falling on it. Formerly called *absorptivity*, the absorptance of a *black body is by definition 1.

absorption 1. The take up of a gas by a solid or liquid, or the take up of a liquid by a solid. Absorption differs from *adsorption in that the absorbed substance permeates the bulk of the absorbing substance. **2.** The conversion of the energy of electromagnetic radiation, sound, streams of particles, etc., into other forms of energy on passing through a medium. A beam of light, for instance, passing through a medium, may lose intensity because of two effects: *scattering of light out of the beam, and absorption of photons by atoms or molecules in the medium. When a photon is absorbed, there is a transition to an excited state.

absorption coefficient *See* Lambert's laws.

absorption spectrum *See* spectrum.

absorptivity *See* absorptance.

abundance 1. The ratio of the total mass of a specified element in the earth's crust to the total mass of the earth's crust, often expressed as a percentage. For example, the abundance of aluminium in the earth's crust is about 8%. **2.** The ratio of the number of atoms of a particular isotope of an element to the total number of atoms of all the isotopes present, often expressed as a percentage. For example, the abundance of uranium-235 in natural uranium is 0.71%. This is the ***natural abundance***, i.e. the abundance as found in nature before any enrichment has taken place.

a.c. *See* alternating current.

acceleration Symbol a. The rate of increase of speed or velocity. It is measured in m s^{-2}. For a body moving linearly with constant acceleration a from a speed of u to a speed of v,

$$a = (v - u)/t = (v^2 - u^2)/2s$$

where t is the time taken and s the distance covered.

If the acceleration is not constant it is given by $dv/dt = d^2s/dt^2$. If the motion is not linear the vector character of displacement, velocity, and acceleration must be considered.

acceleration of free fall Symbol g. The acceleration experienced by any massive object falling freely in the earth's gravitational field. Experimentally this is almost constant for all positions near the earth's surface, independent of the nature of the falling body (provided air resistance is eliminated). This is taken to indicate the strict proportionality of *weight (the force causing the acceleration) and *inertial mass, on the basis of *Newton's second law of motion. There is some variation of g with latitude, because of the earth's rotation and because the earth is not completely spherical. The standard value is taken as 9.806 65 m s^{-2}. The acceleration of free fall is also called the *acceleration due to gravity*.

accelerator An apparatus for increasing the kinetic energies of charged particles, used for research in nuclear and particle physics. *See* cyclotron; synchrocyclotron; linear accelerator.

acceptor A substance that is added as an impurity to a *semiconductor because of its ability to accept electrons from the valence bands, causing p-type conduction by the mobile positive holes left.

accommodation The process by which the focal length of the *lens of the eye is changed so that clear images of objects at a range of distances are displayed on the retina. In man and some other mammals accommodation is achieved by reflex adjustments in the shape of the lens brought about by relaxation and contraction of muscles within the ciliary body.

accumulator (secondary cell; storage battery) A type of *voltaic cell or battery that can be recharged by passing a current through it from an external d.c. supply. The charging current, which is passed in the opposite direction to that in which the cell supplies current, reverses the chemical reactions in the cell. The common types are the *lead–acid accumulator and the *nickel–iron accumulator.

achromatic lens A lens that corrects for chromatic *aberration by using a combination of two lenses, made of different kinds of glass, such that their *dispersions neutralize each other although their *refractions do not. The aberration can be reduced further by using an *apochromatic lens*, which consists of three or more different kinds of glass.

acoustics **1.** The study of sound and sound waves. **2.** The characteristics of a building, especially an auditorium, with regard to its ability to enable speech and music to be heard clearly within it. For this purpose there should be no obtrusive echoes or resonances and the reverberation time should be near the optimum for the hall. Echoes are reduced by avoiding sweeping curved surfaces that could focus the sound and by breaking up large plane surfaces or covering them with sound-absorbing materials. Resonance is avoided by avoiding simple ratios for the main dimensions of the room, so that no one wavelength of sound is a factor of more than one of them. If the reverberation time is too long, speech will sound indistinct and music will be badly articulated, with one note persisting during the next. However, if it is too short, music sounds dead. It is long in a bare room with hard walls, and can be deliberately reduced by carpets, soft furnishings and sound-absorbent ('acoustic') felt. Reverberation times tend to be reduced by the presence of an audience and this must be taken into account in the design of the building.

acoustoelectronic devices (electroacoustic devices) Devices in which electronic signals are converted into acoustic waves. Acoustoelectronic devices are used in constructing *delay lines and also in converting digital data from computers for transmission by telephone lines.

actinic radiation Electromagnetic radiation that is capable of initiating a chemical reaction. The term is used especially of ultraviolet radiation and also to denote radiation that will affect a photographic emulsion.

actinium Symbol Ac. A silvery radioactive metallic element belonging to group IIIB of the periodic table; a.n. 89; mass number of most stable isotope 227 (half-life 21.7 years); m.p. 1050 ± 50°C; b.p. 3300°C (estimated). Actinium–227 occurs in natural uranium to an extent of about 0.715%. Actinium–228 (half-life 6.13 hours) also occurs in nature. There are 22 other artificial isotopes, all radioactive and all with very short half-lives. It has no uses and was discovered by A. Debierne in 1899.

actinium series *See* radioactive series.

actinoid contraction A smooth decrease in atomic or ionic radius with increasing proton number found in the actinoids.

actinometer Any of various instruments for measuring the intensity of electromagnetic radiation. Recent actinometers use the *photoelectric effect but earlier instruments depended either on the fluorescence produced by the radiation on a screen or on the amount of chemical change induced in some suitable substance.

action potential The change in electrical potential that occurs across a cell membrane during the passage of a nerve impulse. As an impulse travels in a wavelike manner along the axon of a nerve, it causes a localized and transient switch in electrical potential across the cell membrane from −60 mV (millivolts) to +45 mV. Nervous stimulation of a muscle fibre has a similar effect.

action spectrum A graphical plot of the efficiency of electromagnetic radiation in producing a photochemical reaction against the wavelength of the radiation used. For example, the action spectrum for photosynthesis using light shows a peak in the region 670–700 nm. This corresponds to a maximum absorption in the absorption spectrum of chlorophylls in this region.

activation analysis An analytical technique that can be used to detect most elements when present in a sample in milligram quantities (or less). In *neutron activation analysis* the sample is exposed to a flux of thermal neutrons in a nuclear reactor. Some of these neutrons are captured by nuclides in the sample to form nuclides of the same atomic number but a higher mass number. These newly formed nuclides emit gamma radiation, which can be used to identify the element present by means of a gamma-ray spectrometer. Activation analysis has also been employed using charged particles, such as protons or alpha particles.

active device **1.** An electronic component, such as a transistor, that is capable of amplification. **2.** An artificial *satellite that receives information and retransmits it after amplification. **3.** A radar device that emits microwave radiation and provides information about a distant body by receiving a reflection of this radiation. *Compare* passive device.

activity **1.** Symbol a. A thermodynamic function used in place of concentration in equilibrium constants for reactions involving nonideal gases and solutions. For example, in a reaction

$$\mathrm{A} \rightleftharpoons \mathrm{B} + \mathrm{C}$$

the true equilibrium constant is given by

$$K = a_B a_C / a_A$$

where a_A, a_B, and a_C are the activities of the components, which function as concentrations (or pressures) corrected for nonideal behaviour. *Activity coefficients* (symbol γ) are defined for gases by $\gamma = a/p$ (where p is pressure) and for solutions by $\gamma = aX$ (where X is the mole fraction). Thus, the equilibrium constant of a gas reaction has the form

$$K_p = \gamma_B p_B \gamma_C p_C / \gamma_A p_A$$

The equilibrium constant of a reaction in solution is

$$K_c = \gamma_B X_B \gamma_C X_C / \gamma_A X_A$$

The activity coefficients thus act as correction factors for the pressures or concentrations. *See* fugacity.

2. Symbol A. The number of atoms of a radioactive substance that disintegrate per unit time. The *specific activity* (a) is the activity per unit mass of a pure radioisotope. *See* radiation units.

additive process *See* colour.

adiabatic demagnetization A technique for cooling a paramagnetic salt, such as potassium chrome alum, to a temperature near *absolute zero. The salt is placed between the poles of an electromagnet and the heat produced during magnetization is removed by liquid helium. The salt is then isolated thermally from the surroundings and the field is switched off; the salt is demagnetized adiabatically and its temperature falls. This is because the demagnetized state, being less ordered, involves more energy than the magnetized state. The extra energy can come only from the internal, or thermal, energy of the substance.

adiabatic process Any process that occurs without heat entering or leaving a system. In general, an adiabatic change involves a fall or rise in temperature of the system. For example, if a gas expands under adiabatic conditions, its temperature falls (work is done against the retreating walls of the container). The *adiabatic equation* describes the relationship between the pressure (p) of an ideal gas and its volume (V), i.e. $pV^\gamma = K$, where γ is the ratio of the principal specific *heat capacities of the gas and K is a constant.

admittance Symbol *Y*. The reciprocal of *impedance. It is measured in siemens.

adsorbate A substance that is adsorbed on a surface.

adsorption The formation of a layer of gas, liquid, or solid on the surface of a solid or, less frequently, of a liquid. There are two types depending on the nature of the forces involved. In *chemisorption* a single layer of molecules, atoms, or ions is attached to the adsorbent surface by chemical bonds. In *physisorption* adsorbed molecules are held by the weaker *van der Waals' forces.

advanced gas-cooled reactor (AGR) *See* nuclear reactor.

aerial (antenna) The part of a radio or television system from which radio waves are transmitted into the atmosphere or space (*transmitting aerial*) or by which they are received (*receiving aerial*). A *directional* or *directive aerial* is one in which energy is transmitted or received more effectively from some directions than others, whereas an *omnidirectional aerial* transmits and receives equally well in all directions.

aerogenerator *See* wind power.

aerosol A colloidal dispersion of a solid or liquid in a gas. The commonly used aerosol sprays contain an inert propellant liquefied under pressure. Halogenated alkanes, such as dichlorodifluoromethane, are commonly used in aerosol cans. This use has been criticized on the grounds that these compounds persist in the atmosphere and may eventually (it is claimed) affect the *ozone layer.

aerospace The earth's atmosphere and the space beyond it.

after-heat Heat produced by a nuclear reactor after it has been shut down. The after-heat is generated by radioactive substances formed in the fuel elements.

AGR Advanced gas-cooled reactor. *See* nuclear reactor.

air *See* earth's atmosphere.

albedo 1. The ratio of the radiant flux reflected by a surface to that falling on it. **2.** The probability that a neutron entering a body of material will be reflected back through the same surface as it entered.

algebraic sum The total of a set of quantities paying due regard to sign, e.g. the algebraic sum of 3 and -4 is -1.

algorithm A method of solving a problem, involving a finite series of steps. In computing practice the algorithm denotes the expression on paper of the proposed computing process (often by means of a flowchart) prior to the preparation of the program. If no algorithm is possible a *heuristic solution has to be sought.

allowed bands *See* energy bands.

alloy A material consisting of two or more metals (e.g. brass is an alloy of copper and zinc) or a metal and a nonmetal (e.g. steel is an alloy of iron and carbon, sometimes with other metals included). Alloys may be compounds, *solid solutions, or mixtures of the components.

alloy steels *See* steel.

Alnico A tradename for a series of alloys, containing iron, aluminium, nickel, cobalt, and copper, used to make permanent magnets.

alpha-iron *See* iron.

alpha particle A helium nucleus emitted by a larger nucleus during the course of the type of radioactive decay known as *alpha decay*. As a helium nucleus consists of two protons and two neutrons bound together as a stable entity the loss of an alpha particle involves a decrease in *nucleon number of 4 and decrease of 2 in the *atomic number, e.g. the decay of a uranium–238 nucleus into a thorium–234 nucleus. A stream of alpha particles is known as an *alpha-ray* or *alpha-radiation*.

alternating current (a.c.) An electric current that reverses its direction with a constant *frequency (f). If a graph of the current against time has the form of a *sine wave, the current is said to be *sinusoidal*. Alternating current, unlike direct current, is therefore continuously varying and its magnitude is either given as its peak value (I_0) or its *root-mean-square value ($I_0/\sqrt{2}$ for a sinusoidal current). This r.m.s. value is more useful as it is comparable to a d.c. value in being a measure of the ability of the current to transmit power. The instantaneous value of a sinusoidal current (I) is given by $I = I_0 \sin 2\pi ft$.

If a direct current is supplied to a circuit the only opposition it encounters is the circuit's *resistance. However, an alternating current is opposed not only by the resistance of the circuit but also by its *reactance. This reactance is caused by *capacitance and *inductance in the circuit. In a circuit consisting of a resistance (R), an inductance (L), and a capacitance (C) all in series, the reactance (X) is equal to $(2\pi fL) - (1/2\pi fC)$. The total opposition to the current, called the *impedance (Z), is then equal to the ratio of the r.m.s. applied p.d. to the r.m.s. current and is given by $\sqrt{(R^2 + X^2)}$.

alternator An *alternating-current generator consisting of a coil or coils that rotate in the magnetic field produced by one or more permanent magnets or electromagnets. The electromagnets are supplied by an independent direct-current source. The frequency of the alternating current produced depends on the speed at which the coil rotates and the number of pairs of magnetic poles. In the large alternators of power stations the electromagnets rotate inside fixed coils; many bicycle dynamos are alternators with rotating permanent magnets inside fixed coils.

altimeter A device used to measure height above sea level. It usually consists of an aneroid *barometer measuring atmospheric pressure. Aircraft are fitted with altimeters, which are set to the atmospheric pressure at a convenient level, usually sea level, before take off. The height of the aircraft can then be read off the instrument as the aircraft climbs and the pressure falls.

AM (amplitude modulation) *See* modulation.

amalgam An alloy of mercury with one or more other metals. Most metals form amalgams (iron and platinum are exceptions), which may be liquid or solid. Some contain definite intermetallic compounds, such as $NaHg_2$.

americium Symbol Am. A radioactive metallic transuranic element belonging to the actinoids; a.n. 95; mass number of most stable isotope 243 (half-life 7.95×10^3 years); r.d. 13.67 (20°C); m.p. 994 ± 4°C; b.p. 2607°C. Ten isotopes are known. The element was discovered by G. T. Seaborg and associates in 1945, who obtained it by bombarding uranium–238 with alpha particles.

ammeter An instrument that measures electric current. The main types are the *moving-coil* ammeter, the *moving-iron* ammeter, and the *thermoammeter*. The moving-coil instrument is a moving-coil *galvanometer fitted with a *shunt to reduce its sensitivity. It can only be used for d.c., but can be adapted for a.c. by using a *rectifier. In moving-iron instruments, a piece of soft iron moves in the magnetic field created when the current to be measured flows through a fixed coil. They can be used with a.c. or d.c. but are less accurate (though more robust) than the moving-coil instruments. In thermoammeters, which can also be used with a.c. or d.c., the current is passed through a resistor, which heats up as the current passes. This is in contact with a thermocouple, which is connected to a galvanometer. This indirect system is mainly used for measuring high frequency a.c. In the *hot-wire* instrument the wire is clamped at its ends and its elongation as it is heated causes a pointer to move over a scale.

ammonia clock A form of atomic clock in which the frequency of a quartz oscillator is controlled by the vibrations of excited ammonia molecules (*see* excitation). The ammonia molecule (NH_3) consists of a pyramid with a nitrogen atom at the apex and one hydrogen atom at each corner of the triangular base. When the molecule is excited, once every 20.9 microseconds the nitrogen atom passes through the base and forms a pyramid the other side: 20.9 microseconds later it returns to its original position. This vibration back and forth has a frequency of 23 870 hertz and ammonia gas will only absorb excitation energy at exactly this frequency. By using a *crystal oscillator to feed energy to the gas and a suitable feedback mechanism, the oscillator can be locked to exactly this frequency.

amorphous Describing a solid that is not crystalline; i.e. one that has no long-range order in its lattice. Many powders that are described as 'amorphous' in fact are composed of microscopic crystals, as can be demonstrated by X-ray diffraction. Glasses are examples of true amorphous solids.

amount of substance Symbol *n*. A measure of the number of entities present in a substance. The specified entity may be an atom, molecule, ion, electron, photon, etc., or any specified group of such entities. The amount of substance of an element, for example, is proportional to the number of atoms present. For all entities, the constant of proportionality is the *Avogadro constant. The SI unit of amount of substance is the *mole.

ampere Symbol A. The SI unit of electric current. The constant current that, maintained in two straight parallel infinite conductors of negligible cross section placed one metre apart in a vacuum, would produce a force between the conductors of 2×10^{-7} N m^{-1}. This definition replaced the earlier international ampere defined as the current required to deposit 0.001 118 00 gram of silver from a solution of silver nitrate in one second. The unit is named after A. M. Ampère (1775–1836).

ampere-hour A practical unit of electric charge equal to the charge flowing in one hour through a conductor passing one ampere. It is equal to 3600 coulombs.

ampere-turn The SI unit of *magnetomotive force equal to the magnetomotive force produced when a current of one ampere flows through one turn of a magnetizing coil.

amplifier A device that increases the strength of an electrical signal by drawing energy from a separate source to that of the signal. The original device used in electronic amplifiers was the *triode valve, in which the cathode–anode current is varied in accordance with the low-voltage signal applied to the valve's control grid. In the more recent *transistor, the emitter–collector current is controlled in much the same way by the signal applied to the transistor's base region. In the most modern devices the complete amplifier circuit is manufactured as a single *integrated circuit. The ratio of the output amplitude (of p.d. or current) of an amplifier (or stage of an amplifier) to the corresponding input amplitude is called the *gain* of the amplifier.

amplitude *See* wave.

amplitude modulation (AM) *See* modulation; radio.

a.m.u. *See* atomic mass unit.

analytical geometry (coordinate geometry) A form of geometry in which points are located in a two-dimensional or three-dimensional space by means of a system of coordinates. Curves are represented by an equation for a set of such points. The geometry of figures can thus be analysed by algebraic methods. *See* Cartesian coordinates; polar coordinates.

anastigmatic lens 1. An objective lens for an optical instrument in which all *aberrations, including *astigmatism, are reduced greatly. **2.** A spectacle lens designed to correct astigmatism. It has different radii of curvature in the vertical and horizontal planes.

anchor ring *See* torus.

and circuit *See* logic circuits.

anechoic Having a low degree of reverberation with little or no reflection of sound. An *anechoic* chamber is one designed for experiments in acoustics. The walls are covered with small pyramids to avoid the formation of stationary waves between facing surfaces and the whole of the interior surface is covered with an absorbent material to avoid reflections.

anemometer An instrument for measuring the speed of the wind or any other flowing fluid. The simple *vane anemometer* consists of a number of cups or blades attached to a central spindle so that the air, or other fluid, causes the spindle to rotate. The instrument is calibrated to give a wind speed directly from a dial. The instrument can be mounted to rotate about a vertical axis and in this form it also gives an indication of the direction of the wind. A *hot-wire anemometer* consists of an electrically heated wire that is cooled by the flow of fluid passing round it. The faster the flow the lower the temperature of the wire and the lower its resistance. Thus the rate of flow can be calculated by measuring the resistance of the wire.

aneroid barometer *See* barometer.

angle modulation *See* modulation.

angle of incidence 1. The angle between a ray falling on a surface and the perpendicular (normal) to the surface at the point at which the ray strikes the surface. **2.** The angle between a wavefront and a surface that it strikes.

angle of reflection 1. The angle between a ray leaving a reflecting surface and the perpendicular (normal) to the surface at the point at which the ray leaves the surface. **2.** The angle between a wavefront and a surface that it leaves.

angle of refraction 1. The angle between a ray that is refracted at a surface between two different media and the perpendicular (normal) to the surface at the point of refraction. **2.** The angle between a wavefront and a surface at which it has been refracted.

angstrom Symbol Å. A unit of length equal to 10^{-10} metre. It was formerly used to measure wavelengths and intermolecular distances but has now been replaced by the nanometre. 1 Å = 0.1 nanometre. The unit is named after the Swedish pioneer of spectroscopy A. J. Ångstrom (1814–74).

angular displacement, velocity, and acceleration *See* rotational motion.

angular frequency (pulsatance) A quantity proportional to the *frequency of a periodic phenomenon but having the dimensions of angular velocity. The angular frequency in radians per second = frequency in hertz $\times 2\pi$ radians per cycle.

angular magnification (magnifying power) *See* magnification.

angular momentum Symbol L. The product of the angular velocity of a body and its *moment of inertia about the axis of rotation, i.e. $L = I\omega$.

anion A negatively charged *ion, i.e. an ion that is attracted to the *anode in *electrolysis. *Compare* cation.

anisotropic Denoting a medium in which certain physical properties are different in different directions. Wood, for instance, is an anisotropic material: its strength along the grain differs from that perpendicular to the grain. Single crystals that are not cubic are anisotropic with respect to some physical properties, such as the transmission of electromagnetic radiation. *Compare* isotropic.

annealing A form of heat treatment applied to a metal to soften it, relieve internal stresses and instabilities, and make it easier to work or machine. It consists of heating the metal to a specified temperature for a specified time, both of which depend on the metal involved, and then allowing it to cool slowly. It is applied to both ferrous and nonferrous metals and a similar process can be applied to other materials, such as glass.

annihilation The destruction of a particle and its *antiparticle as a result of a collision between them. The *annihilation radiation* produced is carried away by *photons or *mesons. For example, in a collision between an electron and a positron the energy produced is carried away by two photons of *gamma radiation, each having an energy of 0.511 MeV, which is equivalent to the rest masses of the annihilated particles plus their kinetic energies. When nucleons annihilate each other the energy is carried away by mesons.

annulus The plane figure formed between two concentric circles of different radii, R and r. Its area is $\pi(R^2 - r^2)$.

anode A positive electrode. In *electrolysis anions are attracted to the anode. In an electronic vacuum tube it attracts electrons from the *cathode and it is therefore from the anode that electrons flow out of the device. In these instances the anode is made positive by external means; however in a *voltaic cell the anode is the electrode that spontaneously becomes positive and therefore attracts electrons to it from the external circuit.

anomaly An angle used to fix the position of a body, such as a planet, in an elliptical orbit. The *true anomaly* of a planet is the angle between the *perihelion, the sun, and the planet in the direction of the planet's motion. The *mean anomaly* is the angle between the perihelion, the sun, and an imaginary planet having the same period as the real planet, but assumed to be moving at constant speed.

antenna *See* aerial.

antiferromagnetism *See* magnetism.

antilogarithm *See* logarithm.

antimatter *See* antiparticle.

antinode *See* stationary wave.

antiparallel vectors Vectors directed along the same line but in opposite directions.

antiparticle A subatomic particle that has the same mass as another particle and equal but opposite values of some other property or properties. For example, the antiparticle of the electron is the positron, which has a positive charge equal in magnitude to the electron's negative charge. The antiproton has a negative charge equal to the proton's positive charge. The neutron and the antineutron have *magnetic moments opposite in sign relative to their *spins. When a particle and its corresponding antiparticle collide *annihilation takes place. *Antimatter* is postulated to consist of matter made up of antiparticles. For example, antihydrogen would consist of an antiproton with an orbiting positron. In principle there seems no reason why matter should not have this form, but, so far, no antimatter has been detected in the universe.

aperture The effective diameter of a lens or mirror. The ratio of the effective diameter to the focal length is called the *relative aperture*, which is commonly known as the aperture, especially in photographic usage. The reciprocal of the relative aperture is called the *focal ratio*. The numerical value of the focal ratio is known as the *f-number* of a lens. For example, a camera lens with a 40 mm focal length and a 10 mm aperture has a relative aperture of 0.25 and a focal ratio of 4. Its f-number would be f/4, often written f4.

The light-gathering power of a telescope depends on the area of the lens, i.e. it is related to the square of the aperture. However, the larger the relative aperture the greater the *aberrations. In microscopy large-aperture objectives (corrected for aberrations) are preferred, since they reduce the blurring caused by *diffraction of light waves.

aperture synthesis *See* radio telescope.

aphelion The point in the orbit of a planet, comet, or artificial satellite in solar orbit at which it is farthest from the sun. The earth is at aphelion on about July 3. *Compare* perihelion.

aplanatic lens A lens that reduces both spherical *aberration and *coma.

apochromatic lens *See* achromatic lens.

apocynthion The point in the orbit around the moon of a satellite launched from the earth that is furthest from the moon. For a satellite launched from the moon the equivalent point is the *apolune*. *Compare* pericynthion.

apogee The point in the orbit of the moon, or an artificial earth satellite, at which it is furthest from the earth. At apogee the moon is 406 700 km from the earth, some 42 000 km further away than at *perigee*, the nearest point to the earth.

apolune *See* apocynthion.

apparent expansivity *See* expansivity.

Appleton layer *See* earth's atmosphere.

apsides The two points in an astronomical orbit that lie closest to (*periapsis*) and farthest from (*apoapsis*) the centre of gravitational attraction. The *line of apsides* is the straight line that joins the two apsides. If the orbit is elliptical the line of apsides is the major axis of the ellipse.

aqueous Describing a solution in water.

arc cos, sin, tan *See* inverse functions.

Archimedes' principle The weight of the liquid displaced by a floating body is equal to the weight of the body. The principle was not in fact stated by the Greek mathematician Archimedes (287–212 BC), though it has some connection with his discoveries. The principle is often stated in the form: when a body is (partially or totally) immersed in a fluid, the upthrust on the body is equal to the weight of fluid displaced.

arc lamp *See* electric lighting.

Argand diagram *See* complex number.

argument 1. A sequence of logical propositions based on a set of premisses and leading to a conclusion. **2.** *See* complex number.

arithmetic average (arithmetic mean) *See* average.

arithmetic series (arithmetic progression) A series or progression of numbers in which there is a common difference between terms, e.g. 3, 9, 15, 21,... is an arithmetic series with a common difference of 6. The general formula for the *n*th term is $[a + (n - 1)d]$ and the sum of *n* terms is $n[2a + (n - 1)d]/2$. *Compare* geometric series.

armature Any moving part in an electrical machine in which a voltage is induced by a magnetic field, especially the rotating coils in an electric motor or generator and the ferromagnetic bar attracted by an electromagnet in a *relay.

associative law The mathematical law stating that the value of an expression is independent of the grouping of the numbers, symbols, or terms in the expression. The *associative law for addition* states that numbers may be added in any order, e.g. $(x + y) + z = (x + z) + y$. The *associative law for multiplication* states that numbers can be multiplied in any order, e.g. $x(yz) = (xy)z$. Subtraction and division are not associative. *Compare* commutative law; distributive law.

astatic galvanometer A sensitive form of moving-magnet *galvanometer in which any effects of the earth's magnetic field are cancelled out. Two small oppositely directed magnets are suspended at the centres of two oppositely wound coils. As its resultant moment on the magnets is zero, the earth's field has no effect and the only restoring torque on the magnets is that provided by the suspending fibre. This makes a sensitive but delicate instrument.

astatine Symbol At. A radioactive halogen element; a.n. 85; r.a.m. 211; m.p. 302°C; b.p. 377°C. It occurs naturally by radioactive decay from uranium and thorium isotopes. Astatine forms at least 20 isotopes, the most stable astatine–210 has a half-life of 8.3 hours. It can also be produced by alpha bombardment of bismuth–200. The existence of At_2 has not yet been established. The element was synthesized by nuclear bombardment in 1940 by D. R. Corson, K. R. MacKenzie, and E. Segré at the University of California.

asteroids (minor planets; planetoids) A number of small bodies that revolve around the sun between the orbits of Mars and Jupiter (in the *asteroid belt*). The size of the bodies varies from the largest, Ceres (with a diameter of about 1000 km), to objects less than 1 km in diameter. It is estimated that there are about 10 bodies with diameters in excess of 300 km and some 200 bodies with diameters over 100 km.

asthenosphere A layer of the earth's mantle (*see* earth) that underlies the lithosphere at a depth of about 70 km. The velocity of seismic waves is considerably reduced in the asthenosphere and it is thought to be a zone of partial melting. It extends to a depth of about 250 km where rocks again become solid.

astigmatism A lens defect in which when rays in one plane are in focus those in another plane are not. In lenses and mirrors it occurs with objects not on the axis and is best controlled by reducing the *aperture to restrict the use of the lens or mirror to its central portion. The eye can also suffer from astigmatism, usually when the cornea is not spherical. It is corrected by using an *anastigmatic lens.

astrometry The branch of astronomy concerned with the measurement of the positions of the celestial bodies on the *celestial sphere.

astronomical telescope *See* telescope.

astronomical unit (AU) The mean distance between the sun and the earth. It is equal to 149 597 870 km (499 light seconds).

astronomy The study of the universe beyond the earth's atmosphere. The main branches are *astrometry, *celestial mechanics, and *astrophysics.

astrophysics The branch of astronomy concerned with the physical processes associated with the celestial bodies and the intervening regions of space. It deals principally with the energy of stellar systems and the relation between this energy and the evolution of the system.

asymmetric atom *See* optical activity.

asymptote A line that a curve approaches but only touches at infinity.

asymptotic series A divergent *series of the type $a_0 + a_1/x + a^2/x^2 + \dots + a_n/x^n$.

atmolysis The separation of a mixture of gases by means of their different rates of diffusion. Usually, separation is effected by allowing the gases to diffuse through the walls of a porous partition or membrane.

atmosphere 1. (atm.) A unit of pressure equal to 101 325 pascals. This is equal to 760.0 mmHg. The actual *atmospheric pressure fluctuates around this value. The unit is usually used for expressing pressures well in excess of standard atmospheric pressure, e.g. in high-pressure chemical processes. **2.** *See* earth's atmosphere.

atmospheric pressure The pressure exerted by the weight of the air above it at any point on the earth's surface. At sea level the atmosphere will support a column of mercury about 760 mm high. This decreases with increasing altitude. The standard value for the atmospheric pressure at sea level in SI units is 101 325 pascals.

atom The smallest part of an element that can exist. Atoms consist of a small dense nucleus of protons and neutrons surrounded by moving electrons. The number of electrons equals the number of protons so the overall charge is zero. The electrons may be thought of as moving in circular or elliptical orbits (*see* Bohr theory) or, more accurately, in regions of space around the nucleus (*see* orbital).

The *electronic structure* of an atom refers to the way in which the electrons are arranged about the nucleus, and in particular the *enegy levels that they occupy. Each electron can be characterized by a set of four quantum numbers, as follows:

(1) The *principal quantum number n* gives the main energy level and has values 1, 2, 3, etc. (the higher the number, the further the electron from the nucleus). Traditionally, these levels, or the orbits corresponding to them, are referred to as *shells* and given letters K, L, M, etc. The K-shell is the one nearest the nucleus.

(2) The *orbital quantum number l*, which governs the angular momentum of the electron. The possible values of l are $(n - 1)$, $(n - 2)$, ... 1, 0. Thus, in the first shell ($n = 1$) the electrons can only have angular momentum zero ($l = 0$). In the second shell ($n = 2$), the values of l can be 1 or 0, giving rise to two *subshells* of slightly different en-

ergy. In the third shell ($n = 3$) there are three subshells, with $l = 2, 1$, or 0. The subshells are denoted by letters $s(l = 0)$, $p(l = 1)$, $d(l = 2)$, $f(l = 3)$. The orbital quantum number is sometimes called the *azimuthal quantum number*.

(3) The *magnetic quantum number m*, which governs the energies of electrons in an external magnetic field. This can take values of $+l$, $+(l - 1)$, ... 1, 0, -1, ... $-(l - 1)$, $-l$. In an *s*-subshell (i.e. $l = 0$) the value of $m = 0$. In a *p*-subshell ($l = 1$), m can have values $+1$, 0, and -1; i.e. there are three *p*-orbitals in the *p*-subshell, usually designated p_x, p_y, and p_z. Under normal circumstances, these all have the same energy level.

(4) The *spin quantum number* m_s, which gives the spin of the individual electrons and can have the values $+½$ or $-½$.

According to the *Pauli exclusion principle, no two electrons in the atom can have the same set of quantum numbers. The numbers define the *quantum state* of the electron, and explain how the electronic structures of atoms occur.

atomic bomb *See* nuclear weapons.

atomic clock An apparatus for measuring or standardizing time that is based on periodic phenomena within atoms or molecules. *See* ammonia clock; caesium clock.

atomic energy *See* nuclear energy.

atomic mass unit (a.m.u.) A unit of mass used to express *relative atomic masses. It is equal to 1/12 of the mass of an atom of the isotope carbon-12 and is equal to $1.660\,33 \times 10^{-27}$ kg. This unit superseded both the physical and chemical mass units based on oxygen-16 and is sometimes called the *unified mass unit* or the *dalton*.

atomic number (proton number) Symbol Z. The number of protons in the nucleus of an atom. The atomic number is equal to the number of electrons orbiting the nucleus in a neutral atom.

atomic orbital *See* orbital.

atomic pile An early form of *nuclear reactor using graphite as a *moderator.

atomic volume The relative atomic mass of an element divided by its density.

atomic weight *See* relative atomic mass.

attenuation **1.** A loss of intensity suffered by sound, radiation, etc., as it passes through a medium. It may be caused by absorption or scattering. **2.** The drop in voltage or current experienced by a signal as it passes through a circuit.

atto- Symbol *a*. A prefix used in the metric system to denote 10^{-18}. For example, 10^{-18} second = 1 attosecond (as).

AU *See* astronomical unit.

audibility The limits of audibility of the human ear are between about 20 hertz (a low rumble) and 20 000 hertz (a shrill whistle). With increased age the upper limit falls quite considerably.

audiofrequency A frequency that is audible to the human ear. *See* audibility.

audiometer An instrument that generates a sound of known frequency and intensity in order to measure an individual's hearing ability.

Auger effect The ejection of an electron from an atom without the emission of an X- or gamma-ray photon, as a result of the de-excitation of an excited electron within the atom. This type of transition occurs in the X-ray region of the emission spectrum. The kinetic energy of the ejected electron, called an *Auger electron*, is equal to the energy of the corresponding X-ray photon minus the binding energy of the Auger electron. The effect was discovered by Pierre Auger (1899–) in 1925.

aurora The luminous phenomena seen in the night sky in high latitudes, occurring most frequently near the earth's geomagnetic poles. The displays of aurora appear as coloured arcs, rays, bands, streamers, and curtains, usually green or red. The aurora is caused by the interaction of the atoms (mainly atomic oxygen) and molecules in the upper atmosphere (above about 100 km) with charged particles streaming from the sun, attracted to the auroral regions by the earth's magnetic field. The aurora is known as the *aurora borealis* (or northern lights) in the northern hemisphere and as the *aurora australis* (or southern lights) in the southern hemisphere.

austenite *See* steel.

autoclave A strong steel vessel used for carrying out chemical reactions, sterilizations, etc., at high temperature and pressure.

autoradiography An experimental technique in which a radioactive specimen is placed in contact with (or close to) a photographic plate, so as to produce a record of the distribution of radioactivity in the specimen. The film is darkened by the ionizing radiation from radioactive parts of the sample. Autoradiography has a number of applications, particularly in the study of living tissues and cells.

avalanche A shower of ionized particles created by a single *ionization as a result of secondary ionizations caused by the original electron and ion being accelerated in an electric field. Each ionization leads to the formation of more electrons and ions, which themselves cause further ionizations. Such avalanches occur in a *Geiger counter.

average (mean) 1. The *arithmetic average* of a set of *n* numbers is the sum of the numbers divided by *n*. **2.** The *geometric average* of a set of *n* numbers is the *n*th root of their product. *See also* root-mean-square value.

Avogadro constant Symbol N_A or L. The number of atoms or molecules in one *mole of substance. It has the value $6.022\,52 \times 10^{23}$. Formerly it was called *Avogadro's number*.

Avogadro's law Equal volumes of all gases contain equal numbers of molecules at the same pressure and temperature. The law, often called *Avogadro's hypothesis*, is true only for ideal gases. It was first proposed in 1811 by Count Amadeo Avogadro (1776–1856).

axis 1. One of a set of reference lines used to locate points on a graph or in a coordinate system. *See* Cartesian coordinates; polar coordinates. **2.** A line about which a figure, curve, or body is symmetrical (*axis of symmetry*) or about which it rotates (*axis of rotation*).

azeotrope (azeotropic mixture; constant-boiling mixture) A mixture of two liquids that boils at constant composition; i.e. the composition of the vapour is the same as that of the liquid. Azeotropes occur because of deviations in Raoult's law leading to a maximum or minimum in the *boiling-point–composition diagram. When the mixture is boiled, the vapour initially has a higher proportion of one component than is present in the liquid, so the proportion of this in the liquid falls with time. Eventually, the maximum and minimum point is reached, at which the two liquids distil together without change in composition. The composition of an azeotrope depends on the pressure.

azimuth *See* polar coordinates.

azimuthal quantum number *See* atom.

B

Babbit metal Any of a group of related alloys used for making bearings. They consist of tin containing antimony (about 10%) and copper (1–2%), and often lead. The original alloy was invented in 1839 by the US inventor Isaac Babbit (1799–1862).

Babo's law The vapour pressure of a liquid is decreased when a solute is added, the amount of the decrease being proportional to the amount of solute dissolved. The law was discovered in 1847 by the German chemist Lambert Babo (1818–99). *See also* Raoult's law.

back e.m.f. An electromotive force that opposes the main current flow in a circuit. For example, when the coils of the armature in an electric motor are rotated a back e.m.f is generated in these coils by their interaction with the field magnet (*see* inductance). Also, in an electric cell, *polarization causes a back e.m.f. to be set up, in this case by chemical means.

background radiation Low intensity *ionizing radiation present on the surface of the earth and in the atmosphere as a result of *cosmic radiation and the presence of radioisotopes in the earth's rocks, soil, and atmosphere. The radioisotopes are either natural or the result of nuclear fallout or waste gas from power stations. Background counts must be taken into account when measuring the radiation produced by a specified source. *See also* microwave background radiation.

balance An accurate weighing device. The simple *beam balance* consists of two pans suspended from a centrally pivoted beam. Known masses are placed on one pan and the substance or body to be weighed is placed in the other. When the beam is exactly horizontal the two masses are equal. An accurate laboratory balance weighs to the nearest hundredth of a milligram. Specially designed balances can be accurate to a millionth of a milligram. More modern *substitution balances* use the substitution principle. In this calibrated weights are removed from the single lever arm to bring the single pan suspended from it into equilibrium with a fixed counter weight. The substitution balance is more accurate than the two-pan device and enables weighing to be carried out more rapidly. In automatic electronic balances, mass is determined not by mechanical deflection but by electronically controlled compensation of an electric force. A scanner monitors the displacement of the pan support generating a current proportional to the displacement. This current flows through a coil forcing the pan support to return to its original position by means of a magnetic force. The signal generated enables the mass to be read from a digital display. The mass of the empty container can be stored in the balance's computer memory and automatically deducted from the mass of the container plus its contents. *See also* spring balance.

ballistic galvanometer A moving-coil *galvanometer designed for measuring charge by detecting a surge of current. It has a heavy coil with minimal damping. When a surge of current is passed through the coil, the initial maximum deflection (the 'throw') is proportional to the total charge that has passed.

ballistic pendulum A device used to measure the velocity of a projectile, such as a bullet. A large mass of relatively soft material is suspended from a horizontal bar and the angle through which this mass is displaced when it is struck by the projectile in flight enables the momentum and hence the velocity of the projectile to be calculated by successive application of the laws of conservation of linear momentum and of energy.

ballistics The study of the flight of projectiles, especially those that have a parabolic flight path from one point on the earth's surface to another.

Balmer series *See* hydrogen spectrum.

band spectrum *See* spectrum.

band theory *See* energy bands.

bandwidth The frequency range over which a radio signal of specified frequency spreads. For example, in a *modulation system it is the range of frequencies occupied by the modulating signal on either side of the carrier wave. In an amplifier, it is the range of frequencies over which the power amplification falls within a specified fraction of the maximum value. In an aerial it is the range of frequencies that an aerial system can handle without mismatch.

bar A c.g.s. unit of pressure equal to 10^6 dynes per square centimetre or 10^5 pascals (approximately 750 mmHg or 0.987 atmosphere). The *millibar* (100 Pa) is commonly used in meteorology.

Barkhausen effect The magnetization of a ferromagnetic substance by an increasing magnetic field takes place in discontinuous steps rather than continuously. The effect results from the orientation of magnetic domains (*see* magnetism). It was first observed by H. Barkhausen (1881–1956) in 1919.

barn A unit of area sometimes used to measure *cross sections in nuclear scattering experiments. It is equal to 10^{-28} square metre. The name comes from the phrase 'side of a barn' (something easy to hit).

barograph A meteorological instrument that records on paper variations in atmospheric pressure over a period. It often consists of an aneroid barometer operating a pen that rests lightly on a rotating drum to which the recording paper is attached.

barometer A device for measuring *atmospheric pressure. The *mercury barometer* in its simplest form consists of a glass tube about 80 cm long sealed at one end and filled with mercury. The tube is then inverted and the open end is submerged in a reservoir of mercury; the mercury column is held up by the pressure of the atmosphere acting on the surface of mercury in the reservoir. This type of device was invented by the Italian scientist Evangelista Torricelli (1608–47), who first noticed the variation in height from day to day, and constructed a barometer in 1644.

In such a device, the force exerted by the atmosphere balanced the weight of the mercury column. If the height of the column is h and the cross-sectional area of the tube is A, then the volume of the mercury in the column is hA and its weight is $hA\rho$, where ρ is the density of mercury. The force is thus $hA\rho g$, where g is the acceleration of free fall and the pressure exerted is this force divided by the area of the tube; i.e. $h\rho g$. Note that the height of the mercury is independent of the diameter of the tube. At standard atmospheric pressure the column is 760 mm high. The pressure is then expressed as 760 mmHg (101 325 pascals).

Mercury barometers of this type, with a reservoir of mercury, are known as *cistern barometers*. A common type is the *Fortin barometer*, in which the mercury is held in a leather bag so that the level in the reservoir can be adjusted. The height is read from a scale along the side of the tube in conjunction with a vernier scale that can be moved up and down. Corrections are made for temperature.

The second main type of barometer is the *aneroid barometer*, in which the cumbersome mercury column is replaced by a metal box with a thin corrugated lid. The air is removed from the box and the lid is supported by a spring. Variations in atmospheric pressure cause the lid to move against the spring. This movement is magnified by a system of delicate levers and made to move a needle around a scale. The aneroid barometer is less accurate than the mercury type but much more robust and convenient, hence its use in *altimeters.

barycentre The *centre of mass of a system.

barye A c.g.s. unit of pressure equal to one dyne per square centimetre (0.1 pascal).

baryon Any of a class of *elementary particles consisting of *nucleons and other particles that have a nucleon in their decay products. Baryons are a subclass of the *hadrons and according to current theory consist of three quarks. Baryons have a quantum number, called the *baryon number*, which is +1 for baryons, −1 for antibaryons, and 0 for other particles. It is conserved in all interactions.

base 1. (in mathematics) **a.** The number of different symbols in a number system. In the decimal system the base is 10; in *binary notation it is 2. **b.** The number that when raised to a certain power has a *logarithm equal to that power. For example if 10 is raised to the power of 3 it is equal to 1000; 3 is then the (common) logarithm of 1000 to the base 10. In natural or Napierian logarithms the base is e. To change the base from common to natural logarithms the formula used is: $\log_{10}y = \log_e y \times \log_{10}e = 0.43429\log_e y$.
2. (in electronics) *See* transistor.

base unit A unit that is defined arbitrarily rather than being defined by simple combinations of other units. For example, the ampere is a base unit in the SI system defined in terms of the force produced between two current-carrying conductors, whereas the coulomb is a *derived unit*, defined as the quantity of charge transferred by one ampere in one second.

basic-oxygen process (BOP process) A high-speed method of making high-grade steel. It originated in the *Linnz–Donnewitz (L–D) process*. Molten pig iron and scrap are charged into a tilting furnace, similar to the Bessemer furnace except that it has no tuyeres. The charge is converted to steel by blowing high-pressure oxygen onto the surface of the metal through a water-cooled lance. The excess heat produced enables up to 30% of scrap to be incorporated into the charge. The process has largely replaced the Bessemer and open-hearth processes.

battery A number of electric cells joined together. The common car battery, or *accumulator, usually consists of six secondary cells connected in series to give a total e.m.f. of 12 volts. A torch battery is usually a dry version of the *Leclanché primary cell, two of which are often connected in series. Batteries may also have cells connected in parallel, in which case they have the same e.m.f. as a single cell, but their capacity is increased, i.e. they will provide more total charge. The capacity of a battery is usually specified in ampere-hours, the ability to supply 1 A for 1 hr, or the equivalent.

BCS theory *See* superconductivity.

beam A group of rays moving in an organized manner. It may consist of particles (e.g. an electron beam) or of electromagnetic radiation (e.g. a radar beam).

beam balance *See* balance.

beam hole A hole through the shielding of a *nuclear reactor to enable a beam of neutrons or other particles to escape for experimental purposes.

beats A periodic increase and decrease in loudness heard when two notes of slightly different frequency are sounded at the same time. If a note of frequency n is heard at the same time as a note of frequency m, the resulting note will have a frequency of about $(n + m)/2$. However the amplitude of this note will vary from the difference to the sum of the amplitudes of the m and n notes and the frequency (called the *beat frequency*) of this variation will be $(m - n)$. The beating sound produced occurs as the waves successively reinforce and oppose each other as they move in and out of phase. *See also* interference.

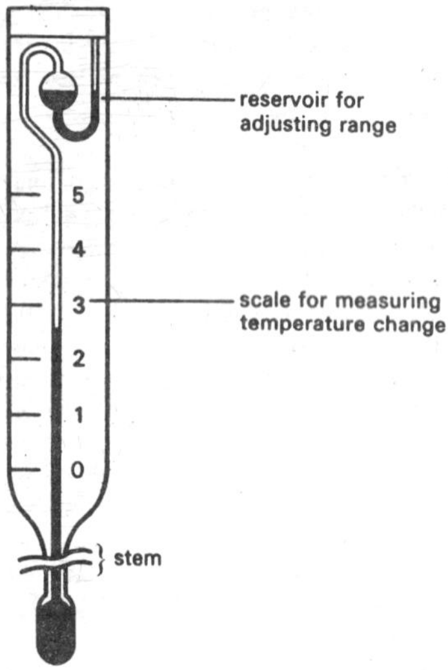

Beckmann thermometer

Beckmann thermometer A thermometer for measuring small changes of temperature. It consists of a mercury-in-glass thermometer with a scale covering only 5 or 6°C calibrated in hundredths of a degree. It has two mercury bulbs, the range of temperature to be measured is varied by running mercury from the upper bulb into the larger lower bulb. It is used particularly for measuring *depression of freezing point or *elevation of boiling point of liquids when solute is added, in order to find relative molecular masses. The instrument was invented by the German chemist E. O. Beckmann (1853–1923).

becquerel Symbol Bq. The SI unit of activity (*see* radiation units). The unit is named after the discoverer of radioactivity A. H. Becquerel (1852–1908).

bel Ten *decibels.

bell metal A type of *bronze used in casting bells. It consists of 60–85% copper alloyed with tin, often with some zinc and lead included.

bending moment (about any point or section of a horizontal beam under load) The algebraic sum of the *moments of all the vertical forces to either side of that point or section.

berkelium Symbol Bk. A radioactive metallic transuranic element belonging to the actinoids; a.n. 97; mass number of the most stable isotope 247 (half-life 1.4×10^3 years); r.d. (calculated) 14. There are eight known isotopes. It was first produced by G. T. Seaborg and associates in 1949 by bombarding americium-241 with alpha particles.

Bernoulli theorem At any point in a pipe through which a fluid is flowing the sum of the pressure energy, the kinetic energy, and the potential energy of a given mass of the fluid is constant. This is equivalent to a statement of the law of the conservation of energy. The law was published in 1738 by the Swiss physicist Daniel Bernoulli (1700–82).

Bessemer process A process for converting *pig iron from a *blast furnace into *steel. The molten pig iron is loaded into a refractory-lined tilting furnace (*Bessemer converter*) at about 1250°C. Air is blown into the furnace from the base and *spiegel is added to introduce the correct amount of carbon. Impurities (especially silicon, phosphorus, and manganese) are removed by the converter lining to form a slag. Finally the furnace is tilted so that the molten steel can be poured off. In the modern VLN (very low nitrogen) version of this process, oxygen and steam are blown into the furnace in place of air to minimize the absorption of nitrogen from the air by the steel. The process is named after the British engineer Sir Henry Bessemer (1813–98), who announced it in 1856. *See also* basic-oxygen process.

beta decay A type of weak interaction (*see* fundamental interactions) in which an unstable atomic nucleus changes into a nucleus of the same mass number but different proton number. The change involves the conversion of a neutron into a proton with the emission of an electron and an antineutrino ($n \rightarrow p + e^- + \bar{\nu}$) or of a proton into a neutron with the emission of a positron and a neutrino ($p \rightarrow n + e^+ + \bar{\nu}$). An example is the decay of carbon–14:

$$^{14}_{6}C \rightarrow ^{14}_{7}N + e^- + \bar{\nu}$$

The electrons or positrons emitted are called *beta particles* and streams of beta particles are known as *beta radiation*.

beta-iron A nonmagnetic allotrope of iron that exists between 768°C and 900°C.

beta particle *See* beta decay.

betatron A particle *accelerator for producing high-energy electrons (up to 340 MeV) for research purposes, including the production of high-energy X-rays. The electrons are accelerated by electromagnetic induction in a doughnut-shaped (toroidal) ring from which the air has been removed. This type of accelerator was first developed by D. W. Kerst (1911–) in 1939; the largest such machine, at the University of Illinois, was completed in 1950.

Bevatron A colloquial name for the proton *synchrotron at the Berkeley campus of the University of California. It produces energies up to 6 GeV.

biaxial crystal *See* double refraction.

biconcave *See* concave.

big-bang theory (superdense theory) The cosmological theory that all the matter and energy in the universe originated in a superdense agglomeration that exploded at a finite moment in the past. This theory successfully explains the *expansion of the universe, the galaxies flying apart like fragments from an exploding bomb; the observed *microwave background radiation, characteristic of *black-body radiation at a temperature of 3 K; and the observed abundance of helium in the universe, formed in the first 100 seconds after the explosion from deuterium at a temperature of 10^9 K. It is now generally considered to be more satisfactory that the rival *steady-state theory. The big-bang theory was first developed in 1927 by A. G. E. Lemaitre (1894–1966) and revived and revised in 1946 by George Gamow (1904–68). Several variants of it have been proposed.

billion 1. (in the UK and Germany) One million million, 10^{12}. **2.** (in the USA and France) One thousand million, 10^9.

bimetallic strip A strip consisting of two metals of different *expansivity riveted or welded together so that the strip bends on heating. If one end is fixed the other end can be made to open and close an electric circuit, as in a *thermostat.

bimorph cell A device consisting of two plates of piezoelectric material, such as Rochelle salt, joined together so that one expands on the application of a potential difference and the other contracts. The cell thus bends as a result of the applied p.d.

The opposite effect is also used, in which the mechanical bending of the cell is used to produce a p.d., as in the crystal microphone and some types of record-player pickups.

binary notation A number system using only two different digits, 0 and 1. Instead of units, tens, hundreds, etc., as used in the decimal system, digits in the binary notation represent units, twos, fours, eights, etc. Thus one in decimal notation is represented by 0001, two by 0010, four by 0100, and eight by 1000. Because 0 and 1 can be made to correspond to off and on conditions in an electric circuit, the binary notation is widely used in computers.

binary stars A pair of stars revolving about a common centre of mass. In a *visual binary* the stars are far enough apart to be seen separately by an optical telescope. In an *astrometric binary* one component is too faint to be seen and its presence is inferred from the perturbations in the motion of the other. In a *spectrosopic binary* the stars cannot usually be resolved by a telescope, but the motions can be detected by different Doppler shifts in the spectrum at each side of the binary, according to whether the components are approaching or receding from the observer.

binding energy The energy equivalent to the *mass defect when nucleons bind together to form an atomic nucleus. When a nucleus is formed some energy is released by the nucleons, since they are entering a more stable lower-energy state. Therefore the energy of a nucleus consists of the energy equivalent of the mass of its individual nucleons minus the binding energy. The binding energy per nucleon plotted against the mass number provides a useful graph showing that up to a mass number of 50–60, the binding energy per nucleon increases rapidly, thereafter it falls slowly. Energy is released both by fission of heavy elements and by fusion of light elements because both processes entail a rearrangement of nuclei in the lower part of the graph to form nuclei in the higher part of the graph.

binoculars Any optical instrument designed to serve both the observer's eyes at once. *Binocular field glasses* consist of two refracting astronomical *telescopes inside each of which is a pair of prisms to increase the effective length and produce an upright image. Simpler binoculars, such as *opera glasses*, consist of two Galilean telescopes which produce upright images without prisms. Commonly, binoculars are specified by a pair of numbers, such as 10 × 50. The first number indicates the angular *magnification produced. The second is the diameter of the objective lens in millimetres, and indicates the amount of light gathered by the instrument. *Binocular microscopes* are used in biology and surgery to enable the observer to obtain a stereoscopic view of small objects or parts.

binocular vision The ability, found only in animals with forward-facing eyes, to produce a focused image of the same object simultaneously on the retinas of both eyes. This permits three-dimensional vision and contributes to distance judgment.

binomial theorem (binomial expansion) A rule for the expansion of a binomial expression (expression consisting of two variables). The general binomial expression $(x + y)^n$ expands to:

$$x^n + nx^{n-1}y + [n(n - 1)/2!]x^{n-2}y^2 + \ldots y^n.$$

bioenergetics The study of the flow and the transformations of energy that occur in living organisms. Typically, the amount of energy that an organism takes in (from food or sunlight) is measured and divided into the amount used for growth of new tissues; that lost through death, wastes, and (in plants) transpiration; and that lost to the environment as heat (through respiration).

bioluminescence The emission of light without heat (*see* luminescence) by living organisms. The phenomenon occurs in glow-worms and fireflies, bacteria and fungi, and in many deep-sea fish (among others); in animals it may serve as a means of protection (e.g. by disguising the shape of a fish) or species recognition or it may provide mating signals. The light is produced during the oxidation of a compound called *luciferin* (the composition of which varies according to the species), the reaction being catalysed by an enzyme, *luciferase*. Bioluminescence may be continuous (e.g. in bacteria) or intermittent (e.g. in fireflies).

biophysics *See* physics.

biprism A glass prism with an obtuse angle that functions as two acute-angle prisms placed base-to-base. A double image of a single object is thus formed; the device was used by Fresnel to produce two coherent beams for interference experiments.

birefringence *See* double refraction.

bistable circuit *See* flip-flop.

bit (binary digit) Either of the digits 0 or 1 as used in the *binary notation. Bits are therefore the basic unit of information in a computer system.

Bitter pattern A microscopic pattern that forms on the surface of a ferromagnetic material that has been painted with a colloidal suspension of small iron particles. The patterns outline the boundaries of the magnetic domains (*see* magnetism). They were first observed by F. Bitter in 1931.

black body A hypothetical body that absorbs all the radiation falling on it. It thus has an *absorptance and an *emissivity of 1. While a true black body is an imaginary concept, a small hole in the wall of an enclosure at uniform temperature is the nearest approach that can be made to it in practice.

Black-body radiation is the electromagnetic radiation emitted by a black body. It extends over the whole range of wavelengths upwards from a lower limit, which is dependent upon the body's temperature. The distribution of energy over this range of wavelengths, however, has a characteristic form with a maximum at a certain wavelength. The radiation emitted by a black body is a function of its temperature only. *See* Stefan's law; Wien's displacement law.

black hole An object in space that has collapsed under its own gravitational forces to such an extent that its *escape velocity is equal to the speed of light. Black holes are thought to originate from *supernovae explosions in which the core of matter left after the explosion is very large (several solar masses). The core of matter then continues to contract and a very large gravitational field is produced. Eventually it reaches a point at which the field is sufficiently large to prevent electromagnetic radiation leaving – the object has then become a black hole.

The boundary marking the point at which the gravitational field reaches the critical value is the *event horizon*, and it is not possible to detect what is happening within this. Observational evidence of objects thought to be black holes comes from their effect on surrounding matter. Thus, if a black hole forms a binary system with another star it will attract and capture matter from this star. The material leaving the star first forms a rotating disc around the black hole, in which the matter becomes compressed and heated to such an extent that it emits X-rays. In the constellation Cygnus there is an X-ray source, Cygnus X-1, which consists of a supergiant revolving around a small invisible companion with a mass ten times that of the sun. The companion is thought to be a black hole. Black holes have also been postulated as the cause of *quasars and *gravitational waves.

The theoretical study of black holes involves use of the general theory of relativity. One problem is the fate of matter within the event horizon, which possibly is compressed to a single point of infinite density (a *singularity*). Theoreticians have also postulated the existence of 'mini' black holes (with masses of about 10^{11} kilogram and radii about 10^{-10} metres). Such entities might have been formed shortly after the big bang when the universe was created. The application of quantum mechanics to gravitational theory indicates that these could lose energy by ejecting particles at their boundary. *See also* Schwarzschild radius; white hole.

blast furnace A furnace for smelting iron ores, such as haematite (Fe_2O_3) or magnetite (Fe_3O_4), to make *pig irən. The furnace is a tall refractory-lined cylindrical structure that is charged at the top with the dressed ore (*see* beneficiation), coke, and a flux, usually limestone. The conversion of the iron oxides to metallic iron is a reduction process in which carbon monoxide and hydrogen are the reducing agents. The overall reaction can be summarized thus:

$$Fe_3O_4 + 2CO + 2H_2 \rightarrow 3Fe + 2CO_2 + 2H_2O$$

The CO is obtained within the furnace by blasting the coke with hot air from a ring of tuyeres about two-thirds of the way down the furnace. The reaction producing the CO is:

$$2C + O_2 \rightarrow 2CO$$

In most blast furnaces hydrocarbons (oil, gas, tar, etc.) are added to the blast to provide a source of hydrogen. In the modern *direct-reduction process* the CO and H_2 may be produced separately so that the reduction process can proceed at a lower temperature. The pig iron produced by a blast furnace contains about 4% carbon and further refining is usually required to produce steel or cast iron.

body-centred cubic (b.c.c.) *See* cubic crystal.

Bohr theory The theory published in 1913 by the Danish physicist Niels Bohr (1885–1962) to explain the line spectrum of hydrogen. He assumed that a single electron of mass m travelled in a circular orbit of radius r, at a velocity v, around a positively charged nucleus. The *angular momentum of the electron would then be mvr. Bohr proposed that electrons could only occupy orbits in which this angular momentum had certain fixed values, $h/2\pi$, $2h/2\pi$, $3h/2\pi$, ... $nh/2\pi$, where h is the Planck constant. This means that the angular momentum is quantized, i.e. can only have certain values, each of which is a multiple of n. Each permitted value of n is associated with an orbit of different radius and Bohr assumed that when the atom emitted or absorbed radiation of frequency ν, the electron jumped from one orbit to another; the energy emitted or absorbed by each jump is equal to $h\nu$. This theory gave good results in predicting the series of lines observed in the *hydrogen spectrum but not with any other spectrum. The idea of quantized values of angular momentum was later explained by the wave nature of the electron. Each orbit has to have a whole number of wavelengths around it; i.e. $n\lambda = 2\pi r$, where λ is the wavelength and n a whole number. The wavelength of a particle is given by h/mv, so $nh/mv = 2\pi r$, which leads to $mvr = nh/2\pi$. Modern atomic theory does not allow subatomic particles to be treated in the same way as large objects, and Bohr's reasoning is somewhat discredited. However, the idea of quantized angular momentum has been retained.

boiling point (b.p.) The temperature at which the saturated vapour pressure of a liquid equals the external atmospheric pressure. As a consequence, bubbles form in the liquid and the temperature remains constant until all the liquid has evaporated. As the boiling point of a liquid depends on the external atmospheric pressure, boiling points are usually quoted for standard atmospheric pressure (760 mmHg = 101 325 Pa).

boiling-point–composition diagram A graph showing how the boiling point and vapour composition of a mixture of two liquids depends on the composition of the mixture. The abscissa shows the range of compositions from 100% A at one end to 100% B at the other. The diagram has two curves: the lower one gives the boiling points (at a fixed pressure) for the different compositions. The upper one is plotted by taking the composition of vapour at each temperature on the boiling-point curve. The two curves would coincide for an ideal mixture, but generally they are different because of deviations from Raoult's law. In some cases, they may show a maximum or minimum and coincide at some intermediate composition, explaining the formation of *azeotropes.

boiling-water reactor *See* nuclear reactor.

bolometer A sensitive instrument used to measure radiant heat. The original form consists of two elements, each comprising blackened platinum strips (about 10^{-3} mm thick) arranged in series on an insulated frame to form a zigzag. The two elements are connected into the adjacent arms of a *Wheatstone bridge; one element is exposed to the radiation, the other is shielded from it. The change in the resistance of the exposed element, as detected by the bridge galvanometer, enables the heat reaching it to be calculated.

Modern semiconductor bolometers are now common, in which the platinum is replaced by a strip of semiconductor: this has a much greater (though usually negative) *temperature coefficient of resistance, and makes the system more sensitive.

Boltzmann constant Symbol k. The ratio of the universal gas constant (R) to the Avogadro constant (N_A). It may be thought of therefore as the gas constant per molecule:

$$k = R/N_A = 1.380\,622 \times 10^{-23}\ \mathrm{J\,K^{-1}}$$

It is named after the Austrian physicist Ludwig Boltzmann (1844–1906).

bomb calorimeter An apparatus used for measuring heats of combustion (e.g. calorific values of fuels and foods). It consists of a strong container in which the sample is sealed with excess oxygen and ignited electrically. The heat of combustion at constant volume can be calculated from the resulting rise in temperature.

Boolean algebra A form of symbolic logic, devised by George Boole (1815–64) in the middle of the 19th century, which provides a mathematical procedure for manipulating logical relationships in symbolic form. For example in Boolean algebra $a + b$ means a or b, while ab means a and b. It makes use of *set theory and is extensively used by the designers of computers to enable the bits 0 and 1, as used in the binary notation, to relate to the logical functions the computer needs in carrying out its calculations.

Born–Haber cycle A cycle of reactions used for calculating the lattice energies of ionic crystalline solids. For a compound MX, the lattice energy is the enthalpy of the reaction

$$M^+(g) + X^-(g) \rightarrow M^+X^-(s)\ \Delta H_L$$

The standard enthalpy of formation of the ionic solid is the enthalpy of the reaction

$$M(s) + \tfrac{1}{2}X_2(g) \rightarrow M^+X^-(s)\ \Delta H_f$$

The cycle involves equating this enthalpy (which can be measured) to the sum of the enthalpies of a number of steps proceeding from the elements to the ionic solid. The steps are:

(1) Atomization of the metal:

$$M(s) \rightarrow M(g)\ \Delta H_1$$

(2) Atomization of the nonmetal:

$$\tfrac{1}{2}X_2(g) \rightarrow X(g)\ \Delta H_2$$

(3) Ionization of the metal:

$$M(g) \rightarrow M^+(g) + e\ \Delta H_3$$

This is obtained from the ionization potential.

(4) Ionization of the nonmetal:

$$X(g) + e \rightarrow X^-(g)\ \Delta H_4$$

This is the electron affinity.

(5) Formation of the ionic solids:

$$M^+(g) + X^-(g) \rightarrow M^+X^-(s)\ \Delta H_L$$

Equating the enthalpies gives:

$$\Delta H_f = \Delta H_1 + \Delta H_2 + \Delta H_3 + \Delta H_4 + \Delta H_L$$

from which ΔH_L can be found. It is named after the German physicist Max Born (1882–1970) and the chemist Fritz Haber (1868–1934).

boron counter A *counter tube containing a *boron chamber*, used for counting slow neutrons. The boron chamber is lined with boron or a boron compound or is filled with the gas boron trifluoride (BF_3). As natural boron contains about 18% of the isotope boron-10, and as this isotope absorbs neutrons with the emission of an alpha particle, the chamber can be coupled with a Geiger counter to count the alpha particles emitted when neutrons enter the chamber.

Bose–Einstein statistics *See* quantum statistics.

boson An *elementary particle with integral spin; i.e. a particle that conforms to Bose–Einstein statistics (*see* quantum statistics), from which it derives its name. *Compare* fermion.

boundary layer The thin layer of fluid formed around a solid body or surface relative to which the fluid is flowing. Adhesion between the molecules of the fluid and that of the body or surface causes the molecules of the fluid closest to the solid to be stationary relative to it. The transfer of heat or mass between a solid and a fluid flowing over it is largely controlled by the nature of the boundary layer.

Bourdon gauge A pressure gauge consisting essentially of a C-shaped or spiral tube with an oval cross section. One end of the tube is connected to the fluid whose pressure is to be measured and the other end is sealed. As the pressure inside the tube is increased, the oval tube tends to become circular and this causes the tube to straighten. The movement of the end of the tube is transferred by a simple mechanism to a needle moving round a dial or to a digital display. With suitable design, Bourdon gauges can be used for high-pressure measurement and also for low pressures.

Boyle's law The volume (V) of a given mass of gas at a constant temperature is inversely proportional to its pressure (p), i.e. pV = constant. This is true only for an *ideal gas. This law was discovered in 1662 by the Irish physicist Robert Boyle (1627–91). On the continent of Europe it is known as *Mariotte's law* after E. Mariotte (1620–84), who discovered it independently in 1676. *See also* gas laws.

Brackett series *See* hydrogen spectrum.

Bragg's law When a beam of X-rays (wavelength λ) strikes a crystal surface in which the layers of atoms or ions are separated by a distance d, the maximum intensity of the reflected ray occurs when $\sin\theta = n\lambda/2d$, where θ (known as the *Bragg angle*) is the complement of the angle of incidence and n is an integer. The law enables the structure of many crystals to be determined. It was discovered in 1912 by Sir Lawrence Bragg (1890–1971).

brass A group of alloys consisting of copper and zinc. A typical yellow brass might contain about 67% copper and 33% zinc.

breakdown The sudden passage of a current through an insulator. The voltage at which this occurs is the *breakdown voltage.*

breaking stress *See* elasticity.

breeder reactor *See* nuclear reactor.

Bremsstrahlung (German: braking radiation) The X-rays emitted when a charged particle, especially a fast electron, is rapidly slowed down, as when it passes through the electric field around an atomic nucleus. The X-rays cover a whole continuous range of wavelengths down to a minimum value, which depends on the energy of the incident particles. Bremsstrahlung are produced by a metal target when it is bombarded by electrons. *See also* X-ray spectrum.

Brewster's law The extent of the polarization of light reflected from a transparent surface is a maximum when the reflected ray is at right angles to the refracted ray. The angle of incidence (and reflection) at which this maximum polarization occurs is called the *Brewster angle* or *polarizing angle.* For this angle i_B, the condition is that $\tan i_B = n$, where n is the refractive index of the transparent medium. The law was discovered in 1811 by the British physicist David Brewster (1781–1868).

Brinell hardness A scale for measuring the hardness of metals introduced around 1900 by the Swedish metallurgist J. A. Brinell (1849–1925). A small chromium-steel ball is pressed into the surface of the metal by a load of known weight. The ratio of the mass of the load in kilograms to the area of the depression formed in square millimetres is the *Brinell number.*

Britannia metal A silvery alloy consisting of 80–90% tin, 5–15% antimony, and sometimes small percentages of copper, lead, and zinc. It is used in bearings and some domestic articles.

British thermal unit (Btu) The Imperial unit of heat, being originally the heat required to raise the temperature of 1 lb of water by 1°F. 1 Btu is now defined as 1055.06 joules.

bronze Any of a group of alloys of copper and tin, sometimes with lead and zinc present. The amount of tin varies from 1% to 30%. The alloy is hard and easily cast and extensively used in bearings, valves, and other machine parts. Various improved bronzes are produced by adding other elements; for instance, *phosphor bronzes* contain up to 1% phosphorus. In addition certain alloys of copper and metals other than tin are called bronzes – *aluminium bronze* is a mixture of copper and aluminium. Other special bronzes include *bell metal and *gun metal.

Brownian movement The continuous random movement of microscopic solid particles (of about 1 micrometre in diameter) when suspended in a fluid medium. First observed by the botanist Robert Brown (1773–1858) in 1827 when studying pollen particles, it was originally thought to be the manifestation of some vital force. It was later recognized to be a consequence of bombardment of the particles by the continually moving molecules of the liquid. The smaller the particles the more extensive is the motion. The effect is also visible in particles of smoke suspended in a still gas.

brush An electrical contact to a moving commutator on a motor or generator. It is made of a specially prepared form of carbon and is kept in contact with the moving part by means of a spring.

brush discharge A luminous discharge from a conductor that takes the form of luminous branching threads that penetrate into the surrounding gas. It is a form of *corona and it occurs when the electric field near the surface of the conductor exceeds a certain value but is not sufficiently high for a spark to appear.

bubble chamber A device for detecting ionizing radiation. It consists of a chamber containing a liquid, often hydrogen, kept at slightly above its boiling point at its preliminary pressure. Immediately before the passage of the ionizing particles the pressure is reduced, and the particles then act as centres for the formation of bubbles, which can be photographed to obtain a record of the particles' tracks. The device was invented in 1952 by the US physicist D. A. Glaser (1926–). *Compare* cloud chamber.

bulk modulus *See* elastic modulus.

bumping Violent boiling of a liquid caused by superheating so that bubbles form at a pressure above atmospheric pressure. It can be prevented by putting pieces of porous pot in the liquid to enable bubbles of vapour to form at the normal boiling point.

Bunsen burner A laboratory gas burner having a vertical metal tube into which the gas is led, with a hole in the side of the base of the tube to admit air. The amount of air can be regulated by a sleeve on the tube. When no air is admitted the flame is luminous and smoky. With air, it has a faintly visible hot outer part (the oxidizing part) and an inner blue cone where combustion is incomplete (the cooler reducing part of the flame). The device is named after the German chemist Robert Bunsen (1811–99), who used a similar device (without a regulating sleeve) in 1855.

Bunsen cell A *primary cell consisting of a zinc cathode immersed in dilute sulphuric acid and a carbon anode immersed in concentrated nitric acid. The electrolytes are separated by a porous pot. The cell gives an e.m.f. of about 1.9 volts.

buoyancy The upward thrust on a body immersed in a fluid. This force is equal to the weight of the fluid displaced (*see* Archimedes' principle).

byte A subdivision of a *word in a computer, it usually consists of eight *bits.

C

cadmium cell *See* Weston cell.

caesium clock An *atomic clock that depends on the energy difference between two states of the caesium–133 nucleus when it is in a magnetic field. In one type, atoms of caesium–133 are irradiated with *radio-frequency radiation, whose frequency is chosen to correspond to the energy difference between the two states. Some caesium nuclei absorb this radiation and are excited to the higher state. These atoms are deflected by a further magnetic field, which causes them to hit a detector. A signal from this detector is fed back to the radio-frequency oscillator to prevent it drifting from the resonant frequency of 9 192 631 770 hertz. In this way the device is locked to this frequency with an accuracy better than 1 part in 10^{13}. The caesium clock is used in the *SI unit definition of the second.

calculus A series of mathematical techniques developed independently by Isaac Newton (1642–1727) and Gottfried Leibniz (1646–1716). *Differential calculus* treats a continuously varying quantity as if it consisted of an infinitely large number of infinitely small changes. For example, the velocity v of a body at a particular instant can be regarded as the infinitesimal distance, written ds, that it travels in the vanishingly small time interval, dt; the instantaneous velocity v is then ds/dt, which is called the *derivative* of s with respect to t. If s is a known function of t, v at any instant can be calculated by the process of *differentiation. The differential calculus is a powerful technique for solving many problems concerned with rate processes, maxima and minima, and similar problems.

Integral calculus is the opposite technique. For example, if the velocity of a body is a known function of time, the infinitesimal distance ds travelled in the brief instant dt is given by $ds = vdt$. The measurable distance s travelled between two instants t_1 and t_2 can then be found by a process of summation, called *integration, i.e.

$$s = \int_{t_1}^{t_2} v\mathrm{d}t$$

The technique is used for finding areas and volumes and other problems involving the summation of infinitesimals.

californium Symbol Cf. A radioactive metallic transuranic element belonging to the actinoids; a.n. 98; mass number of the most stable isotope 251 (half-life about 700 years). Nine isotopes are known; californium-252 is an intense neutron source, which makes it useful in neutron *activation analysis and potentially useful as a radiation source in medicine. The element was first produced by G. T. Seaborg and associates in 1950.

caloric theory A former theory concerning the nature of heat, which was regarded as a weightless fluid (called *caloric*). It was unable to account for the fact that friction could produce an unlimited quantity of heat and it was abandoned when Joule showed that heat is a form of energy.

calorie The quantity of heat required to raise the temperature of 1 gram of water by 1°C (1 K). The calorie, a c.g.s. unit, is now largely replaced by the *joule, an *SI unit. 1 calorie = 4.186 8 joules.

Calorie (kilogram calorie; kilocalorie) 1000 calories. This unit is still in limited use in estimating the energy value of foods, but is obsolescent.

calorific value The heat per unit mass produced by complete combustion of a given substance. Calorific values are used to express the energy values of fuels; usually these are expressed in megajoules per kilogram ($\mathrm{MJ\,kg^{-1}}$). They are also used to measure the energy content of foodstuffs; i.e. the energy produced when the food is oxidized in the body. The units used in this context are kilojoules per gram ($\mathrm{kJ\,g^{-1}}$), although Calories (kilocalories) are often still used in nontechnical contexts. Calorific values are measured using a *bomb calorimeter.

calorimeter Any of various devices used to measure thermal properties such as *calorific value, *specific heat capacity, *specific latent heat, etc. *See* bomb calorimeter.

calx A metal oxide formed by heating an ore in air.

camera 1. An optical device for obtaining still photographs or for exposing cinematic film. It consists of a light-proof box with a lens at one end and a plate or film at the other. To make an exposure the shutter is opened and an image of the object to be photographed is formed on the light-sensitive film. The length of the exposure is determined by the intensity of light available, the film speed, and the aperture of the lens (*see* f-number). In the simpler cameras the shutter speed and aperture are controlled manually, but in automatic cameras the iris over the lens or the shutter is adjusted on the basis of information provided by a built-in *exposure meter. In ciné cameras the shutter automatically opens as the film comes to rest behind the lens for each frame; the film passes through the camera so that a set number (commonly 16, 18, or

24) of frames are exposed every second. **2.** The part of a television system that converts optical images into electronic signals. It consists of a lens system, which focuses the image to be televised on the photosensitive mosaic of the camera tube, causing localized discharge of those of its elements that are illuminated. This mosaic is scanned from behind by an electron beam so that the beam current is varied as it passes over areas of light and shade. The signal so picked up by the scanning beam is preamplified in the camera and passed to the transmitter with sound and synchronization signals. In *colour television three separate camera tubes are used, one for each *primary colour.

Canada balsam A yellow-tinted resin used for mounting specimens in optical microscopy. It has similar optical properties to glass.

canal rays Streams of positive ions produced in a *discharge tube by boring holes (canals) in the cathode. The positive ions attracted to the cathode pass through the holes and emerge on the other side as positive rays.

candela Symbol Cd. The *SI unit of luminous intensity equal to the luminous intensity in the perpendicular direction of the *black-body radiation from an area of 1/600 000 square metre at the temperature of freezing platinum under a pressure of 101 325 pascals.

candle power Luminous intensity as formerly expressed in terms of the international candle but now expressed in candela.

capacitance The property of a conductor or system of conductors that describes its ability to store electric charge. The capacitance (C) is given by Q/V, where Q is stored charge on one conductor and V the potential difference between the two conductors (or between a single conductor and earth); it is measured in farads.

An isolated sphere has a capacitance of $4\pi\varepsilon r$, where r is the radius and ε the *permittivity of the medium surrounding it. Capacitance is more commonly applied to systems of conductors (or semiconductors) separated by insulators (*see* capacitor).

capacitor An arrangement of conductors separated by an insulator (dielectric) used to store charge or introduce *reactance into an alternating-current circuit. The earliest form was the *Leyden jar. Capacitors used as circuit elements have two conducting plates separated by the dielectric. The dielectric may be air, paper impregnated with oil or wax, plastic film, or ceramic. The simplest form has two parallel rectangular conducting plates (area A) separated by a dielectric (thickness d, permittivity ε). The capacitance of such a capacitor is $A\varepsilon/d$. *Electrolytic capacitors* are devices in which the dielectric is formed by electrolysis; for example, it may be an oxide film on one of the electrodes. In these the electrolysis commonly continues during the charging process and is reversed during discharge.

capacitor microphone A microphone consisting of a *capacitor with a steady voltage applied across its parallel plates. One plate is fixed, the other is a thin diaphragm that is moved by the pressure of the sound waves. The movements of the diaphragm cause a variation in the spacing and therefore in the *capacitance of the device. This variation in capacitance is, in turn, reflected in a similar variation in the charge carried by each plate. The consequent current to and from one plate is carried by a resistor, the varying potential difference across which constitutes the device's output signal. It was formerly known as a *condenser microphone*.

capillarity *See* surface tension.

capillary A tube of small diameter, such as the narrowest type of blood vessel in the vertebrate circulatory system.

capture Any of various processes in which a system of particles absorbs an extra particle. There are several examples in atomic and nuclear physics. For instance, a positive ion may capture an electron to give a neutral atom or molecule. Similarly, a neutral atom or molecule capturing an electron becomes a negative ion. An atomic nucleus may capture a neutron to produce a different (often unstable) nucleus. Another type of nuclear capture is the process in which the nucleus of an atom absorbs an electron from the innermost orbit (the K shell) to transform into a different nucleus. In this process (called *K capture*) the atom is left in an excited state and generally decays by emission of an X-ray photon.

Radiative capture is any such process in which the capture results in an excited state that decays by emission of photons. A common example is neutron capture to yield an excited nucleus, which decays by emission of a gamma ray.

carat 1. A measure of fineness (purity) of gold. Pure gold is described as 24-carat gold. 14-carat gold contains 14 parts in 24 of gold, the remainder usually being copper. **2.** A unit of mass equal to 0.200 gram, used to measure the masses of diamonds and other gemstones.

carbon cycle A series of nuclear reactions in which four hydrogen nuclei combine to form a helium nucleus with the liberation of energy and two positrons. The process is believed to be the source of energy in many stars and to take place in six stages. In this series carbon-12 acts as if it were a catalyst, being reformed at the end of the series:

$$^{12}_{6}C + ^{1}_{1}H \rightarrow ^{13}_{7}N + \gamma$$
$$^{13}_{7}N \rightarrow ^{13}_{6}C + e^{+}$$
$$^{13}_{6}C + ^{1}_{1}H \rightarrow ^{14}_{7}N + \gamma$$
$$^{14}_{7}N + ^{1}_{1}H \rightarrow ^{15}_{8}O + \gamma$$
$$^{15}_{8}O \rightarrow ^{15}_{7}N + e^{+}$$
$$^{15}_{7}N + ^{1}_{1}H \rightarrow ^{12}_{6}C + ^{4}_{2}He.$$

See stellar evolution.

carbon dating (radiocarbon dating) A method of estimating the ages of archaeological specimens of biological origin. As a result of *cosmic radiation a small number of atmospheric nitrogen nuclei are continuously being transformed by neutron bombardment into radioactive nuclei of carbon-14:

$$^{14}_{7}N + n \rightarrow ^{14}_{6}C + p$$

Some of these radiocarbon atoms find their way into living trees and other plants in the form of carbon dioxide, as a result of *photosynthesis. When the tree is cut down photosynthesis stops and the ratio of radiocarbon atoms to stable carbon atoms begins to fall as the radiocarbon decays. The ratio $^{14}C/^{12}C$ in the specimen can be measured and enables the time that has elapsed since the tree was cut down to be calculated. The method has been shown to give consistent results for specimens up to some 40 000 years old, though its accuracy depends upon assumptions concerning the past intensity of the cosmic radiation. The technique was developed by Willard F. Libby (1908-80) and his coworkers in 1946-47.

carbon fibres Fibres of carbon in which the carbon has an oriented crystal structure. Carbon fibres are made by heating textile fibres and are used in strong composite materials for use at high temperatures.

Carnot cycle The most efficient cycle of operations for a reversible *heat engine. Published in 1824 by the French physicist N. L. S. Carnot (1746-1832), it consists of four operations on the working substance in the engine:

a. Isothermal expansion at thermodynamic temperature T_1 with heat Q_1 taken in.

b. Adiabatic expansion with a fall of temperature to T_2.

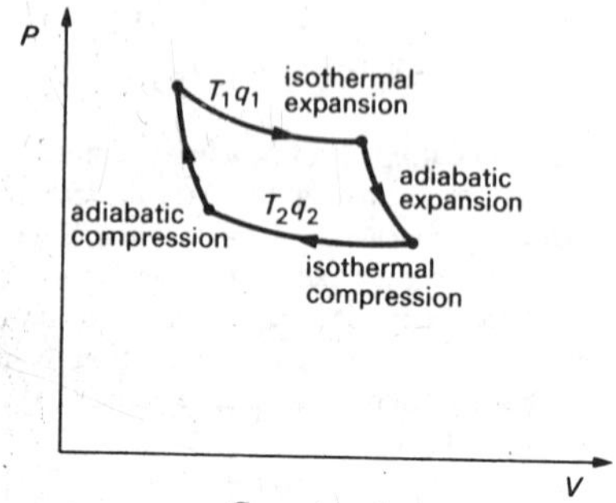

Carnot cycle

c. Isothermal compression at temperature T_2 with heat Q_2 given out.
d. Adiabatic compression with a rise of temperature back to T_1.
According to the *Carnot principle*, the efficiency of any reversible heat engine depends only on the temperature range through which it works, rather than the properties of the working substances. In any reversible engine, the efficiency (η) is the ratio of the work done (W) to the heat input (Q_1), i.e. $\eta = W/Q_1$. As, according to the first law of *thermodynamics, $W = Q_1 - Q_2$, it follows that $\eta = (Q_1 - Q_2)/Q_1$. As the system is reversible, $Q_1/Q_2 = T_1/T_2$, therefore $\eta = (T_1 - T_2)/T_1$. For maximum efficiency T_1 should be as high as possible and T_2 as low as possible.

carrier **1.** *See* carrier wave. **2.** *See* charge carrier.

carrier gas The gas that carries the sample in *gas chromatography.

carrier wave An electromagnetic wave of specified frequency and amplitude that is emitted by a radio transmitter in order to carry information. The information is superimposed onto the carrier by making small changes in either its amplitude or its frequency (*see* modulation).

Cartesian coordinates A system used in analytical geometry to locate a point *P*,

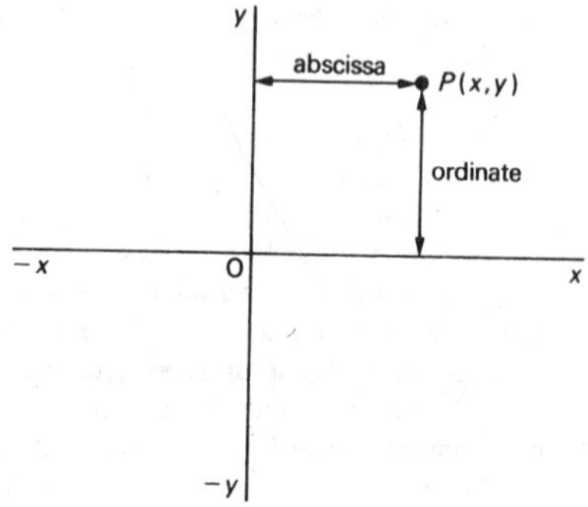

two-dimensional system

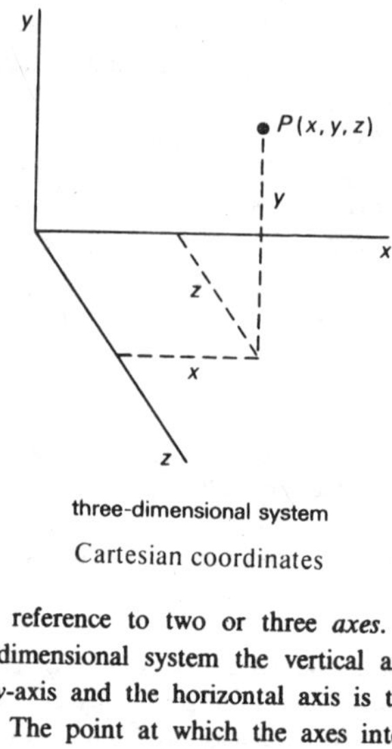

three-dimensional system

Cartesian coordinates

with reference to two or three *axes*. In a two-dimensional system the vertical axis is the *y*-axis and the horizontal axis is the *x*-axis. The point at which the axes intersect each other is called the *origin, O*. Values of $y < O$ fall on the *y*-axis below the origin, values of $x < O$ fall on the *x*-axis to the left of the origin. Any point *P* is located by its perpendicular distances from the two axes. The distance from the *x*-axis is called the *ordinate*; the distance from the *y*-axis is the *abscissa*. The position is indicated numerically by enclosing the values of the abscissa and the ordinate in a bracket and separating the two by means of a comma, e.g. (x,y). The system can also be used to locate a point with reference to a third, z-axis. It is named after René Descartes (1596–1650).

cascade liquefier An apparatus for liquefying a gas of low *critical temperature. Another gas, already below its critical temperature, is liquified and evaporated at a reduced pressure in order to cool the first gas to below its critical temperature. In practice a series of steps is often used, each step enabling the critical temperature of the next gas to be reached.

cascade process Any process that takes place in a number of steps, usually because the single step is too inefficient to produce the desired result. For example, in various uranium-enrichment processes the separation of the desired isotope is only poorly achieved in a single stage; to achieve better separation the process has to be repeated a number of times, in a series, with the enriched fraction of one stage being fed to the succeeding stage for further enrichment. Another example of cascade process is that operating in a *cascade liquefier.

case hardening The hardening of the surface layer of steel, used for tools and certain mechanical components. The commonest method is to carburize the surface layer by heating the metal in a hydrocarbon or by dipping the red hot metal into molten sodium cyanide. Diffusion of nitrogen into the surface layer to form nitrides is also used.

Cassegrainian telescope *See* telescope.

cast iron A group of iron alloys containing 1.8 to 4.5% of carbon. It is usually cast into specific shapes ready for machining, heat treatment, or assembly. It is sometimes produced direct from the *blast furnace or it may be made from remelted *pig iron.

catenary A curve formed when a chain or rope of uniform density hangs from two fixed points. If the lowest point on the curve passes through the origin, the equation is $y = c(\cosh x/c)$, where c is the distance between the x-axis and the directrix.

cathetometer A telescope or microscope fitted with crosswires in the eyepiece and mounted so that it can slide along a graduated scale. Cathetometers are used for accurate measurement of lengths without mechanical contact. The microscope type is often called a *travelling microscope*.

cathode A negative electrode. In *electrolysis cations are attracted to the cathode. In vacuum electronic devices electrons are emitted by the cathode and flow to the *anode. It is therefore from the cathode that electrons flow into these devices. However, in a primary or secondary cell the cathode is the electrode that spontaneously becomes negative during discharge, and from which therefore electrons emerge.

cathode-ray oscilloscope (CRO) An instrument based on the *cathode-ray tube that provides a visual image of electrical signals. The horizontal deflection is usually provided by an internal time base, which causes the beam to sweep across the screen at a specified rate. The signal to be investigated is fed to the vertical deflection plates after amplification. Thus the beam traces a graph of the signal against time.

cathode rays Streams of electrons emitted at the cathode in an evacuated tube containing a cathode and an anode. They were first observed in gas *discharge tubes operated at low pressure. Under favourable conditions electrons produced in the region near the cathode are accelerated down the tube to the anode. In such devices as the *cathode-ray tube the electrons are produced by *thermionic emission from a hot cathode in a vacuum.

CATHODE-RAY TUBE

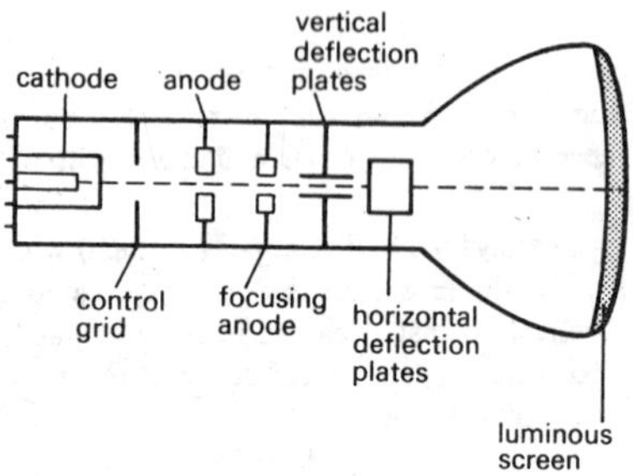

Cathode-ray tube

cathode-ray tube (CRT) The device that provides the viewing screen in the television tube, the radar viewer, and the *cathode-ray oscilloscope. The cathode-ray tube consists of an evacuated tube containing a heated cathode and two or more ring-shaped anodes through which the cathode rays can pass so that they strike the enlarged end of the tube. This end of the tube is coated with fluorescent material so that it provides a screen. Any point on the screen that is struck by the cathode ray becomes luminous. A *control grid between the cathode and the anode enables the intensity of the beam to be varied, thus controlling the brightness of the illumination on the screen. The assembly of cathode, control grid, and anode is called the *electron gun. The beam emerging from the electron gun is focused and deflected by means of plates providing an electric field or coils providing a magnetic field. This enables the beam to be focused to a small point of light and deflected to produce the illusion of an illuminated line as this point sweeps across the tube.

The television tube is a form of cathode-ray tube in which the beam is made to scan the screen 625 times to form a frame, with 25 new frames being produced every second. (These are the figures for standard television tubes in the UK). Each frame creates a picture by variations in the intensity of the beam as it forms each line.

cathodic protection *See* sacrificial protection.

cation A positively charged ion, i.e. an ion that is attracted to the cathode in *electrolysis. *Compare* anion.

cationic dye *See* dyes.

CAT scanner (computerized axial tomography scanner) *See* tomography.

caustic (in optics) The curve or surface formed by the reflection of parallel rays of light in a large-aperture concave mirror. The apex of the caustic lies at the principal focus of the mirror. Such a curve can sometimes be seen on the surface of the liquid in a cup as a result of reflection by the curved walls of the cup. A similar curve is formed by a convex lens with spherical surfaces refracting parallel rays of light.

cavitation The formation of gas- or vapour-filled cavities in liquids in motion when the pressure is reduced to a critical value while the ambient temperature remains constant. If the velocity of the flowing liquid exceeds a certain value, the pressure can be reduced to such an extent that *Bernoulli's theorem breaks down. It is at this point that cavitation occurs, causing a restriction on the speed at which hydraulic machinery can be run without noise, vibration, erosion of metal parts, or loss of efficiency.

cavity resonator *See* resonant cavity.

celestial equator *See* equator.

celestial mechanics The study of the motions of and forces between the celestial bodies. It is based on *Newton's laws of motion and *Newton's law of gravitation. Refinements based on the general theory of *relativity are also included in the study, although the differences between the two theories are only important in a few cases.

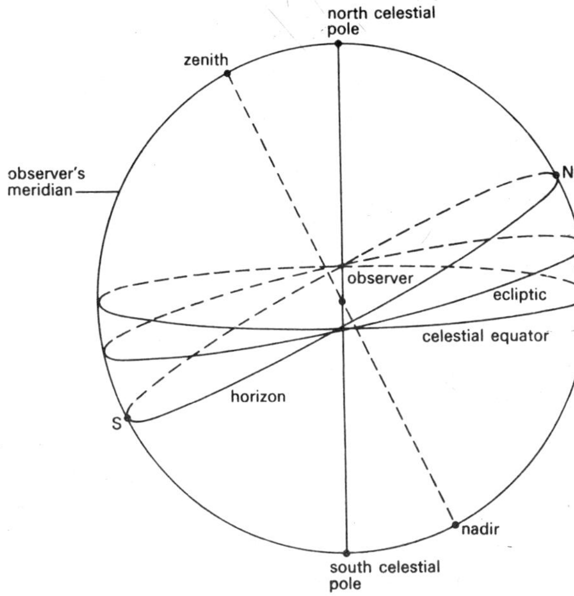

Celestial sphere

celestial sphere The imaginary sphere of infinite radius within which celestial bodies appear to lie. The earth, and the observer, are visualized as being at the centre of the sphere and the sphere as rotating once every sidereal *day. The sphere is used to describe the position of celestial bodies with respect to the earth.

cell 1. A system in which two electrodes are in contact with an electrolyte. The electrodes are metal or carbon plates or rods or, in some cases, liquid metals (e.g. mercury). In an *electrolytic cell a current from an outside source is passed through the electrolyte to produce chemical change (*see* electrolysis). In a *voltaic cell, spontaneous reactions between the electrodes and electrolyte(s) produce a potential difference between the two electrodes.

Voltaic cells can be regarded as made up of two *half cells, each composed of an electrode in contact with an electrolyte. For instance, a zinc rod dipped in zinc sulphate solution is a $Zn|Zn^{2+}$ half cell. In such a system zinc atoms dissolve as zinc ions, leaving a negative charge on the electrode

$$Zn(s) \rightarrow Zn^{2+}(aq) + 2e$$

The solution of zinc continues until the charge build-up is sufficient to prevent further ionization. There is then a potential difference between the zinc rod and its solution. This cannot be measured directly, since measurement would involve making contact with the electrolyte, thereby introducing another half cell (*see* electrode potential). A rod of copper in copper sulphate solution comprises another half cell. In this case the spontaneous reaction is one in which copper ions in solution take electrons from the electrode and are deposited on the electrode as copper atoms. In this case, the copper acquires a positive charge.

The two half cells can be connected by using a porous pot for the liquid junction (as in the *Daniell cell) or by using a salt bridge. The resulting cell can then supply current if the electrodes are connected through an external circuit. The cell is written

$Zn(s)|Zn^{2+}(aq)|Cu^{2+}(aq)|Cu \quad E = 1.10\ V$

Here, E is the e.m.f. of the cell equal to the potential of the right-hand electrode minus that of the left-hand electrode for zero current. Note that 'right' and 'left' refer to the cell as written. Thus, the cell could be written

$$Cu(s)|Cu^{2+}(aq)|Zn^{2+}(aq)|Zn(s) \quad E = -1.10\ V.$$

The overall reaction for the cell is

$$Zn(s) + Cu^{2+}(aq) \rightarrow Cu(s) + Zn^{2+}(aq)$$

This is the direction in which the cell reaction occurs for a positive e.m.f.

The cell above is a simple example of a *chemical cell*; i.e. one in which the e.m.f. is produced by a chemical difference. *Concentration cells* are cells in which the e.m.f. is caused by a difference of concentration. This may be a difference in concentration of the electrolyte in the two half cells. Alternatively, it may be an electrode concentration difference (e.g. different concentrations of metal in an amalgam, or different pressures of gas in two gas electrodes). Cells are also classified into cells *without transport* (having a single electrolyte) and *with transport* (having a liquid junction across which ions are transferred). Various types of voltaic cell exist, used as sources of current, standards of potential, and experimental set-ups for studying electrochemical reactions.

2. *See* photoelectric cell.
3. *See* solar cell.
4. *See* Kerr effect (for Kerr cell).

Celsius scale A *temperature scale in which the fixed points are the temperatures at standard pressure of ice in equilibrium with water (0°C) and water in equilibrium with steam (100°C). The scale, between these two temperatures, is divided in 100 degrees. The degree Celsius (°C) is equal in magnitude to the *kelvin. This scale was formerly known as the *centigrade scale*; the name was officially changed in 1948 to avoid confusion with a hundredth part of a grade. It is named after the Swedish astronomer Anders Celsius (1701–44), who devised the inverted form of this scale (ice point 100°, steam point 0°) in 1742.

cementation Any metallurgical process in which the surface of a metal is impregnated by some other substance, especially an obsolete process for making steel by heating bars of wrought iron to red heat for several days in a bed of charcoal. *See also* case hardening.

cementite *See* steel.

centi- Symbol c. A prefix used in the metric system to denote one hundredth. For example, 0.01 metre = 1 centimetre (cm).

centigrade scale *See* Celsius scale.

central processing unit (CPU) *See* computer.

centre of curvature The centre of the sphere of which a *lens surface or curved *mirror forms a part. The *radius of curvature* is the radius of this sphere.

centre of gravity *See* centre of mass.

centre of mass The point at which the whole mass of a body may be considered to be concentrated. This is the same as the *centre of gravity*, the point at which the whole weight of a body may be considered to act, if the body is situated in a uniform gravitational field.

centrifugal force *See* centripetal force.

centrifugal pump *See* pump.

centrifuge A device in which solid or liquid particles of different densities are separated by rotating them in a tube in a horizontal circle. The denser particles tend to move along the length of the tube to a greater radius of rotation, displacing the lighter particles to the other end.

centripetal force A force acting on a body causing it to move in a circular path. If the mass of the body is *m*, its constant speed *v*, and the radius of the circle *r*, the magnitude of the force is mv^2/r and it is directed towards the centre of the circle. Even though the body is moving with a constant speed *v*, its velocity is changing, because its direction is constantly changing. There is therefore an acceleration v^2/r towards the centre of the circle. For example, when an object is tied to a string and swung in a horizontal circle there is a tension in the string equal to mv^2/r. If the string breaks, this restraining force disappears and the object will move off in a straight line along the tangent to the circle in which it was previously moving.

In the case of a satellite (mass *m*) orbiting the earth (mass *M*), the centripetal force holding the satellite in orbit is the *gravitational force, GmM/d^2, where *G* is the gravitational constant and *d* is the height of the satellite above the centre of the earth. Therefore $GmM/d^2 = mv^2/d$ This equation enables the height of the orbit to be calculated for a given orbital velocity.

Another way of looking at this situation, which was once popular, is to assume that the centripetal force is balanced by an equal and opposite force, acting away from the centre of the circle, called the *centrifugal force*. One could then say that the satellite stays in orbit when the centrifugal force balances the gravitational force. This is, however, a confusing argument because the centrifugal force is fictitious – it does not exist. The gravitational force is not balanced by the centrifugal force: it *is* the centripetal force.

Another example is that of a car rounding a bend. To an observer in the car, a tennis ball lying on the back shelf will roll across the shelf as if it was acted on by an outward centrifugal force. However, to an observer outside the car it can be seen that the ball, because of its almost frictionless contact with the car, is continuing in its straight line motion, uninfluenced by the centripetal force. Occasionally the concept of a centrifugal force can be useful, as long as it is recognized as a fictitious force. A true centrifugal force is exerted, as a *reaction, by the rotating object on whatever is providing its centripetal force.

centroid The point within an area or volume at which the centre of mass would be if the surface or body had a uniform density. For a symmetrical area or volume it coincides with the centre of mass. For a nonsymmetrical area or volume it has to be found by integration.

Cerenkov counter (Cerenkov detector) A type of *counter for detecting and counting high-energy charged particles. The particles pass through a liquid and the light emitted as *Cerenkov radiation is registered by a *photomultiplier tube.

Cerenkov radiation Electromagnetic radiation, usually bluish light, emitted by a beam of high-energy charged particles passing through a transparent medium at a speed greater than the speed of light in that medium. It was discovered in 1934 by the Russian physicist Pavel Cerenkov (1904–). The effect is similar to that of a *sonic boom when an object moves faster than the speed of sound; in this case the radiation is a shock wave set up in the electromagnetic field. Cerenkov radiation is used in the *Cerenkov counter.

cermet A composite material consisting of a ceramic in combination with a sintered metal, used when a high resistance to temperature, corrosion, and abrasion is needed.

CERN (Conseil Européen pour la Recherche Nucléaire) The European Organization for Nuclear Research, which is situated close to Geneva in Switzerland and is supported by twelve European nations. It runs the *super proton synchrotron* (SPS), which has a 7-kilometre underground tunnel, enabling protons to be accelerated to 400 GeV.

cetane number A number that provides a measure of the ignition characteristics of a Diesel fuel when it is burnt in a standard Diesel engine. It is the percentage of cetane (hexadecane) in a mixture of cetane and 1-methylnaphthalene that has the same ignition characteristics as the fuel being tested. *Compare* octane number.

c.g.s. units A system of *units based on the centimetre, gram, and second. Derived from the metric system, it was badly adapted to use with thermal quantities (based on the inconsistently defined *calorie) and with electrical quantities (in which two systems, based respectively on unit permittivity and unit permeability of free space, were used). For scientific purposes c.g.s. units have now been replaced by *SI units.

chain reaction A reaction that is self-sustaining as a result of the products of one step initiating a subsequent step.

In nuclear chain reactions the succession depends on production and capture of neutrons. Thus, one nucleus of the isotope uranium-235 can disintegrate with the production of two or three neutrons, which cause similar fission of adjacent nuclei. These in turn produce more neutrons. If the total amount of material exceeds a *critical mass, the chain reaction may cause an explosion.

Chemical chain reactions usually involve free radicals as intermediates. An example is the reaction of chlorine with hydrogen initiated by ultraviolet radiation. A chlorine molecule is first split into atoms:

$$Cl_2 \rightarrow Cl\cdot + Cl\cdot$$

These react with hydrogen as follows

$$Cl\cdot + H_2 \rightarrow HCl + H\cdot$$

$$H\cdot + Cl_2 \rightarrow HCl + Cl\cdot \text{ etc.}$$

Combustion and explosion reactions involve similar free-radical chain reactions.

change of phase (change of state) A change of matter in one physical *phase (solid, liquid, or gas) into another. The change is invariably accompanied by the evolution or absorption of energy, even if it takes place at constant temperature (*see* latent heat).

channel 1. The region between the source and the drain in a field-effect *transistor. The conductivity of the channel is controlled by the voltage applied to the gate. **2.** A path, or a specified frequency band, along which signals, information, or data flow.

characteristic *See* logarithm.

charge A property of some *elementary particles that gives rise to an interaction between them and consequently to the host of material phenomena described as electrical. Charge occurs in nature in two forms, conventionally described as *positive* and *negative* in order to distinguish between the two kinds of interaction between particles. Two particles that have similar charges (both negative or both positive) interact by repelling each other; two particles that have dissimilar charges (one positive, one negative) interact by attracting each other. The size of the interaction is determined by *Coulomb's law.

The natural unit of negative charge is the charge on an *electron, which is equal but opposite in effect to the positive charge on the proton. Large-scale matter that consists of equal numbers of electrons and protons is electrically neutral. If there is an excess of electrons the body is negatively charged; an excess of protons results in a positive charge. A flow of charged particles, especially a flow of electrons, constitutes an electric current. Charge is measured in coulombs, the charge on an electron being 1.602×10^{-19} coulombs.

charge carrier The entity that transports electric charge in an electric current. The nature of the carrier depends on the type of conductor: in metals, the charge carriers are electrons; in *semiconductors the carriers are electrons (*n*-type) or positive *holes (*p*-type); in gases the carriers are positive ions and electrons; in electrolytes they are positive and negative ions.

charge density **1.** The electric charge per unit volume of a medium or body (*volume charge density*). **2.** The electric charge per unit surface area of a body (*surface charge density*).

Charles' law The volume of a fixed mass of gas at constant pressure expands by a constant fraction of its volume at 0°C for each Celsius degree or kelvin its temperature is raised. For any *ideal gas the fraction is approximately 1/273. This can be expressed by the equation $V = V_0(1 + t/273)$, where V_0 is the volume at 0°C and V is its volume at t°C. This is equivalent to the statement that the volume of a fixed mass of gas at constant pressure is proportional to its thermodynamic temperature, $V = kT$, where k is a constant. The law resulted from experiments begun around 1787 by the French scientist J. A. C. Charles (1746–1823) but was properly established only by the more accurate results published in 1802 by the French scientist Joseph Gay-Lussac (1778–1850). Thus the law is also known as *Gay-Lussac's law*. An equation similar to that given above applies to pressures for ideal gases: $p = p_0(1 + t/273)$, a relationship known as *Charles' law of pressures*. *See also* gas laws.

charm A hypothetical property of certain *elementary particles that is expressed as a *quantum number and is used in the quark hypothesis. It was originally suggested to account for the unusually long lifetime of the *psi particle. In this theory the three original quark–antiquark pairs were supplemented by a fourth pair – the charmed quark and its antiquark. The psi particle itself is a meson having zero charm as it consists of the charmed pair. However, there is some evidence that some charmed *hadrons exist. Charm is thought to be conserved in strong and electromagnetic interactions.

chemical cell *See* cell.

chemical dating An absolute *dating technique that depends on measuring the chemical composition of a specimen. Chemical dating can be used when the specimen is known to undergo slow chemical change at a known rate. For instance, phosphate in buried bones is slowly replaced by fluoride ions from the ground water. Measurement of the proportion of fluorine present gives a rough estimate of the time that the bones have been in the ground. Another, more accurate, method depends on the fact that amino acids in living organisms are L- optical isomers. After death, these racemize and the age of bones can be estimated by measuring the relative amounts of D- and L-amino acids present.

chemical potential Symbol: μ. For a given component in a mixture, the coefficient $\partial G/\partial n$, where G is the Gibbs free energy and n the amount of substance of the component. The chemical potential is the change in Gibbs free energy with respect to change in amount of the component, with pressure, temperature, and amounts of other components being constant. Components are in equilibrium if their chemical potentials are equal.

chemiluminescence *See* luminescence.

chemisorption *See* adsorption.

chip *See* silicon chip.

chirality The property of existing in left- and right-handed structural forms. *See* optical activity.

Chiron A minor planet discovered in 1977. It has an orbit of 50.68 years that, unlike other known minor planets, lies almost entirely outside that of Saturn. Its diameter is uncertain, but appears to be of the order of 300 km.

choke A coil of wire with high inductance and low resistance. It is used in radio circuits to impede the passage of audio-frequency or radio-frequency currents or to smooth the output of a rectifying circuit.

cholesteric crystal *See* liquid crystal.

chromatic aberration *See* aberration.

chromaticity An objective description of the colour quality of a visual stimulus that does not depend on its luminance but which, together with its luminance, completely specifies the colour. The colour quality is defined in terms of *chromaticity coordinates*, x, y, and z, where

$$x = X/(X + Y + Z)$$

$$y = Y/(X + Y + Z)$$

$$\text{and } z = Z/(X + Y + Z)$$

X, Y, and Z are the *tristimulus values* of a light, i.e. they are the amounts of three reference stimuli needed to match exactly the light under consideration in a trichromatic system.

chromium steel Any of a group of *stainless steels containing 8–25% of chromium. A typical chromium steel might contain 18% of chromium, 8% of nickel, and 0.15% of carbon. Chromium steels are highly resistant to corrosion and are used for cutlery, chemical plant, ball bearings, etc.

chromosphere The layer of the *sun's atmosphere immediately above the *photosphere. The chromosphere is normally only visible when the photosphere is totally eclipsed by the moon. The chromosphere is about 10 000 kilometres thick and the temperature in it rises from 4000 K, where it merges with the photosphere, to about 50 000 K, where it reaches the transition region below the *corona.

chronon A hypothetical 'particle' of time defined as the time taken for electromagnetic radiation to cross an electron; i.e. the diameter of an electron divided by the speed of light. This value is approximately 2×10^{-23} second.

circle A closed curve every point on which is a fixed distance (the *radius*) from a point (the *centre*) within the curve. The *diameter* is a line that joins two points on the *circumference* and passes through the centre: the diameter is twice the radius and equal to the circumference divided by 3.141 592, a constant known by the Greek letter π (pi), i.e. $c = \pi d = 2\pi r$. The area of a circle is equal to πr^2. In analytical geometry the equation of a circle, centred at the origin, is $x^2 + y^2 = r^2$.

circular measure A method of measuring angles by treating them as the angle formed by a sector of a circle at the circle's centre. The unit of measure is the *radian*, the angle subtended at the centre of a circle by an arc of equal length to the radius. Since an arc of length r subtends an angle of 1 radian, the whole circumference, length $2\pi r$, will subtend an angle of $2\pi r/r = 2\pi$ radians. Thus, $360° = 2\pi$ radians; 1 radian $= 57.296°$.

circular polarization *See* polarization of light.

circumnutation *See* nutation.

cladding **1.** A thin coating of an expensive metal rolled on to a cheaper one. **2.** A thin covering of a metal around a fuel element in a nuclear reactor to prevent corrosion of the fuel elements by the coolant.

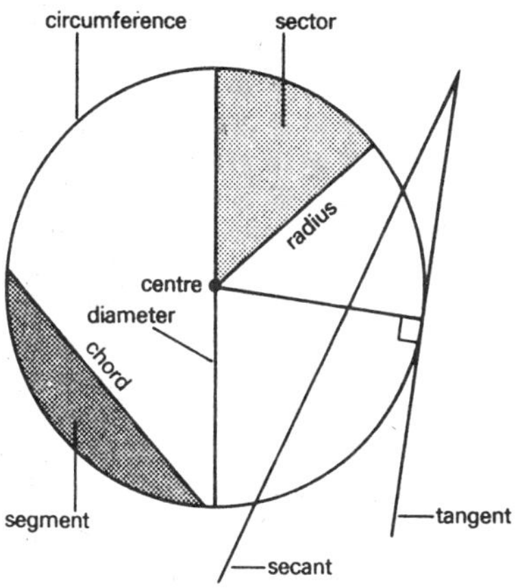

A circle

Clark cell A type of *voltaic cell consisting of an anode made of zinc amalgam and a cathode of mercury both immersed in a saturated solution of zinc sulphate. The Clark cell was formerly used as a standard of e.m.f.; the e.m.f. at 15°C is 1.4345 volts. It is named after the British scientist Hosiah Clark (*d.* 1898).

classical physics Theoretical physics up to approximately the end of the 19th century, before the concepts of *quantum theory (1900) and special *relativity (1905). Classical physics relied largely on Newton's mechanics and James Clerk Maxwell's theory of electromagnetism. It may still be applied with high precision to large-scale phenomena involving no very rapid relative motion.

Claude process A process for liquefying air on a commercial basis. Air under pressure is used as the working substance in a piston engine, where it does external work and cools adiabatically. This cool air is fed to a counter-current *heat exchanger, where it reduces the temperature of the next intake of high-pressure air. The same air is re-compressed and used again, and after several cycles eventually liquefies. The process was perfected in 1902 by the French scientist Georges Claude (1870–1960).

close packing The packing of spheres so as to occupy the minimum amount of space. In a single plane, each sphere is surrounded by six close neighbours in a hexagonal arrangement. The spheres in the second plane fit into depressions in the first layer, and so on. Each sphere has 12 other touching spheres. There are two types of close packing. In *hexagonal close packing* the spheres in the third layer are directly over those in the first, etc., and the arrangement of planes is ABAB.... In *cubic close packing* the spheres in the third layer occupy a different set of depressions than those in the first. The arrangement is ABCABC.... *See also* cubic crystal.

cloud chamber A device for making visible the paths of particles of *ionizing radiation. The *Wilson (expansion) cloud chamber* consists of a container containing air and ethanol vapour, which is cooled suddenly by adiabatic expansion, causing the vapour to become supersaturated. The excess moisture in the vapour is then deposited in drops on the tracks of ions created by the passage of the ionizing radiation. The resulting row of droplets can be photographed. If the original moving particle was being deflected by electric or magnetic fields, the extent of the deflection provides information on its mass and charge. This device was invented in 1911 by C. T. R. Wilson (1869–1961).

A simpler version of this apparatus is the *diffusion cloud chamber*, developed by Cowan, Needels, and Nielsen in 1950, in which supersaturation is achieved by placing a row of felt strips soaked in a suitable alcohol at the top of the chamber. The lower part of the chamber is cooled by solid carbon dioxide. The vapour continuously diffuses downwards, and that in the centre (where it becomes supersaturated) is almost continuously sensitive to the presence of ions created by the radiation.

Clusius column A device for separating isotopes by *thermal diffusion. One form consists of a vertical column some 30 metres high with a heated electric wire running along its axis. The lighter isotopes in a gaseous mixture of isotopes diffuse faster than the heavier isotopes. Heated by the axial wire, and assisted by natural convection, the lighter atoms are carried to the top of the column, where a fraction rich in lighter isotopes can be removed for further enrichment.

cluster A group of stars, which are sufficiently close to each other to be physically associated. In current theories of stellar evolution it is suggested that stars are formed in clusters rather than singly. A *globular cluster* is a roughly spherical group of up to one million stars, which are thought to share a common origin. An *open cluster* is a diffuse group of stars, often only twenty to several hundred strong, that disperses much more easily than a globular cluster.

coagulation The process in which colloidal particles come together to form larger masses. Coagulation can be brought about by adding ions to neutralize the charges stabilizing the colloid. Ions with a high charge are particularly effective (e.g. alum, containing Al^{3+}, is used in styptics to coagulate blood). Another example of ionic coagulation is in the formation of river deltas, which occurs when colloidal silt particles in rivers are coagulated by ions in sea water. Heating is another way of coagulating certain colloids (e.g. boiling an egg coagulates the albumin).

coaxial cable A cable consisting of a central conductor surrounded by an insulator, which is in turn contained in an earthed sheath of another conductor. The central conductor and the outer conductor are coaxial (i.e. have the same axis). They are used to transmit high-frequency signals as they produce no external fields and are not influenced by them.

cobalt steel Any of a group of *alloy steels containing 5–12% of cobalt, 14–20% of tungsten, usually with 4% of chromium and 1–2% of vanadium. They are very hard but somewhat brittle. Their main use is in high-speed tools.

Cockcroft–Walton generator The first proton accelerator; a simple *linear accelerator producing a potential difference of some 800 kV (d.c.) from a circuit of rectifiers and capacitors fed by a lower (a.c.) voltage. The experimenters, Sir John Cockcroft (1897–1967) and E. T. S. Walton (1903–), used this device in 1932 to achieve the first artificially induced nuclear reaction by bombarding lithium with protons to produce helium:

$$^{1}_{1}H + ^{7}_{3}Li = ^{4}_{2}He + ^{4}_{2}He$$

coefficient 1. (in mathematics) A number or other known factor by which a variable quantity is multiplied, e.g. in $ax^2 + bx + c = 0$, a is the coefficient of x^2 and b is the coefficient of x. **2.** (in physics) A measure of a specified property of a particular substance under specified conditions, e.g. the coefficient of *friction of a substance.

coefficient of expansion *See* expansivity.

coefficient of friction *See* friction.

coelostat A device that enables light from the same area of the sky to be continuously reflected into the field of view of an astronomical telescope or other instrument. It consists of a plane mirror driven by a clockwork or electrical mechanism so that it rotates from east to west to compensate for the apparent west-to-east rotation of the *celestial sphere.

coercive force (coercivity) The magnetizing force necessary to reduce the flux density in a magnetic material to zero. *See* hysteresis.

coherent radiation Electromagnetic radiation in which two or more sets of waves have a constant phase relationship, i.e. with peaks and troughs always similarly spaced.

coherent units A system of *units of measurement in which derived units are obtained by multiplying or dividing base units without the use of numerical factors. *SI units form a coherent system; for example the unit of force is the newton, which is equal to 1 kilogram metre per second squared ($kg\,m\,s^{-2}$), the kilogram, metre, and second all being base units of the system.

coincidence circuit An electronic logic device that gives an output only if two input signals are fed to it simultaneously or within a specified time of each other. A *coincidence counter* is an electronic counter incorporating such a device.

cold emission The emission of electrons by a solid without the use of heat (thermal emission), either as a result of field emission (*see* field-emission microscope) or *secondary emission.

collector *See* transistor.

colligative properties Properties that depend on the concentration of particles (molecules, ions, etc.) present in a solution, and not on the nature of the particles. Examples of colligative properties are osmotic pressure (*see* osmosis), *lowering of vapour pressure, *depression of freezing point, and *elevation of boiling point.

collimator 1. Any device for producing a parallel beam of radiation. A common arrangement used for light consists of a convex achromatic lens fitted to one end of a tube with an adjustable slit at the other end, the slit being at the principal focus of the lens. Light rays entering the slit leave the lens as a parallel beam. Collimators for particle beams and other types of electromagnetic radiation utilize a system of slits or apertures. **2.** A small fixed telescope attached to a large astronomical telescope to assist in lining up the large one onto the desired celestial body.

collision density The number of collisions that occur in unit volume in unit time when a given neutron flux passes through matter.

colloids Colloids were originally defined by Thomas Graham in 1861 as substances, such as starch or gelatin, which will not diffuse through a membrane. He distinguished them from *crystalloids* (e.g. inorganic salts), which would pass through membranes. Later it was recognized that colloids were distinguished from true solutions by the presence of particles that were too small to be observed with a normal microscope yet were much larger than normal molecules. Colloids are now regarded as systems in which there are two or more phases, with one (the *dispersed phase*) distributed in the other (the *continuous phase*). Moreover, at least one of the phases has small dimensions (in the range 10^{-9}–10^{-6} m). Colloids are classified in various ways.

Sols are dispersions of small solid particles in a liquid. The particles may be macromolecules or may be clusters of small molecules. *Lyophobic sols* are those in which there is no affinity between the dispersed phase and the liquid. An example is silver chloride dispersed in water. In such colloids the solid particles have a surface charge, which tends to stop them coming together. Lyophobic sols are inherently unstable and in time the particles aggregate and form a precipitate. *Lyophilic sols*, on the other hand, are more like true solutions in which the solute molecules are large and have an affinity for the solvent. Starch in water is an example of such a system. *Association colloids* are systems in which the dispersed phase consists of clusters of molecules that have lyophobic and lyophilic parts. Soap in water is an association colloid (*see* micelle).

Emulsions are colloidal systems in which the dispersed and continuous phases are both liquids, e.g. oil-in-water or water-in-oil. Such systems require an emulsifying agent to stabilize the dispersed particles.

Gels are colloids in which both dispersed and continuous phases have a three-dimensional network throughout the material, so that it forms a jelly-like mass. Gelatin is a common example. One component may sometimes be removed (e.g. by heating) to leave a rigid gel (e.g. silica gel).

Other types of colloid include *aerosols* (dispersions of liquid or solid particles in a gas, as in a mist or smoke) and foams (dispersions of gases in liquids or solids).

cologarithm The logarithm of the reciprocal of a number.

colorimeter Any instrument for comparing or reproducing colours. Monochromatic colorimeters match a *colour with a mixture of monochromatic and white lights. Trichromatic colorimeters use a mixture of three *primary colours.

colour The sensation produced when light of different wavelengths falls on the human eye. Although the visible spectrum covers a continuously varying range of colours from red to violet it is usually split into seven colours (the *visible spectrum*) with the following approximate wavelength ranges:
red 740–620 nm
orange 620–585 nm
yellow 585–575 nm
green 575–500 nm
blue 500–445 nm
indigo 445–425 nm
violet 425–390 nm
A mixture of all these colours in the proportions encountered in daylight gives white light; other colours are produced by varying the proportions or omitting components.

A coloured light has three attributes: its *hue*, depending on its wavelength; its *saturation*, depending on the degree to which it departs from white light; and its *luminosity. Coloured objects that owe their colour to pigments or dyes absorb some components of white light and reflect the rest. For example, a red book seen in white light absorbs all the components except the red, which it reflects. This is called a *subtractive process* as the final colour is that remaining after absorption of the rest. This is the basis of the process used in *colour photography. Combining coloured lights, on the other hand, is an *additive process* and this is the method used in *colour television. *See also* primary colours.

colour charge *See* elementary particles.

colour photography Any of various methods of forming coloured images on film or paper by photographic means. One common process is a subtractive reversal system that utilizes a film with three layers of light-sensitive emulsion, one responding to each of the three *primary colours. On development a black image is formed where the scene is blue. The white areas are dyed yellow, the *complementary colour of blue, and the blackened areas are bleached clean. A yellow filter between this emulsion layer and the next keeps blue light from the second emulsion, which is green-sensitive. This is dyed magenta where no green light has fallen. The final emulsion is red-sensitive and is given a cyan (blue-green) image on the negative after dying. When white light shines through the three dye layers the cyan dye subtracts red where it does not occur in the scene, the magenta subtracts green, and the yellow subtracts blue. The light projected by the negative therefore reconstructs the original scene either as a transparency or for use with printing paper.

colour television A television system in which the camera filters the light from the scene into the three *primary-colour components, red, blue, and green, which are detected by separate camera tubes. The separate information so obtained relating to the colour of the image is combined with the sound and synchronization signals and transmitted using one of three systems, the American, British, or French. At the receiver, the signal is split again into red, blue, and green components, each being fed to a separate *electron gun in the cathode-ray tube of the receiver. By an additive process (*see* colour) the picture is reconstituted by the beam from each gun activating a set of phosphor dots of that colour on the screen.

colour temperature The temperature of a non-black body as indicated by the temperature of a black body having approximately the same spectral distribution.

coma 1. A nebulous cloud of gas and dust that surrounds the nucleus of a *comet. **2.** An *aberration of a lens or mirror in which the image of a point lying off the axis has a comet-shaped appearance.

combinations *See* permutations and combinations.

combustion A chemical reaction in which a substance reacts rapidly with oxygen with the production of heat and light. Such reactions are often free-radical chain reactions. *See also* flame.

comet A small body that travels around the sun in an eccentric orbit. *Short-period comets* have orbital periods of less than 150 years. The others have very long periods, some exceeding 100 000 years. Typical comets have three components: the *nucleus* of ice and dust, the *coma of gas and dust, and the *comet tail*, which only appears when the comet is near the sun (it, too, consists of gas and dust). The nuclei of most comets are thought to be 'dirty snowballs' about one kilometre in diameter, although the solar system has a few comets with nuclei exceeding 10 km in diameter. The coma may be 10^4–10^5 km in diameter, and the tail can be 10^7 km in length. *See also* Halley's comet.

common logarithm *See* logarithm.

communication satellite An unmanned artificial satellite sent by rocket into a geostationary orbit (*see* synchronous orbit) around the earth to enable television broadcasts and telephone communications to be made between points on the earth's surface that could not otherwise communicate by radio owing to the earth's curvature. Modulated *microwaves are transmitted to the satellite, which amplifies them and retransmits them at a different frequency to the receiving station. The satellites are powered by *solar cells. Three or more satellites in equatorial orbits can provide a world-wide communications linkage. The satellites are placed well above the *ionosphere and therefore the carrier waves used have to be in the microwave region of the spectrum in order to pass through the ionosphere.

commutative law The mathematical law stating that the value of an expression is independent of the order of combination of the numbers, symbols, or terms in the expression. The *commutative law* for addition applies if $x + y = y + x$. The commutative law of multiplication applies if $x \times y = y \times x$. Subtraction and division are not commutative. *Compare* associative law; distributive law.

commutator The part of the armature of an electrical motor or generator through which connections are made to external circuits. It consists of a cylindrical assembly of insulated copper conductors, each of which is connected to one point in the armature winding. Spring-loaded carbon brushes are positioned around the commutator to carry the current to or from it.

compass A small magnet pivoted at its central point to revolve in a horizontal plane. In the earth's magnetic field the magnet (called the compass needle) aligns itself so that its north-seeking end points to the earth's magnetic north pole. A scale (called a compass card) is placed below the needle for use in navigation. In some navigation compasses the entire card is pivoted, indicating direction by a fixed mark on the casing. Such compasses are often filled with alcohol to provide damping. Magnetic compasses suffer from being affected by magnetic metals in their vicinity and to a large extent they have been replaced by *gyrocompasses.

complementarity The concept that a single model may not be adequate to explain all the observations made of atomic or subatomic systems in different experiments. For example, *electron diffraction is best explained by assuming that the electron is a wave (*see* de Broglie wavelength), whereas the *photoelectric effect is described by assuming that it is a particle. The idea of two different but complementary concepts to treat quantum phenomena was first put forward by the Danish physicist Niels Bohr (1855–1962) in 1927. *See also* light.

complementary colours A pair of coloured lights of specific hue (*see* colour) that produce the sensation of white when mixed in appropriate intensities. There is an infinite number of such pairs, an example (with wavelengths) is orange (608 nm) and blue (490 nm).

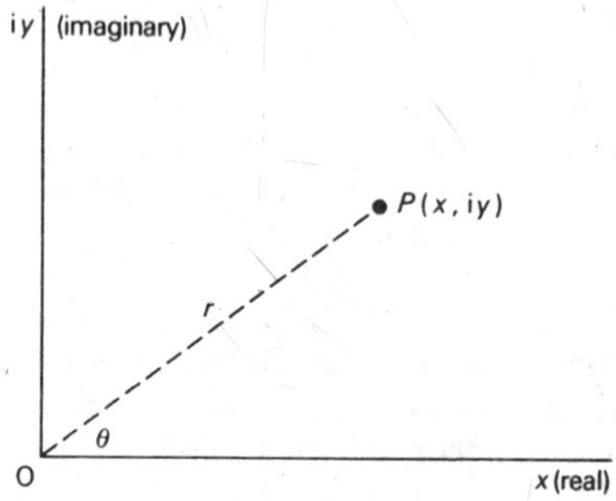

Argand diagram

complex number A number that has a real part, x, and an imaginary part, iy, where $i = \sqrt{-1}$ and x and y are real (x can also equal 0). The complex number therefore has the form $x + iy$, which can also be written in the polar form $r\cos\theta$ + i$r\sin\theta$, where r is the *modulus* and θ is the *argument* (or *amplitude*). A complex number can be represented on an *Argand diagram*, devised by J. R. Argand (1768–1822), in which the horizontal axis represents the real part of the number and the vertical axis the imaginary part. In the polar form the modulus is the line joining the origin to the point representing the complex number and the argument is the angle between the modulus and the x-axis.

component vectors Two or more vectors that produce the same effect as a given vector; the vectors that combine to produce the effect of a resultant vector. A component vector in a given direction is the projection of the given vector (V) along that direction, i.e. $V\cos\theta$, where θ is the angle between the given vector and the direction.

compound microscope *See* microscope.

compressibility The reciprocal of bulk modulus (*see* elastic modulus). The compressibility (k) is given by $-dV/Vdp$, where dV/dp is the rate of change of volume (V) with pressure.

compression ratio The ratio of the total volume enclosed in the cylinder of an *internal-combustion engine at the beginning of the compression stroke to the volume enclosed at the end of the compression stroke. For petrol engines the compression ratio is 8.5–9:1, with a recent tendency to the lower end of the range in order to make use of unleaded petrols. For Diesel engines the compression ratio is in the range 12–18:1.

Compton effect A reduction in the energy of X-rays or gamma-ray photons when they are scattered by free electrons. The phenomenon, first observed in 1923 by the US physicist A. H. Compton (1892–1902), occurs when the photon collides with an electron; some of the photon's energy is transferred to the electron and consequently the photon loses energy $h(\nu_1 - \nu_2)$, where h is the Planck constant and ν_1 and ν_2 are the frequencies before and after collision. As $\nu_1 > \nu_2$, the wavelength of the radiation increases after the collision.

computer An electronic device that processes information according to a set of instructions, called the *program*. The most versatile type of computer is the *digital computer*, in which the input is in the form of characters, represented within the machine in *binary notation. The three basic components of a digital computer are the peripheral input and output devices, the memory, and the *central processing unit* (CPU). The input usually consists of a keyboard, tape drive, or a punched-card or tape reader. The output is usually a printer providing information on paper, a tape drive that provides it on magnetic tape, or a visual-display unit (VDU) providing it on a cathode-ray screen. The memory consists of one or more storage devices, such as magnetic disks, on which both data and programs are stored. The CPU controls the whole system and performs arithmetical and logical operations on the data. Computers range in size from the *microprocessor* with a few thousand logic elements, to the large *mainframe computer* with millions of logic circuits.

The *analog computer* is used in scientific experiments, industrial control, etc. In this type of device the input and output are continuously varying quantities, such as a voltage, rather than the discrete digits of the more commercially useful digital device. *Hybrid computers* combine the properties of both digital and analog devices. Input is usually in analog form, but processing is carried out digitally in a CPU.

Computer *hardware* consists of the actual electronic or mechanical devices used in the system; the *software* consists of the programs and data.

concave Curving inwards. A *concave mirror* is one in which the reflecting surface is formed from the interior surface of a sphere or paraboloid. A *concave lens* has at least one face formed from the interior surface of a sphere. A *biconcave lens* has both faces concave and is therefore thinnest at its centre. The *plano-concave lens* has one plane face and one concave face. The *concavo-convex lens* (also called a *meniscus*) has one concave face and one *convex face. *See* lens.

concavo-convex *See* concave.

concentration cell *See* cell.

conchoidal fracture Fracture of a solid in which the surface of the material is curved and marked by concentric rings. It occurs particularly in amorphous materials.

condensation The change of a vapour or gas into a liquid. The change of phase is accompanied by the evolution of heat (*see* latent heat).

condensation pump *See* diffusion pump.

condensed-matter physics *See* solid-state physics.

condenser 1. A mirror or set of lenses used in optical instruments, such as a microscope or film projector, to concentrate the light diverging from a compact source. A common form consists of two plano-convex lenses with the plane faces pointing outwards. **2.** A device used to cool a vapour to cause it to condense to a liquid. In a steam engine the condenser acts as a reservoir that collects the part of the steam's internal energy that has not been used in doing work on the piston. The cooling water passed through the condenser is warmed and is used as fresh feedwater for the boiler. **3.** *See* capacitor.

condenser microphone *See* capacitor microphone.

conductance The reciprocal of electrical resistance in a direct-current circuit. The ratio of the resistance to the square of the *impedance in an alternating-current circuit. The SI unit is the siemens, formerly called the mho or reciprocal ohm.

conduction 1. (thermal conduction) The transmission of heat through a substance from a region of high temperature to a region of lower temperature. In gases and most liquids, the energy is transmitted mainly by collisions between atoms and molecules with those possessing lower kinetic energy. In solid and liquid metals, heat conduction is predominantly by migration of fast-moving electrons, followed by collisions between these electrons and ions. In solid insulators the absence of *free electrons restricts heat transfer to the vibrations of atoms and molecules within crystal lattices. *See* conductivity. **2. (electrical conduction)** The passage of electric charge through a substance under the influence of an electric field. *See also* charge carrier; energy bands.

conduction band *See* energy bands.

conductivity 1. (thermal conductivity) A measure of the ability of a substance to conduct heat. For a block of material of cross section A, the energy transferred per unit time E/t, between faces a distance, l, apart is given by $E/t = \lambda A(T_2 - T_1)/l$, where λ is the conductivity and T_2 and T_1 are the temperatures of the faces. This equation assumes that the opposite faces are parallel and that there is no heat loss through the sides of the block. The SI unit is therefore $\mathrm{J\,s^{-1}\,m^{-1}\,K^{-1}}$. **2. (electrical conductivity)** The reciprocal of the *resistivity of a material. It is measured in siemens per metre in SI units. When a fluid is involved the electrolytic conductivity is given by the ratio of the current density to the electric field strength.

conductivity water *See* distilled water.

conductor 1. A substance that has a high thermal *conductivity. Metals are good conductors on account of the high concentration of *free electrons they contain. Energy is transmitted through a metal predominantly by means of collisions between electrons and ions. Most nonmetals are poor conductors (good *thermal insulators*) because there are relatively few free electrons. **2.** A substance that has a high electrical conductivity. Again conduction results from the movement of free electrons. *See* energy bands.

cone 1. (in optics) A type of light-sensitive receptor cell, found in the *retinas of all diurnal vertebrates. Cones are specialized to transmit information about colour and they function best in bright light. They are not evenly distributed on the retina, being concentrated in the fovea and absent on the margin of the retina. *Compare* rod. **2.** (in mathematics) A solid figure generated by a line (the *generator*) joining a point on the perimeter of a closed plane curve (the *directrix*) to a point (the *vertex*) outside this plane, as the line moves round the directrix.

If the directrix is a circle, the figure is a *circular cone* standing on a circular *base*. If the line joining the vertex to the centre of the base (the *axis*) is perpendicular to the base the figure is a *right circular cone*, which has a volume $\pi r^2 h/3$, where r is the radius of the base and h is the height of the vertex above the base. If the axis of the cone is not perpendicular to the base, the figure is an *oblique cone*. In general, the volume of any cone is one third of its base area multiplied by the perpendicular distance of the vertex from the base.

configuration **1.** The arrangement of atoms or groups in a molecule. **2.** The arrangement of electrons about the nucleus of an *atom.

conic A figure formed by the intersection of a plane and a *cone. If the intersecting plane is perpendicular to the axis of a right circular cone, the figure formed is a *circle. If the intersecting plane is inclined to the axis at an angle in excess of half the apex angle of the cone it is an *ellipse. If the plane is parallel to the sloping side of the cone, the figure is a *parabola. If the plane cuts both halves of the cone a *hyperbola is formed.

A conic can be defined as a plane curve in which for all points on the curve the ratio of the distance from a fixed point (the *focus*) to the perpendicular distance from a straight line (the *directrix*) is a constant called the *eccentricity* e. For a parabola $e = 1$, for an ellipse $e < 1$, and for a hyperbola $e > 1$.

conjugate points Two points in the vicinity of a *lens or *mirror such that a bright object placed at one will form an image at the other.

conjunction The alignment of two celestial bodies within the solar system so that they have the same longitude as seen from the earth. A planet that orbits between the sun and the earth (Venus and Mercury) is in *superior conjunction* when it is in line with the sun and the earth but on the opposite side of the sun to the earth. It is in *inferior conjunction* when it lies between the earth and the sun. Conjunction may also occur between two planets or a moon and a planet.

conservation law A law stating that the total magnitude of a certain physical property of a system, such as its mass, energy, or charge, remain unchanged even though there may be exchanges of that property between components of the system. For example, imagine a table with a bottle of salt solution (NaCl), a bottle of silver nitrate solution ($AgNO_3$), and a beaker standing on it. The mass of this table and its contents will not change even when some of the contents of the bottles are poured into the beaker. As a result of the reaction between the chemicals two new substances (silver chloride and sodium nitrate) will appear in the beaker:

$$NaCl + AgNO_3 \rightarrow AgCl + NaNO_3,$$

but the total mass of the table and its contents will not change. This *conservation of mass* is a law of wide and general applicability, which is true for the universe as a whole, provided that the universe can be considered a closed system (nothing escaping from it, nothing being added to it). According to Einstein's mass–energy relationship, every quantity of energy (E) has a mass (m), which is given by E/c^2, where c is the speed of light. Therefore if mass is conserved, the law of *conservation of energy* must be of equally wide application. The laws of *conservation* of *linear momentum* and *angular momentum* also are believed to be universally true.

Because no way is known of either creating or destroying electric charge, the law of *conservation of charge* is also a law of universal application. Other quantities are also conserved in reactions between elementary particles.

conservative field A field of force in which the work done in moving a body from one point to another is independent of the path taken. The force required to move the body between these points in a conservative field is called a *conservative force.*

consolute temperature The temperature at which two partially miscible liquids become fully miscible as the temperature is increased.

constant 1. A component of a relationship between variables that does not change its value, e.g. in $y = ax + b$, b is a constant. **2.** A fixed value that has to be added to an indefinite integral. Known as the *constant of integration*, it depends on the limits between which the integration has been performed. **3.** *See* fundamental constants.

constantan An alloy having an electrical resistance that varies only very slightly with temperature (over a limited range around normal room temperatures). It consists of copper (50–60%) and nickel (40–50%) and is used in resistance wire, thermocouples, etc.

constant-boiling mixture *See* azeotrope.

contact potential difference The potential difference that occurs between two electrically connected metals or between the base regions of two semiconductors. If two metals with work functions ϕ_1 and ϕ_2 are brought into contact, their Fermi levels will coincide. If $\phi_1 > \phi_2$ the first metal will acquire a positive surface charge with respect to the other at the area of contact. If the metals are brought close to each other at some other point there will be a potential difference between them.

containment 1. The prevention of the escape of radioactive materials from a *nuclear reactor. **2.** The process of preventing the plasma in a *thermonuclear reactor from touching the walls of the vessel by means of magnetic fields.

continuous function A function $f(x)$ is continuous at $x = a$ if the limit of $f(x)$ as x approaches a is $f(a)$. A function that does not satisfy this condition is said to be a *discontinuous function.*

continuous phase *See* colloid.

continuous spectrum *See* spectrum.

continuous wave A wave that is transmitted continuously rather than in pulses.

continuum A system of axes that form a *frame of reference. The three dimensions of space and the dimension of time together can be taken to form a four-dimensional continuum; this was suggested by Minkowski in connection with special *relativity.

control grid A wire-mesh electrode placed between the cathode and anode in a *thermionic valve or a *cathode-ray tube to control the flow of electrons from one to the other. A fluctuating potential signal fed to the control grid produces at the anode a current signal with similar but amplified fluctuations. It thus forms the basis of the electronic valve amplifier. In a cathode-ray tube the grid controls the intensity of the electron beam and hence the brightness of the image on the screen.

control rod One of a number of rods of a material, such as boron or cadmium, that absorbs neutrons. Control rods can be moved into or out of the core of a *nuclear reactor to control the rate of the reaction taking place within it.

convection A process by which heat is transferred from one part of a fluid to another by movement of the fluid itself. In *natural convection* the movement occurs as a result of gravity; the hot part of the fluid expands, becomes less dense, and is displaced by the colder denser part of the fluid as this drops below it. This is the process that occurs in most domestic hot-water systems between the boiler and the hot-water cylinder. A natural convection current is set up transferring the hot water from the boiler up to the cylinder (always placed above the boiler) so that the cold water from the cylinder can move down into the boiler to be heated. In some modern systems, where small-bore pipes are used or it is inconvenient to place the cylinder above the boiler, the circulation between boiler and hot-water cylinder relies upon a pump. This is an example of *forced convection*, where hot fluid is transferred from one region to another by a pump or fan.

conventional current A 19th-century convention, still in use, that treats any electrical current as a flow of positive charge from a region of positive potential to one of negative potential. The real motion, however, in the case of electrons flowing through a metal conductor, is in the opposite direction, from negative to positive. In semiconductors *hole conduction is in the direction of the conventional current; electron conduction is in the opposite direction.

convergent series A series in which the difference between consecutive terms decreases as the number of terms increases, e.g. $1 + 1/2 + 1/4 + 1/8 + \ldots + 1/2^n$. The sum of such a series approaches a limit as n tends to infinity (the limit is 2 in the example given). If a series is not convergent it is said to be a *divergent series*, in which case the sum does not approach a limit as n tends to infinity. Its sum is plus or minus infinity or it may oscillate. In a divergent series the difference between consecutive terms is constant or increases as n increases.

converging lens or mirror A lens or mirror that can refract or reflect a parallel beam of light so that it converges at a point (the principal focus). Such a mirror is concave; a converging lens is thicker at its centre than at its edges (i.e. it is biconvex, plano-convex, or convexo-concave). *Compare* diverging lens or mirror.

conversion electron *See* internal conversion.

converter 1. An electrical machine for converting alternating current into direct current, or less frequently, vice versa. **2.** The reaction vessel in the *Bessemer process or some similar steel-making process. **3.** A computer device for converting information coded in one form into some other form.

converter reactor A *nuclear reactor that converts fertile material (e.g. thorium–232) into *fissile material (e.g. uranium–233). A converter reactor can also be used to produce electrical power.

convex Curving outwards. A *convex mirror* is one in which the reflecting surface is formed from the exterior surface of a sphere or paraboloid. A *convex lens* has at least one face formed from the exterior surface of a sphere. A *biconvex lens* has both faces convex and is therefore thickest at its centre. The *plano-convex lens* has one plane face and one convex face. The *convexo-concave lens* (also called a *meniscus*) has one convex face and one *concave face. *See* lens.

coolant A fluid used to remove heat from a system by *convection (usually forced), either to control the temperature or to extract energy. In a water-cooled car engine the coolant is water (or water and antifreeze), which is pumped around the engine and cooled in the radiator. In a *nuclear reactor the coolant is used to transfer the heat of the reaction from the core to a heat exchanger or to the steam-raising plant. In gas-cooled reactors the coolant is usually carbon dioxide. Pressurized water or boiling water is used as both coolant and *moderator in several types of reactor. In fast reactors, liquid sodium is used as the coolant.

Cooper pairs *See* superconductivity.

coordinate *See* Cartesian coordinates; polar coordinates.

coordinate geometry *See* analytical geometry.

coordination number The number of groups, molecules, atoms, or ions surrounding a given atom or ion in a complex or crystal. For instance, in a square-planar complex the central ion has a coordination number of four. In a close-packed crystal (*see* close packing) the coordination number is twelve.

Copernican astronomy The system of astronomy proposed by the Polish canon Nicolaus Copernicus (1473–1543) in his book *De revolutionibus orbium coelestium*, which was published in the month of his death and first seen by him on his deathbed. It used some elements of *Ptolemaic astronomy, but rejected the notion, then current, that the earth was a stationary body at the centre of the universe. Instead, Copernicus proposed the apparently unlikely concept that the sun was at the centre of the universe and that the earth was hurtling through space in a circular orbit about it. Galileo's attempts, some 70 years later, to convince the Catholic church that in spite of scriptural authority to the contrary, the Copernican system was correct, resulted in *De revolutionibus* being placed on the Index of forbidden books, where it remained until 1835.

core 1. A rod or frame of magnetic material that increases the inductance of a coil through which it passes. Cores are used in transformers, electromagnets, and the rotors and stators of electrical machines. It may consist of laminated metal, ferrite, or compressed ferromagnetic particles in a matrix of an insulating binder (*dust core*). **2.** The inner part of a *nuclear reactor in which the nuclear reaction takes place. **3.** The devices that make up the memory in certain types of computer. **4.** The central region of a star or planet.

Coriolis force A fictitious force sometimes used to simplify calculations involving rotating systems, such as the movement of air, water, and projectiles over the surface of the rotating earth. The concept was first used in 1835 by Gaspard de Coriolis (1792–1843), a French physicist. The daily rotation of the earth means that in 24 hours a point on its equator moves a distance of some 40 000 kilometres, giving it a tangential velocity of about 1670 kilometres per hour. A point at the latitude of, say, Rome, travels a shorter distance in the same time and therefore has a lower tangential velocity – about 1340 km/hr. Air over the equator has the full tangential velocity of 1670 km/hr and as it travels north, say, it will retain this velocity; to an observer outside the earth this would be clear. However, to an observer in Rome it appears to be moving eastwards, because the earth at that point is moving eastwards more slowly than the air. The Coriolis force (which is quite fictitious) is the force that a naive observer thinks is needed to push the air eastwards.

corona 1. The outer part of the sun's atmosphere. Its two main components are the K-corona (or inner corona), with a temperature of about 2×10^6 K at a height of some 75 000 km, and the F-corona (or outer corona), which is considerably cooler and extends for several million kilometres into space. **2.** A glowing region of the air surrounding a conductor when the potential gradient near it exceeds a critical value. It is caused by ionization of the air and may be accompanied by hissing sounds. *Corona dis-*

charge (or *point discharge*) occurs at sharp points where the surface charge density is high by the attraction, charging, and consequent repulsion of air molecules.

corpuscular theory *See* light.

corrosion Chemical or electrochemical attack on the surface of a metal. *See also* electrolytic corrosion; rusting.

cosine rule In any triangle, with sides of length a, b, and c, $c^2 = a^2 + b^2 - 2ab\cos\theta$, where θ is the angle between sides a and b.

cosmic radiation High-energy particles that fall on the earth from space. *Primary cosmic rays* consist of nuclei of the most abundant elements, with *protons (hydrogen nuclei) forming by far the highest proportion; electrons, positrons, neutrinos, and gamma-ray photons are also present. The particle energies range from 10^{-11} J to 10 J (10^8 to 10^{20} eV) and as they enter the earth's atmosphere they collide with oxygen and nitrogen nuclei producing *secondary cosmic rays*. The secondary rays consist of elementary particles and gamma-ray photons. A single high-energy primary particle can produce a large *shower* of secondary particles. The sources of the primary radiation are not all known, although the sun is believed to be the principal source of particles with energies up to about 10^{10} eV. It is believed that all particles with energies of less than 10^{18} eV originate within the Galaxy.

cosmology The study of the nature, origin, and evolution of the universe. Various theories concerning the origin and evolution of the universe exist. *See* big-bang theory; steady-state theory.

Cottrell precipitator An electrostatic precipitator used to remove dust particles from industrial waste gases, by attracting them to charged grids or wires.

coudé system *See* telescope.

coulomb Symbol C. The *SI unit of electric charge. It is equal to the charge transferred by a current of one ampere in one second. The unit is named after Charles de Coulomb (1736–1806), a French physicist.

Coulomb's law The force (sometimes called the *Coulomb force*) between two charged particles, regarded as point charges Q_1 and Q_2 a distance d apart, is proportional to the product of the charges and inversely proportional to the square of the distance between them. The law is now usually stated in the form $F = Q_1Q_2/4\pi\varepsilon d^2$, where ε is the absolute *permittivity of the intervening medium. $\varepsilon = \varepsilon_r\varepsilon_0$, where ε_r is the relative permittivity (the dielectric constant) and ε_0 is the electric constant. The electric field surrounding a point charge is called the *Coulomb field* and the scattering of charged particles by the Coulomb field surrounding an atomic nucleus is called *Coulomb scattering*. The law was first published by Charles de Coulomb in 1785.

counter Any device for detecting and counting objects or events, often incident charged particles or photons. The latter devices usually work by allowing the particle to cause ionization, which creates a current or voltage pulse. The pulses are then counted electronically. *See* Cerenkov counter; crystal counter; Geiger counter; proportional counter; scintillation counter; semiconductor counter; spark counter. These names are often applied merely to the actual detectors; the ancillary counting mechanism is then called a *scaler.

couple Two equal and opposite parallel forces applied to the same body that do not act in the same line. The forces create a torque, the *moment of which is equal to the product of the force and the perpendicular distance between them.

creep The continuous deformation of a solid material, usually a metal, under a constant stress that is well below its *yield point. It usually only occurs at high temperatures and the creep characteristics of any material destined to be used under conditions of high stress at high temperatures must be investigated.

critical angle *See* total internal reflection.

critical damping *See* damping.

critical mass The minimum mass of fissile material that will sustain a nuclear *chain reaction. For example, when a nucleus of uranium–235 disintegrates two or three neutrons are released in the process, each of which is capable of causing another nucleus to disintegrate, so creating a chain reaction. However, in a mass of U–235 less than the critical mass, too many neutrons escape from the surface of the material for the chain reaction to proceed. In the atom bomb, therefore, two or more subcritical masses have to be brought together to make a mass in excess of the critical mass before the bomb will explode.

critical pressure The pressure of a fluid in its *critical state; i.e. when it is at its critical temperature and critical volume.

critical reaction A nuclear *chain reaction in which, on average, one transformation causes exactly one other transformation so that the chain reaction is self-sustaining. If the average number of transformations caused by one transformation falls below one, the reaction is *subcritical* and the chain reaction ceases; if it exceeds one the reaction is *supercritical* and proceeds explosively.

critical state The state of a fluid in which the liquid and gas phases both have the same density. The fluid is then at its *critical temperature, *critical pressure, and *critical volume.

critical temperature **1.** The temperature above which a gas cannot be liquefied by an increase of pressure. *See also* critical state. **2.** *See* transition temperature.

critical volume The volume of a fixed mass of a fluid in its *critical state; i.e. when it is at its critical temperature and critical pressure. The *critical specific volume* is its volume per unit mass in this state: in the past this has often been called the critical volume.

CRO *See* cathode-ray oscilloscope.

cross section **1.** A plane surface formed by cutting a solid, especially by cutting at right angles to its longest axis. **2.** The area of such a surface. **3.** A measure of the probability that a collision will occur between a beam of radiation and a particular particle, expressed as the effective area presented by the particle in that particular process. It is measured in square metres or *barns.

crucible A dish or other vessel in which substances can be heated to a high temperature.

cryogenic pump A *vacuum pump in which pressure is reduced by condensing gases on surfaces maintained at about 20 K by means of liquid hydrogen or at 4 K by means of liquid helium. Pressures down to 10^{-8} mmHg (10^{-6} Pa) can be maintained; if they are used in conjunction with a *diffusion pump, pressures as low as 10^{-15} mmHg (10^{-13} Pa) can be reached.

cryogenics The study of very low temperatures and the techniques for producing them. Objects are most simply cooled by placing them in a bath containing liquefied gas maintained at a constant pressure. In general, a liquefied gas can provide a constant bath temperature from its triple point to its critical temperature and the bath temperature can be varied by changing the pressure above the liquid. The lowest practical temperature for a liquid bath is 0.3 K. Refrigerators (*see* refrigeration) consist essentially of devices operating on a repeated cycle, in which a low-temperature reservoir is a continuously replenished liquid bath. Above 1 K they work by compressing and expanding suitable gases. Below this temperature liquids or solids are used and by *adiabatic demagnetization it is possible to reach 10^{-6} K.

cryohydrate A eutectic mixture of ice and some other substance (e.g. an ionic salt) obtained by freezing a solution.

cryometer A thermometer designed to measure low temperatures. *Thermocouples can be used down to about 1 K and *resistance thermometers can be used at 0.01 K. Below this magnetic thermometers (0.001 K) and nuclear-resonance thermometers (3×10^{-7} K) are required.

cryoscopic constant *See* depression of freezing point.

cryostat A vessel enabling a sample to be maintained at a very low temperature. The *Dewar flask is the most satisfactory vessel for controlling heat leaking in by radiation, conduction, or convection. Cryostats usually consist of two or more Dewar flasks nesting in each other. For example, a liquid nitrogen bath is often used to cool a Dewar flask containing a liquid helium bath.

cryotron A switch that relies on *superconductivity. It consists of a coil of wire of one superconducting material surrounding a straight wire of another superconducting material; both are immersed in a liquid-helium bath. A current passed through the coil creates a magnetic field, which alters the superconducting properties of the central wire, switching its resistance from zero to a finite value. Cryotron switches can be made very small and take very little current.

crystal A solid with a regular polyhedral shape. All crystals of the same substance grow so that they have the same angles between their faces. However, they may not have the same external appearance because different faces can grow at different rates, depending on the conditions. The external form of the crystal is referred to as the *crystal habit*. The atoms, ions, or molecules forming the crystal have a regular arrangement and this is the *crystal structure*.

crystal counter A type of solid-state *counter in which a potential difference is applied across a crystal; when the crystal is struck by an elementary particle or photon, the electron–ion pairs created cause a transient increase in conductivity. The resulting current pulses are counted electronically.

crystal habit *See* crystal.

crystal lattice The regular pattern of atoms, ions, or molecules in a crystalline substance. A crystal lattice can be regarded as produced by repeated translations of a *unit cell* of the lattice. *See also* crystal system.

crystallography The study of crystal form and structure. *See also* X-ray crystallography.

crystalloids *See* colloids.

crystal microphone A microphone in which the sound waves fall on a plate of Rochelle salt or similar material with piezoelectric properties, the variation in pressure being converted into a varying electric field by the *piezoelectric effect. Crystal microphones have strong high-frequency response and are nondirectional; they are now rarely used except when their cheapness in important.

crystal oscillator An oscillator of precisely determined frequency in which a piezoelectric crystal is used to produce the oscillation in a tuned circuit or is coupled to a tuned circuit to control its frequency. An alternating electric field applied to two metallic films sputtered onto the parallel faces of a crystal, usually of quartz, causes it to vibrate at its natural frequency; this frequency can be in the kilohertz or megahertz range, depending on how the crystal is cut. The mechanical vibrations in turn create an alternating electric field across the crystal that does not suffer from frequency drift. The device is widely used in *quartz clocks and watches.

crystal pick-up A pick-up in a record player in which the mechanical vibrations produced by undulations in the record groove are transmitted to a piezoelectric crystal, which produces a varying electric field of the same frequency as the sound. This signal is amplified and fed to loudspeakers in order to recreate the sound.

crystal structure *See* crystal.

crystal system A method of classifying crystalline substances on the basis of their unit cell. There are seven crystal systems. If the cell is a parallelopiped with sides a, b, and c and if α is the angle between b and c, β the angle between a and c, and γ the angle between a and b, the systems are:

(1) *cubic* $a=b=c$ and $\alpha=\beta=\gamma=90°$
(2) *tetragonal* $a=b=c$ and $\alpha=\beta=\gamma=90°$
(3) *rhombic* (or *orthorhombic*) $a=b=c$ and $\alpha=\beta=\gamma=90°$
(4) *hexagonal* $a=b=c$ and $\alpha=\beta=\gamma=90°$
(5) *trigonal* $a=b=c$ and $\alpha=\beta=\gamma=90°$
(6) *monoclinic* $a=b=c$ and $\alpha=\gamma=90°=\beta$
(7) *triclinic* $a=b=c$ and $\alpha=\beta=\gamma$

cubic crystal A crystal in which the unit cell is a cube (*see* crystal system). There are three possible packings for cubic crystals: *simple cubic*, *face-centred cubic*, and *body-centred cubic*. See illustration.

cubic equation An equation in which the highest power of the variable is three. It has the general form $ax^3 + bx^2 + cx + d = 0$ and, in general, is satisfied by three values of x.

cubic expansivity *See* expansivity.

curie The former unit of *activity (*see* radiation units). It is named after the Polish-born French physicist Marie Curie (1867–1934).

Curie point (Curie temperature) The temperature at which a ferromagnetic substance loses its ferromagnetism and becomes only paramagnetic. For iron the Curie point is 760°C and for nickel 356°C.

Curie's law The susceptibility (χ) of a paramagnetic substance is proportional to the thermodynamic temperature (T), i.e. $\chi = C/T$, where C is the Curie constant. A modification of this law, the *Curie–Weiss law*, is more generally applicable. It states that $\chi = C/(T - \theta)$, where θ is the Weiss constant, a characteristic of the material. The law was first proposed by the French physicist Pierre Curie (1859–1906) and modified by another French physicist, Pierre-Ernest Weiss (1865–1940).

curium Symbol Cm. A radioactive metallic transuranic element belonging to the actinoids; a.n. 96; mass number of the most stable isotope 247 (half-life 1.64×10^7 years); r.d. (calculated) 13.51; m.p. 1340±40°C. There are nine known isotopes. The element was first identified by G. T. Seaborg and associates in 1944 and first produced by L. B. Werner and I. Perlman in 1947 by bombarding americium-241 with neutrons.

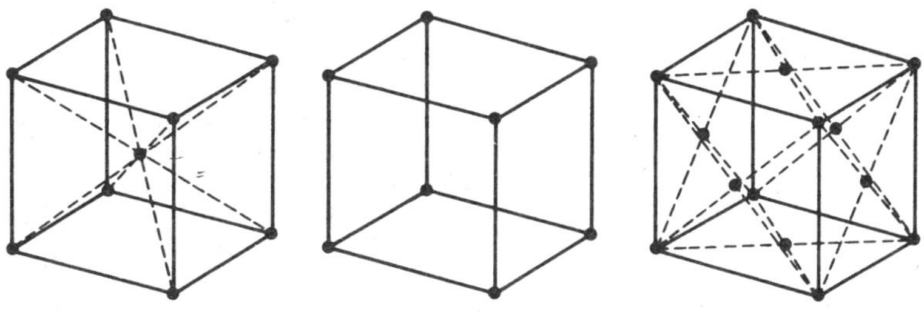

Cubic crystal structures

current Symbol *I*. A flow of electric charge through a conductor. The current at a particular cross section is the rate of flow of charge. The charge may be carried by electrons, ions, or positive holes (*see* charge carrier). The unit of current is the ampere. *See also* conventional current.

current balance An instrument used to measure a current absolutely, on the basis of the definition of the ampere. An accurate form consists of a beam balance with similar coils attached to the ends of the balance arms. Fixed coils are situated above and below these two coils. The six coils are then connected in series so that a current passing through them creates a torque on the beam, which is restored to the horizontal by means of a rider. From the position and weight of the rider, and the geometry of the system, the current can be calculated.

current density **1.** The current flowing through a conductor per unit cross-sectional area, measured in amperes per square metre. **2.** The current flowing through an electrolyte per unit area of electrode.

cusp A point at which two arcs of a curve intersect.

cycle A regularly repeated set of changes to a system that brings back all its parameters to their original values once in every set of changes. The duration of one cycle is called its *period and the rate of repetition of cycle, called the *frequency, is measured in *hertz. *See* simple harmonic motion.

cycloid The curve traced by a point on the circumference of a circle as it rolls without slipping along a straight line. The length of the arc formed by one revolution of the circle is $8r$, where r is the radius of the circle. The horizontal distance between cusps is $2\pi r$.

cyclotron A cyclic particle *accelerator in which charged particles fed into the centre of the device are accelerated in an outward spiral path inside two hollow D-shaped conductors placed to form a split circle. A magnetic field is applied at right-angles to the plane of the dees and an alternating potential difference is applied between them. The frequency of the alternating p.d. is arranged so that the particles are accelerated each time they reach the evacuated gap between the dees. The magnetic field makes them follow curved paths. After several thousand revolutions inside the dees the particles reach the perimeter of the dees, where a deflecting field directs them onto the target. In this device protons can achieve an energy of 10^{-12} J (10 MeV). The first working cyclotron was produced in 1931 by the US physicist E. O. Lawrence (1901–58). *See also* synchrocyclotron.

cylindrical polar coordinates *See* polar coordinates.

D

dalton *See* atomic mass unit.

Dalton's atomic theory A theory of chemical combination, first stated by the British chemist John Dalton (1766–1844) in 1803. It involves the following postulates:

(1) Elements consist of indivisible small particles (atoms).

(2) All atoms of the same element are identical; different elements have different types of atom.

(3) Atoms can neither be created nor destroyed.

(4) 'Compound elements' (i.e. compounds) are formed when atoms of different elements join in simple ratios to form 'compound atoms' (i.e. molecules).

Dalton also proposed symbols for atoms of different elements (later replaced by the present notation using letters).

Dalton's law The total pressure of a mixture of gases or vapours is equal to the sum of the partial pressures of its components, i.e. the sum of the pressures that each component would exert if it were present alone and occupied the same volume as the mixture of gases. Strictly speaking, the principle is true only for ideal gases.

damping A decrease in the amplitude of an oscillation as a result of energy being drained from the oscillating system to overcome frictional or other resistive forces. For example, a pendulum soon comes to rest unless it is supplied with energy from an outside source; in a pendulum clock, energy is supplied through an *escapement from a wound spring or a falling mass to compensate for the energy lost through friction. Damping is introduced intentionally in measuring instruments of various kinds to overcome the problem of taking a reading from an oscillating needle. A measuring instrument is said to be *critically damped* if the system just fails to oscillate and the system comes to rest in the shortest possible time. If it is *underdamped* it will oscillate repeatedly before coming to rest; if it is *overdamped* it will not oscillate but it will take longer to come to rest than it would if it was critically damped. An instrument, such as a galvanometer, that is critically damped is often called a *deadbeat* instrument.

database A large collection of information that has been coded and stored in a computer in such a way that it can be extracted under a number of different category headings.

dating techniques Methods of estimating the age of rocks, palaeontological specimens, archaeological sites, etc. *Relative dating techniques* date specimens in relation to one another; for example, stratigraphy is used to establish the succession of fossils. *Absolute* (or *chronometric*) *techniques* give an absolute estimate of the age and fall into two main groups. The first depends on the existence of something that develops at a seasonally varying rate, as in dendrochronology and varve dating. The other uses some measurable change that occurs at a known rate, as in *chemical dating, *radioactive* (or *radiometric*) *dating* (*see* carbon dating; fission-track dating; potassium–argon dating; rubidium–strontium dating; uranium–lead dating), and *thermoluminescence.

daughter 1. A nuclide produced by radioactive *decay of some other nuclide (the *parent*). **2.** An ion or free radical produced by dissociation or reaction of some other (*parent*) ion or radical.

day The time taken for the earth to complete one revolution on its axis. The *solar day* is the interval between two successive returns of the sun to the *meridian. The *mean solar day* of 24 hours is the average value of the solar day for one year. The *sidereal day* is measured with respect to the fixed stars and is 4.09 minutes shorter than the mean solar day as a result of the imposition of the earth's orbital motion on its rotational motion.

d.c. *See* direct current.

deadbeat *See* damping.

de Broglie wavelength The wavelength of the wave associated with a moving particle. The wavelength (λ) is given by $\lambda = h/mv$, where h is the Planck constant, m is the mass of the particle, and v its velocity. The *de Broglie wave* was first suggested by the French physicist Louis de Broglie (1892–) in 1924 on the grounds that electromagnetic waves can be treated as particles (*photons) and one could therefore expect particles to behave in some circumstances like waves (*see* complementarity). The subsequent observation of *electron diffraction substantiated this argument and the de Broglie wave became the basis of *wave mechanics.

debye A unit of electric *dipole moment in the electrostatic system, used to express dipole moments of molecules. It is the dipole moment produced by two charges of opposite sign, each of 1 statcoulomb and placed 10^{-18} cm apart, and has the value 3.335 64 $\times 10^{-30}$ coulomb metre.

Debye–Huckel theory A theory to explain the nonideal behaviour of electrolytes, published in 1923 by Peter Debye (1884–1966) and Erich Huckel (1896–). It assumes that electrolytes in solution are fully dissociated and that nonideal behaviour arises because of electrostatic interactions between the ions. The theory shows how to calculate the extra free energy per ion resulting from such interactions, and consequently the activity coefficient. It gives a good description of nonideal electrolyte behaviour for very dilute solutions, but cannot be used for more concentrated electrolytes.

deca- Symbol da. A prefix used in the metric system to denote ten times. For example, 10 coulombs = 1 decacoulomb (daC).

decay The spontaneous transformation of one radioactive nuclide into a daughter nuclide, which may be radioactive or may not, with the emission of one or more particles or photons. The decay of N_0 nuclides to give N nuclides after time t is given by $N = N_0 \exp(-\gamma t)$, where γ is called the *decay constant* or the *disintegration constant*. The reciprocal of the decay constant is the *mean life*. The time required for half the original nuclides to decay (i.e. $N = \frac{1}{2}N_0$) is called the *half-life* of the nuclide. The same terms are applied to elementary particles that spontaneously transform into other particles. For example, a free neutron decays into a proton and an electron (*see* beta decay). *See also* alpha particle.

deci- Symbol d. A prefix used in the metric system to denote one tenth. For example, 0.1 coulomb = 1 decicoulomb (dC).

decibel A unit used to compare two power levels, usually applied to sound or electrical signals. Although the decibel is one tenth of a *bel*, it is the decibel, not the bel, that is invariably used. Two power levels P and P_0 differ by n decibels when $n = 10\log_{10} P/P_0$. If P is the level of sound intensity to be measured, P_0 is a reference level, usually the intensity of a note of the same frequency at the threshold of audibility.

The logarithmic scale is convenient as human audibility has a range of 1 (just audible) to 10^{12} (just causing pain) and one decibel, representing an increase of some

26%, is about the smallest change the ear can detect.

decimal system A number system based on the number 10; the number system in common use. All rational numbers can be written as a *finite decimal* (e.g. $\frac{1}{4} = 0.25$) or a *repeating decimal* (e.g. $5/27 = 0.185\,185\,185\ldots$). An *irrational number can be written to any number of decimal places, but can never be given exactly (e.g. $\sqrt{3} = 1.732\,050\,8\ldots$).

declination 1. The angle between the magnetic meridian and the geographic meridian at a point on the surface of the earth. *See* geomagnetism. **2.** The angular distance of a celestial body north (positive) or south (negative) of the celestial *equator.

defect 1. A discontinuity in a crystal lattice. A *point defect* consists either of a missing atom or ion creating a *vacancy* in the lattice (a vacancy is sometimes called a *Schottky defect*) or an extra atom or ion between two normal lattice points creating an *interstitial*. A *Frenkel defect* consists of a vacancy in which the missing atom or ion has moved to an interstitial position. If more than one adjacent point defect occurs in a crystal there may be a slip along a surface causing a *line defect* (or *dislocation*). Defects are caused by strain or, in some cases, by irradiation. All crystalline solids contain an equilibrium number of point defects above absolute zero; this number increases with temperature. The existence of defects in crystals is important in the conducting properties of *semiconductors. **2.** *See* mass defect.

definite integral *See* integration.

degaussing The process of neutralizing the magnetization in an object that has inadvertently become magnetized. For example, ferromagnetic components of TV sets may become magnetized and misdirect the electron beams. A degaussing coil is often provided and fed with a diminishing alternating current each time the set is switched on. Ships can be degaussed by surrounding them with current-carrying cables that set up an equal and opposite field. This prevents the ships from detonating magnetic mines. Degaussing is used to protect scientific and other electronic devices from strong magnetic fields; usually a system of coils is designed to neutralize such fields over the important region or the equipment is surrounded by a shield of suitable alloy (e.g. Mumetal).

degenerate 1. Denoting a state of matter in which all the electrons have been stripped from the atoms as a result of extremely high temperatures. This is believed to occur in certain classes of stars, such as *white dwarfs and *neutron stars. In this state, matter consists of a mass of extremely dense closely packed nuclei and electrons. **2.** Denoting a *semiconductor in which the *Fermi level is located either in the valence band or the conduction band (*see* energy levels), causing the material to behave as a metal. **3.** Denoting energy levels for different quantum states in an atom, ion, molecule, etc., that have the same energy. **4.** Denoting a gas in which the concentration of particles causes a significant departure from the classical *Maxwell–Boltzmann distribution.

degree 1. A unit of plane angle equal to 1/360th of a complete revolution. **2.** A division on a *temperature scale. **3.** The power to which a variable is raised. If one expression contains several variables the overall degree of the expression is the sum of the powers. For example, the expression $p^2q^3r^4$ has a degree of 9 overall (it is a second-degree expression in p). The degree of a polynomial is the degree of the variable with the highest power, e.g. $ax^5 + bx^4 + c$ has a degree of 5. **4.** The highest power to which the derivative of the highest order is raised in a *differential equation. For example, $(d^2y/dx^2)^3 + dy/dx = c$ is a differential equation of the third degree (but second order).

degrees of freedom **1.** The number of independent parameters required to specify the configuration of a system. This concept is applied in the *kinetic theory to specify the number of independent ways in which an atom or molecule can take up energy. There are however various sets of parameters that may be chosen, and the details of the consequent theory vary with the choice. For example, in a monatomic gas each atom may be allotted three degrees of freedom, corresponding to the three coordinates in space required to specify its position. The mean energy per atom for each degree of freedom is the same, according to the principle of the *equipartition of energy, and is equal to $kT/2$ for each degree of freedom (where k is the *Boltzmann constant and T is the thermodynamic temperature). Thus for a monatomic gas the total molar energy is $3LkT/2$, where L is the Avogadro constant (the number of atoms per mole). As $k = R/L$, where R is the molar gas constant, the total molar energy is $3RT/2$.

In a diatomic gas the two atoms require six coordinates between them, giving six degrees of freedom. Commonly these are interpreted as six independent ways of storing energy: on this basis the molecule has three degrees of freedom for different directions of *translational motion, and in addition there are two degrees of freedom for rotation of the molecular axis and one vibrational degree of freedom along the bond between the atoms. The rotational degrees of freedom each contribute their share, $kT/2$, to the total energy; similarly the vibrational degree of freedom has an equal share of kinetic energy and must on average have as much potential energy (*see* simple harmonic motion). The total energy per molecule for a diatomic gas is therefore $3kT/2$ (for translational energy of the whole molecule) plus $2kT/2$ (for rotational energy of each atom) plus $2kT/2$ (for vibrational energy), i.e. a total of $7kT/2$.

2. The least number of independent variables required to define the state of a system in the *phase rule. In this sense a gas has two degrees of freedom (e.g. temperature and pressure).

dekatron A neon-filled tube with a central anode surrounded by ten cathodes and associated transfer electrodes. As voltage pulses are received by the tube a glow discharge moves from one set of electrodes to the next, enabling the device to be used as a visual counting tube in the decimal system. The tube can also be used for switching.

delayed neutrons The small proportion of neutrons that are emitted with a measurable time delay in a nuclear fission process. *Compare* prompt neutrons.

delay line A component in an electronic circuit that is introduced to provide a specified delay in transmitting the signal. Coaxial cable or inductor-capacitor networks can be used to provide a short delay but for longer delays an *acoustic delay line* is required. In this device the signal is converted by the *piezoelectric effect into an acoustic wave, which is passed through a liquid or solid medium, before reconversion to an electronic signal.

deliquescence The absorption of water from the atmosphere by a hygroscopic solid to such an extent that a concentrated solution of the solid eventually forms.

delta-iron *See* iron.

demagnetization The removal of the ferromagnetic properties of a body by disordering the domain structure (*see* magnetism). One method of achieving this is to insert the body within a coil through which an alternating current is flowing; as the magnitude of the current is reduced to zero, the domains are left with no predominant direction of magnetization.

demodulation The process of extracting the information from a modulated carrier wave (*see* modulation; radio). The device used is called a *demodulator* or a *detector*.

denature **1.** To add a poisonous or unpleasant substance to ethanol to make it unsuitable for human consumption (*see* methylated spirits). **2.** To produce a structural change in a protein or nucleic acid that results in the reduction or loss of its biological properties. Denaturation is caused by heat, chemicals, and extremes of pH. The differences between raw and boiled eggs are largely a result of denaturation. **3.** To add another isotope to a fissile material to make it unsuitable for use in a nuclear weapon.

densitometer An instrument used to measure the *photographic density of an image on a film or photographic print. Densitometers work by letting the specimen transmit or reflect a beam of light and monitoring the transmitted or reflected intensity. They originally consisted of visual *photometers but most instruments are now photoelectric. The simplest transmission densitometer consists of a light source, a photosensitive cell, and a microammeter: the density is measured in terms of the meter readings with and without the sample in place. They have a variety of uses, including detecting the sound track on a cinematic film, measuring intensities in spectrographic records, and checking photographic prints.

density **1.** The mass of a substance per unit of volume. In *SI units it is measured in $kg\,m^{-3}$. *See also* relative density; vapour density. **2.** *See* charge density. **3.** *See* photographic density.

depleted Denoting a material that contains less of a particular isotope than it normally contains, especially a residue from a nuclear reactor or isotope-separation plant containing fewer fissile atoms than natural uranium.

depletion layer A region in a *semiconductor that has a lower-than-usual number of mobile charge carriers. A depletion layer forms at the interface between two dissimilar regions of conductivity (e.g. a *p–n* junction). *See* semiconductor diode.

depolarization The prevention of *polarization in a *primary cell. For example, maganese(IV) oxide (the *depolarizer*) is placed around the positive electrode of a *Leclanché cell to oxidize the hydrogen released at this electrode.

depression of freezing point The reduction in the freezing point of a pure liquid when another substance is dissolved in it. It is a *colligative property – i.e. the lowering of the freezing point is proportional to the number of dissolved particles (molecules or ions), and does not depend on their nature. It is given by $\Delta t = K_f C_m$, where C_m is the molar concentration of dissolved solute and K_f is a constant (the *cryoscopic constant*) for the solvent used. Measurements of freezing-point depression (using a Beckmann thermometer) can be used for finding relative molecular masses of unknown substances.

depth of field The range of distance in front of and behind an object that is being focused by an optical instrument, such as a microscope or camera, within which other objects will be in focus. The *depth of focus* is the amount by which the distance between the camera and the film can be changed without upsetting the sharpness of the image.

derivative *See* differentiation; calculus.

derived unit *See* base unit.

desorption The removal of adsorbed atoms, molecules, or ions from a surface.

detector 1. *See* demodulation. 2. *See* counter.

deuterated compound A compound in which some or all of the hydrogen-1 atoms have been replaced by deuterium atoms.

deuterium (heavy hydrogen) Symbol D. The isotope of hydrogen that has a mass number 2 (r.a.m. 2.0144). Its nucleus contains one proton and one neutron. The abundance of deuterium in natural hydrogen is about 0.015%. It is present in water as the oxide, deuterium oxide (*see* heavy water), from which it is usually obtained by electrolysis or fractional distillation. Its chemical behaviour is almost identical to hydrogen although deuterium compounds tend to react rather more slowly than the corresponding hydrogen compounds. Its physical properties are slightly different from those of hydrogen, e.g. b.p. 23.6 K (hydrogen 20.4 K).

deuterium oxide *See* heavy water.

deuteron A nucleus of a deuterium atom, consisting of a proton and a neutron bound together.

deviation 1. (angle of deviation) The angle formed between a ray of light falling on a surface or transparent body and the ray leaving it. **2.** The difference between one of an observed set of values and the true value, usually represented by the mean of all the observed values. The *mean deviation* is the mean of all the individual deviations of the set. *See* standard deviation.

dew *See* precipitation.

Dewar flask A vessel for storing hot or cold liquids so that they maintain their temperature independently of the surroundings. Heat transfer to the surroundings is reduced to a minimum: the walls of the vessel consist of two thin layers of glass (or, in large vessels, steel) separated by a vacuum to reduce conduction and convection; the inner surface of a glass vessel is silvered to reduce radiation; and the vessel is stoppered to prevent evaporation. It was devised around 1872 by the British physicist Sir James Dewar (1842–1923) and is also known by its first trade name *Thermos flask*. *See also* cryostat.

dew point The temperature at which the water vapour in the air is saturated. As the temperature falls the dew point is the point at which the vapour begins to condense as droplets of water.

dew-point hygrometer *See* hygrometer.

dextrorotatory Denoting a compound that rotates the plane of polarization of plane-polarized light to the right (clockwise as observed by someone facing the oncoming radiation). *See* optical activity.

***d*-form** *See* optical activity.

diamagnetism *See* magnetism.

diamond The hardest known mineral (with a hardness of 10 on Mohs' scale). It is an allotropic form of pure carbon that has crystallized in the cubic system, usually as octahedra or cubes, under great pressure. Diamond crystals may be colourless and transparent or yellow, brown, or black. They are highly prized as gemstones but also have extensive uses in industry, mainly for cutting and grinding tools. Industrial diamonds are increasingly being produced synthetically.

diaphragm An opaque disc with a circular aperture at its centre. Diaphragms of different sizes are used to control the total light flux passing through an optical system or to reduce aberration by restricting the light passing through a system to the central portion. An *iris diaphragm* consists of a number of overlapping crescent-shaped discs arranged so that the central aperture can be continuously varied in diameter.

diastereoisomers Stereoisomers that are not identical and yet not mirror images. For instance, the *d*-form of tartaric acid and the meso form constitute a pair of diastereoisomers. *See* optical activity.

dichroism The property of some crystals, such as tourmaline, of selectively absorbing light vibrations in one plane while allowing light vibrations at right angles to this plane to pass through. Polaroid is a synthetic dichroic material. *See* polarization.

dielectric A nonconductor of electric charge in which an applied electric field causes a *displacement of charge but not a flow of charge. Electrons within the atoms of a dielectric are, on average, displaced by an applied field with respect to the nucleus, giving rise to a dipole that has an electric moment in the direction of the field. The resulting stress within the dielectric is known as the *electric polarization* (P) and is defined by $P = D - E\varepsilon_0$, where D is the displacement, E is the electric field strength, and ε_0 is the electric constant.

The *dielectric constant* is now called the relative *permittivity. The *dielectric strength* is the maximum potential gradient that can be applied to a material without causing it to break down. It is usually expressed in volts per millimetre. *See also* capacitor.

dielectric constant *See* permittivity.

dielectric heating The heating of a dielectric material, such as a plastic, by applying a radio-frequency electric field to it. The most common method is to treat the material as the dielectric between the plates of a capacitor. The heat produced is proportional to $V^2fA\phi/t$, where V is the applied potential difference, f its frequency, A is the area of the dielectric, t its thickness, and ϕ is the loss factor of the material (related to its *permittivity).

Diesel engine *See* internal-combustion engine.

differential calculus *See* calculus.

differential equation An equation in which a derivative of y with respect to x appears as well as the dependent variables x and y. The *order* of a differential equation is the order of its highest derivative. The *degree* of the equation is the highest power present of the highest order derivative. There are many types of differential equation, each having its own method of solution. The simplest type has separable variables, enabling each side of the equation to be integrated separately.

differentiation The process of finding the *derivative* of a function in differential *calculus. If $y = f(x)$, the derivative of y, written dy/dx or $f'(x)$, is equal to the limit as $\Delta x \rightarrow 0$ of $[f(x + \Delta x) - f(x)]/\Delta x$. In general, if $y = x^n$, then $dy/dx = nx^{n-1}$. On a graph of $y = f(x)$, the derivative dy/dx is the gradient of the tangent to the curve at the point x.

diffraction The spreading or bending of waves as they pass through an aperture or round the edge of a barrier. The diffracted waves subsequently interfere with each other (*see* interference) producing regions of reinforcement and weakening. First noticed as occurring with light by Francesco Grimaldi (1618–63), the phenomenon gave considerable support to the wave theory of light. Diffraction also occurs with streams of particles. *See also* Fresnel diffraction; Fraunhofer diffraction; electron diffraction.

diffraction grating A device for producing spectra by diffraction and interference. The usual grating consists of a glass or speculum-metal sheet with a very large number of equidistant parallel lines ruled on it (usually of the order of 1000 per mm). Diffracted light after transmission through the glass or reflection by the speculum produces maxima of illumination (spectral lines) according to the equation $m\lambda = d(\sin i + \sin\theta)$, where d is the distance between grating lines, λ is the wavelength of the light, i is the angle of incidence, θ the direction of the diffracted maximum, and m is the 'order' of the spectral line. Reflection gratings are also used to produce spectra in the ultraviolet region of the electromagnetic spectrum.

diffusion **1.** The process by which different substances mix as a result of the random motions of their component atoms, molecules, and ions. In gases, all the components are perfectly miscible with each other and mixing ultimately becomes nearly uniform, though slightly affected by gravity (*see also* Graham's law). The diffusion of a solute through a solvent to produce a solution of uniform concentration is slower, but otherwise very similar to the process of gaseous diffusion. In solids, however, diffusion occurs very slowly at normal temperatures. **2.** The scattering of a beam of light by reflection at a rough surface or by transmission through a translucent (rather than transparent) medium, such as frosted glass. **3.** The passage of elementary particles through matter when there is a high probability of scattering and a low probability of capture.

diffusion cloud chamber *See* cloud chamber.

diffusion pump (condensation pump) A *vacuum pump in which oil or mercury vapour is diffused through a jet, which entrains the gas molecules from the container in which the pressure is to be reduced. The diffused vapour and entrained gas molecules are condensed on the cooled walls of the pump. Pressures down to 10^{-7} Pa can be reached by sophisticated forms of the diffusion pump.

digit A symbol used to represent a single number. For example, the number 479 consists of three digits.

digital computer *See* computer.

digital display A method of indicating a reading of a measuring instrument, clock, etc., in which the appropriate numbers are generated on a fixed display unit by the varying parameter being measured rather than fixed numbers on a scale being indicated by a moving pointer or hand. *See* digitron; light-emitting diode; liquid-crystal display.

digital recording A method of recording or transmitting sound in which the sound itself is not transmitted or recorded. Instead the pressure in the sound wave is sampled at least 30 000 times per second and the successive values represented by numbers, which are then transmitted or recorded. Afterwards they are restored to analogue form in the receiver or player. This method is used for very high fidelity recordings as no distortion or interference occurs during transmission or in the recording process.

digitron An electronic gas-discharge tube that provides a *digital display in calculators, counters, etc. It usually has 10 cold cathodes shaped into the form of the digits 0–9. The cathode selected receives a voltage pulse causing a glow discharge to illuminate the digit. It has now largely been superseded by *light-emitting diodes and *liquid-crystal displays.

dihedral (dihedron) An angle formed by the intersection of two planes (e.g. two faces of a polyhedron). The *dihedral angle* is the angle formed by taking a point on the line of intersection and drawing two lines from this point, one in each plane, perpendicular to the line of intersection.

dilatancy *See* Newtonian fluid.

dilation (*or* **dilatation**) **1.** An increase in volume. **2.** *See* time dilation.

dilatometer A device for measuring the cubic *expansivities of liquids. It consists of a bulb of known volume joined to a graduated capillary tube, which is closed at the top to prevent evaporation. A known mass of liquid is introduced into the device, which is submerged in a bath maintained at different temperatures t_1 and t_2. The two volumes corresponding to these temperatures, V_1 and V_2, are read off the calibrated stem. The value of the cubic expansivity (γ) is then given by

$$\gamma = (V_2 - V_1)/V_1(t_2 - t_1).$$

dimensional analysis A method of checking an equation or a solution to a problem by analysing the dimensions in which it is expressed. It is also useful for establishing the form, but not the numerical coefficients, of an empirical relationship. If the two sides of an equation do not have the correct dimensions, the equation is wrong. If it does have the correct dimensions it can still be wrong, but the error is likely to be in the arithmetic rather than the method of solution.

dimensions The product or quotient of the basic physical quantities, raised to the appropriate powers, in a derived physical quantity. The basic physical quantities of a mechanical system are usually taken to be mass (M), length (L), and time (T). Using these dimensions, the derived physical quantity velocity will have the dimensions L/T and acceleration will have the dimensions L/T^2. As force is the product of a mass and an acceleration (*see* Newton's law of motion), force has the dimensions MLT^{-2}. In electrical work, current, I, can be regarded as dimensionally independent and the dimensions of other electrical units can be found from standard relationships. Charge, for example, is measured as the product of current and time. It therefore has the dimension IT. Potential difference is given by the relationship $P = VI$, where P is power. As power is force $\times$ distance $\div$ time ($MLT^{-2} \times L \times T^{-1} = ML^2T^3$), voltage V is given by $V = ML^2T^{-3}I^{-1}$.

diode An electronic device with two electrodes. In the obsolescent *thermionic diode* a heated cathode emits electrons, which flow across the intervening vacuum to the anode when a positive potential is applied to it. The device permits flow of current in one direction only as a negative potential applied to the anode repels the electrons. This property of diodes was made use of in the first thermionic radios, in which the diode was used to demodulate the transmitted signal (*see* modulation). In the *semiconductor diode*, a *p–n* junction performs a similar function. The forward current increases potential difference above a certain threshold, whereas the reverse current is very small indeed. *See* semiconductor; transistor.

dioptre A unit for expressing the power of a lens or mirror equal to the reciprocal of its focal length in metres. Thus a lens with a focal length of 0.5 metre has a power of 1/0.5 = 2 dioptres. The power of a converging lens is usually taken to be positive and that of a diverging lens negative. Because the power of a lens is a measure of its ability to cause a beam to converge, the dioptre is now sometimes called the radian per metre.

dip *See* geomagnetism.

dipole 1. Two equal and opposite charges that are separated by a distance. The *dipole* *moment* is the product of either charge and the distance between them. Some molecules behave as dipoles and measurement of the dipole moments can often provide information regarding the configuration of the molecule. **2.** An aerial commonly used for frequencies below 30 megahertz. It consists of a horizontal rod, fed or tapped at its centre. It may be half a wavelength or a full wavelength long.

direct current (d.c.) An electric current in which the net flow of charge is in one direction only. *Compare* alternating current.

direct-current motor *See* electric motor.

direct motion 1. The apparent motion of a planet from west to east as seen from the earth against the background of the stars. **2.** The anticlockwise rotation of a planet, as seen from its north pole. *Compare* retrograde motion.

directrix 1. A plane curve defining the base of a *cone. **2.** A straight line from which the distance to any point on a *conic is in a constant ratio to the distance from that point to the focus.

discharge 1. The conversion of the chemical energy stored in a *secondary cell into electrical energy. **2.** The release of electric charge from a capacitor in an external circuit. **3.** The passage of charge carriers through a gas at low pressure in a *discharge tube*. A potential difference applied between cathode and anode creates an electric field that accelerates any free electrons and ions to their appropriate electrodes. Collisions between electrons and gas molecules create more ions. Collisions also produce excited ions and molecules (*see* excitation), which decay with emission of light in certain parts of the tube.

discontinuous function *See* continuous function.

disintegration Any process in which an atomic nucleus breaks up spontaneously into two or more fragments in a radioactive decay process or breaks up as a result of a collision with a high-energy particle or nuclear fragment.

disintegration constant *See* decay.

dislocation *See* defect.

disperse phase *See* colloids.

dispersion The splitting up of a ray of light of mixed wavelengths by refraction into its components. Dispersion occurs because the *deviation for each wavelength is different on account of the different speeds at which waves of different wavelengths pass through the refracting medium. If a ray of white light strikes one face of a prism and passes out of another face, the white light will be split into its components and the full visible spectrum will be formed. The *dispersive power* of a prism (or other medium) for white light is defined by

$$(n_b - n_r)/(n_y - 1),$$

where n_b, n_r, and n_y are the *refractive indexes for blue, red, and yellow light respectively. The term is sometimes applied to the separation of wavelengths produced by a *diffraction grating.

dispersion forces *See* van der Waals' force.

dispersive power *See* dispersion.

displacement (electric flux density) Symbol D. The ratio of the charge displaced in a medium subjected to an electric field and the area of an imagined plane perpendicular to the potential gradient.

distortion The extent to which a system fails to reproduce the characteristics of its input in its output. It is most commonly applied to electronic amplifiers and to optical systems. *See* aberration.

distributive law The mathematical law stating that one operation is independent of being carried out before or after another operation. For example, multiplication is distributive with respect to addition and subtraction, i.e. $x(y + z) = xy + xz$. *Compare* associative law; commutative law.

diurnal Daily; denoting an event that happens once every 24 hours.

div (divergence) Symbol ∇. The expression

$$\partial u_1/\partial x + \partial u_2/\partial y + \partial u_3/\partial z$$

where u_1, u_2, and u_3 are the x, y, and z components of the vector $\boldsymbol{u}$. In physics, $\nabla.\boldsymbol{F}$ is used to describe an excess flux leaving a volume in space, where $\boldsymbol{F}$ is a three-dimensional vector function. *See also* Laplace equation.

divergent series *See* convergent series.

diverging lens or mirror A lens or mirror that can refract or reflect a parallel beam of light into a diverging beam. A diverging lens is predominantly concave; a diverging mirror is convex. *Compare* converging lens or mirror.

***dl*-form** *See* optical activity; racemic mixture.

D-lines Two close lines in the yellow region of the visible spectrum of sodium, having wavelengths 589.0 and 589.6 nm. As they are prominent and easily recognized they are used as a standard in spectroscopy.

domain *See* magnetism.

donor *See* semiconductor.

doping *See* semiconductor.

Doppler effect The apparent change in the observed frequency of a wave as a result of relative motion between the source and the observer. For example, the sound made by a low-flying aircraft as it approaches appears to fall in pitch as it passes and flies away. In fact, the frequency of the aircraft engine remains constant but as it is approaching more sound waves per second impinge on the ear and as it recedes fewer sound waves per second impinge on the ear. The apparent frequency, F, is given by

$$F = f(c - u_o)/(c - u_s),$$

where f is the true frequency, c is the speed of sound, and u_o and u_s are the speeds of the observer and the source, respectively.

Although the example of sound is most commonly experienced, the effect was suggested by Christian Johann Doppler (1803–53), an Austrian physicist, as an attempt to explain the coloration of stars. In fact the Doppler effect cannot be observed visually in relation to the stars, although the effect does occur with electromagnetic radiation and the *red shift of light from receding stars can be observed spectroscopically. The Doppler effect is also used in radar to distinguish between stationary and moving targets and to provide information regarding the speed of moving targets by measuring the frequency shift between the emitted and reflected radiation.

For electromagnetic radiation, the speed of light, c, features in the calculation and as there is no fixed medium to provide a frame of reference, relativity has to be taken into account, so that

$$F = f\sqrt{[(1 - v/c)/(1 + v/c)]},$$

where v is the speed at which source and observer are moving apart. If v^2/c^2 is small compared to 1, i.e. if the speed of separation is small compared to the speed of light, this equation simplifies to

$$F = f(1 - v/c).$$

***d*-orbital** *See* orbital.

dose A measure of the extent to which matter has been exposed to *ionizing radiation. The *absorbed dose* is the energy per unit mass absorbed by matter as a result of such exposure. The SI unit is the gray, although it is often measured in rads (1 rad = 0.01 gray; *see* radiation units). The *maximum permissible dose* is the recommended upper limit of absorbed dose that a person or organ should receive in a specified period according to the International Commission on Radiological Protection. *See also* linear energy transfer.

dosimeter Any device used to measure absorbed *dose of ionizing radiation. Methods used include the *ionization chamber, photographic film, or the rate at which certain chemical reactions occur in the presence of ionizing radiation.

double refraction The property, possessed by certain crystals (notably calcite), of forming two refracted rays from a single incident ray. The *ordinary ray* obeys the normal laws of refraction. The other refracted ray, called the *extraordinary ray*, follows different laws. The light in the ordinary ray is polarized at right angles to the light in the extraordinary ray. Along an *optic axis the ordinary and extraordinary rays travel with the same speed. Some crystals, such as calcite, quartz, and tourmaline, have only one optic axis; they are *uniaxial crystals*. Others, such as mica and selenite, have two optic axes; they are *biaxial crystals*. The phenomenon is also known as *birefringence* and the double-refracting crystal as a *birefringent crystal*. *See also* polarization.

doublet 1. A pair of optical lenses of different shapes and made of different materials used together so that the chromatic aberration produced by one is largely cancelled by the reverse aberration of the other. **2.** A pair of associated lines in certain spectra, e.g. the two lines that make up the sodium D line.

drain *See* transistor.

dry cell A primary or secondary cell in which the electrolytes are in the form of a paste. Many torch, radio, and calculator batteries are *Leclanché cells in which the electrolyte is an ammonium chloride paste and the container is the negative zinc electrode (with an outer plastic wrapping).

dry ice Solid carbon dioxide used as a refrigerant. It is convenient because it sublimes at −78°C (195 K) at standard pressure rather than melting.

ductility The ability of certain metals, such as copper, to retain their strength when their shape is changed, especially the ability of such metals to be drawn into a thin wire without cracking or breaking.

Dulong and Petit's law For a solid element the product of the relative atomic mass and the specific heat capacity is a constant equal to about 25 J mol^{-1} K^{-1}. Formulated in these terms in 1819 by the French scientists Pierre Dulong (1785–1838) and A. T. Petit (1791–1820), the law in modern terms states: the molar heat capacity of a solid element is approximately equal to $3R$, where R is the *gas constant. The law is only approximate but applies with fair accuracy at normal temperatures to elements with a simple crystal structure.

dust core *See* core.

dwarf star A small star that lies on the main sequence in a *Hertzsprung–Russell diagram. *See also* white dwarf.

dynamic equilibrium *See* equilibrium.

dynamics (kinetics) The branch of mechanics concerned with the motion of bodies under the action of forces. Time intervals, distances, and masses are regarded as fundamental and bodies are assumed to possess *inertia. Bodies in motion have an attribute called *momentum (*see* Newton's laws of motion), which can only be changed by the application of a force. *Compare* statics.

dynamo An electric *generator, especially one designed to provide *direct current. Alternating-current generators can be called dynamos but are more often called alternators.

dynamometer 1. An instrument used to measure a force, often a spring balance. **2.** A device used to measure the output power of an engine or motor. **3.** (**current dynamometer**) A variety of *current balance, for measuring electric current.

dyne The unit of force in the *c.g.s. system; the force required to give a mass of one gram an acceleration of $1\ cm\ s^{-2}$. 1 dyne = 10^{-5} newton.

dystectic mixture A mixture of substances that has a constant maximum melting point.

E

e The irrational number defined as the limit as n tends to infinity of $(1 + 1/n)^n$. It has the value 2.718 28 . . . . It is used as the base of natural *logarithms and occurs in the *exponential function, e^x.

earth The planet that orbits the sun between the planets Venus and Mars at a mean distance from the sun of 149 600 000 km. It has a mass of about 5.976×10^{24} kg and an equatorial diameter of 12 756 km. The earth consists of three layers: the gaseous atmosphere (*see* earth's atmosphere), the liquid hydrosphere, and the solid lithosphere. The solid part of the earth also consists of three layers: the *crust* with a mean thickness of about 32 km under the land and 10 km under the seas; the *mantle*, which extends some 2900 km below the crust; and the *core*, part of which is believed to be liquid. The crust has a relative density of about 3 and consists largely of sedimentary rocks overlaying igneous rocks. The composition of the crust is: oxygen 47%, silicon 28%, aluminium 8%, iron 4.5%, calcium 3.5%, sodium and potassium 2.5% each, and magnesium 2.2%. Hydrogen, carbon, phosphorus, and sulphur are all present to an extent of less than 1%. The mantle reaches a relative density of about 5.5 at its maximum depth and is believed to consist mainly of silicate rocks. The core is believed to have a maximum relative density of 13 and a maximum temperature of 6400 K. *See also* geomagnetism.

earthquake A sudden movement or fracturing within the earth's lithosphere, followed by the series of shocks generated by this movement. This may range from a mild tremor to a large-scale earth movement causing extensive damage over a wide area. The point at which the earthquake originates is known as the *seismic focus*; the point on the earth's surface directly above this is the *epicentre* (or *hypocentre*). Earthquakes result from a build-up of stresses within the rocks until they are strained to the point beyond which they will fracture. They occur in narrow continuous belts of activity, which correspond with the junction of lithospheric plates, including the circum-Pacific belt, the Alpine–Himalayan belt, and the mid-ocean ridges. The scale of the shock of an earthquake is known as the magnitude; the most commonly used scale for comparing the magnitude of earthquakes is the logarithmic *Richter scale (8.9 is the highest recorded magnitude on the scale).

earth's atmosphere The gas that surrounds the earth. The composition of dry air at sea level is: nitrogen 78.08%, oxygen 20.95%, argon 0.93%, carbon dioxide 0.03%,

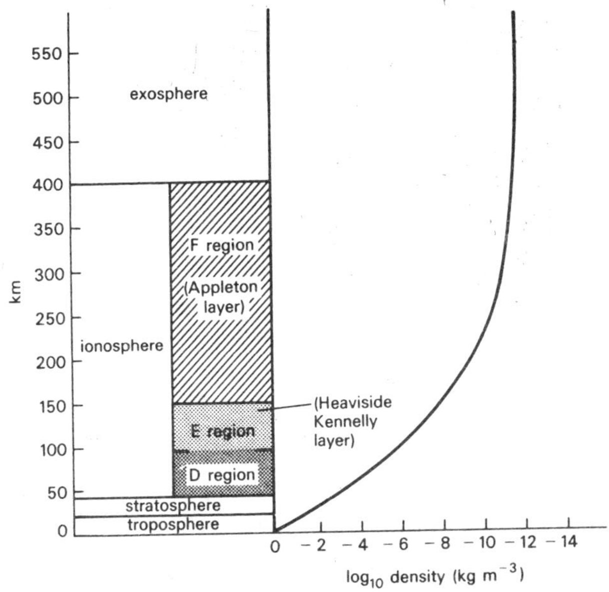

Earth's atmosphere

neon 0.0018%, helium 0.0005%, krypton 0.0001%, and xenon 0.00001%. In addition to water vapour, air in some localities contains sulphur compounds, hydrogen peroxide, hydrocarbons, and dust particles.

The lowest level of the atmosphere, in which most of the weather occurs, is called the *troposphere*. Its thickness varies from about 7 km at the poles to 28 km at the equator and in this layer temperature falls with increasing height. The next layer is the *stratosphere*, which goes up to about 50 km. Here the temperature remains approximately constant. Above this is the *ionosphere*, which extends to about 1000 km, with the temperature rising and the composition changing substantially. At about 100 km and above most of the oxygen has dissociated into atoms; at above 150 km the percentage of nitrogen has dropped to nil. In the ionosphere the gases are ionized by the absorption of solar radiation. This enables radio transmissions to be made round the curved surface of the earth as the ionized gas acts as a reflector for certain wavelengths. The ionosphere is divided into three layers. The D-layer (50–90 km) contains a low concentration of free electrons and reflects low-frequency radio waves. The E-layer (90–150 km) is also called the *Heaviside layer* or *Heaviside–Kennelly layer* as its existence was predicted independently by Oliver Heaviside (1850–1925) and Arthur E. Kennelly (1861–1939). This layer reflects medium-frequency waves. The F-layer (150–1000 km) is also called the *Appleton layer* after its discoverer Sir Edward Appleton (1892–1965). It has the highest concentration of free electrons and is the most useful for radio transmission. Wavelengths between 8 mm and 20 m are not reflected by the ionosphere but escape into space. Therefore television transmissions, which utilize this range, require artificial *satellites for reflection (or reception, amplification, and retransmission). From about 400 km, the outermost region of the atmosphere is also called the *exosphere*.

earthshine Sunlight reflected from the surface of the earth. An observer in space may see nearby objects dimly illuminated by earthshine, as things on earth may be illuminated by moonlight. Under certain conditions near new moon the dark disc of the moon can be seen faintly illuminated by earthshine – a phenomenon called 'the old moon in the new moon's arms'.

earth's magnetic field *See* geomagnetism.

ebullioscopic constant *See* elevation of boiling point.

eccentricity *See* conic.

ECG *See* electrocardiogram.

echelon A form of *interferometer consisting of a stack of glass plates arranged stepwise with a constant offset. It gives a high resolution and is used in spectroscopy to study hyperfine line structure. In the *transmission echelon* the plates are made equal in optical thickness to introduce a constant delay between adjacent parts of the wavefront. The *reflecting echelon* has the exposed steps metallized and acts like an exaggerated *diffraction grating.

echo The reflection of a wave by a surface or object so that a weaker version of it is detected shortly after the original. The delay between the two is an indication of the distance of the reflecting surface. An *echo sounder* is an apparatus for determining the depth of water under a ship. The ship sends out a sound wave and measures the time taken for the echo to return after reflection by the sea bottom. *Sonar* (*so*und *na*vigation *r*anging) is a technique for locating underwater objects by a similar method. Echoes also occur with radio waves; reflection of waves causes an echo in radio transmission and ghosts in television pictures. *See also* radar.

echolocation 1. *See* radar; echo. **2.** A method used by some animals (such as bats, dolphins, and certain birds) to detect objects in the dark. The animal emits a series of high-pitched sounds that echo back from the object and are detected by the ear or some other sensory receptor. From the direction of the echo and from the time between emission and reception of the sounds the object is located, often very accurately.

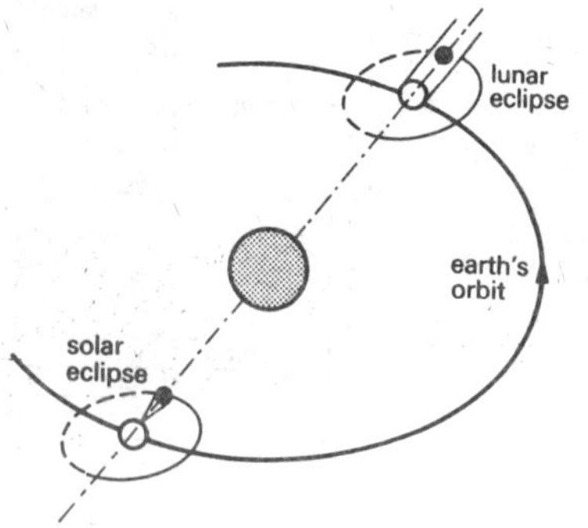

Solar and lunar eclipses

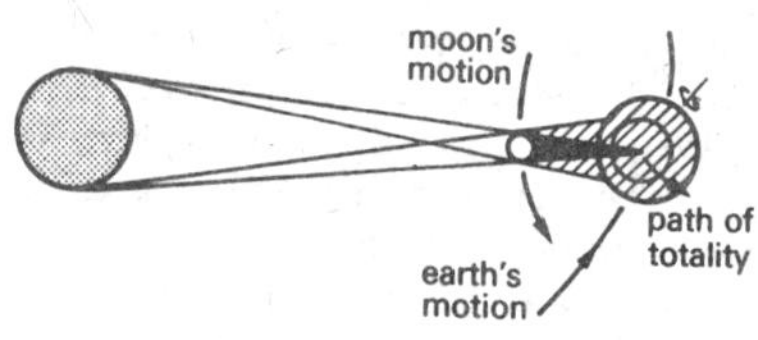

Moon's shadow in solar eclipse

eclipse The total (*total eclipse*) or partial (*partial eclipse*) obscuring of light from a celestial body as it passes behind or through the shadow of another body. A *lunar eclipse* occurs when the sun, earth, and moon are in a straight line and the shadow of the earth falls on the moon. A *solar eclipse* occurs when the shadow of the moon falls on the earth.

ecliptic The *great circle in which the plane of the earth's orbit round the sun intersects the *celestial sphere. It is thus the sun's apparent annual path across the sky.

eddy current A current induced in a conductor situated in a changing magnetic field or moving in a fixed one. Any imagined circuit within the conductor will change its magnetic flux linkage, and the consequent induced e.m.f. will drive current around the circuit. In a substantial block of metal the resistance will be small and the current therefore large. Eddy currents occur in the cores of transformers and other electrical machines and represent a loss of useful energy (the *eddy-current loss*). To reduce this loss to a minimum metal cores are made of insulated sheets of metal, the resistance between these laminations reducing the current. In high-frequency circuits *ferrite cores can be used. Eddy currents in a moving conductor interact with the magnetic field producing them to retard the motion of the conductor. This enables some electrical instruments (moving-coil type) to utilize eddy currents to create damping. Eddy currents are also used in *induction heating.

Edison cell *See* nickel–iron accumulator.

EEG *See* electroencephalogram.

effective temperature *See* luminosity.

effective value *See* root-mean-square value.

efficiency A measure of the performance of a machine, engine, etc., being the ratio of the energy or power it delivers to the energy or power fed to it. In general, the efficiency of a machine varies with the conditions under which it operates and there is usually a load at which it operates with the highest efficiency. The *thermal efficiency* of a heat engine is the ratio of the work done by the engine to the heat supplied by the fuel. For a *reversible heat engine this efficiency is equal to $(T_1 - T_2)/T_1$, where T_1 is the thermodynamic temperature at which all the heat is taken up and T_2 is the thermodynamic temperature at which it is given out (*see* Carnot cycle). For real engines it is always less than this.

effusion The flow of a gas through a small aperture. The relative rates at which gases effuse, under the same conditions, is approximately inversely proportional to the square roots of their densities.

eigenfunction An allowed *wave function of a system in quantum mechanics. The associated energies are *eigenvalues*.

Einstein equation **1.** The mass–energy relationship announced by Einstein in 1905 in the form $E = mc^2$, where E is a quantity of energy, m its mass, and c is the speed of light. It presents the concept that energy possesses mass. *See also* relativity. **2.** The relationship $E_{max} = hf - W$, where E_{max} is the maximum kinetic energy of electrons emitted in the photoemissive effect, h is the Planck constant, f the frequency of the incident radiation, and W the *work function of the emitter. This is also written $E_{max} = hf - \phi e$, where e is the electronic charge and ϕ a potential difference, also called the work function. (Sometimes W and ϕ are distinguished as *work function energy* and *work function potential*.) The equation can also be applied to photoemission from gases, when it has the form: $E = hf - I$, where I is the ionization potential of the gas.

einsteinium Symbol Es. A radioactive metallic transuranic element belonging to the actinoids; a.n. 99; mass number of the most stable isotope 254 (half-life 270 days). Eleven isotopes are known. The element was first identified by A. Ghiorso and associates in debris from the first hydrogen bomb explosion in 1952. Microgram quantities of the element did not become available until 1961.

Einstein shift *See* redshift.

elastance The reciprocal of *capacitance. It is measured in farad^{-1} (sometimes called a 'daraf').

elastic collision A collision in which the total kinetic energy of the colliding bodies after collision is equal to their total kinetic energy before collision. Elastic collisions occur only if there is no conversion of kinetic energy into other forms, as in the collision of atoms. In the case of macroscopic bodies this will not be the case as some of the energy will become heat. In a collision between polyatomic molecules, some kinetic energy may be converted into vibrational and rotational energy of the molecules, but otherwise molecular collisions appear to be elastic.

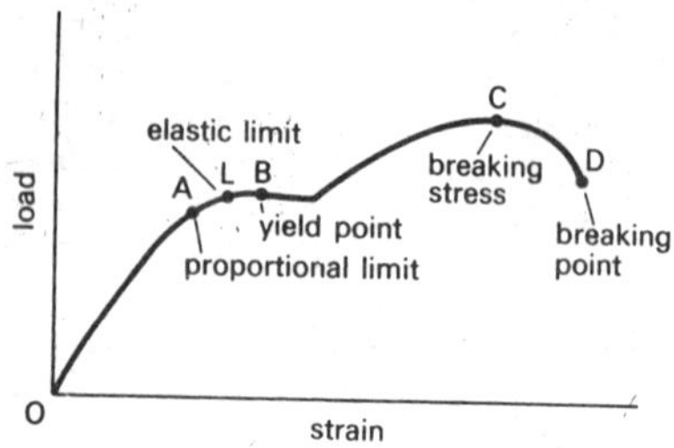

Elasticity

elasticity The property of certain materials that enables them to return to their original dimensions after an applied *stress has been removed. In general, if a stress is applied to a wire, the *strain will increase in proportion (*see OA* on the illustration) until a certain point called the *limit of proportionality* is reached. This is in accordance with *Hooke's law. Thereafter there is at first a slight increase in strain with increased load until a point *L* is reached. This is the *elastic limit*; up to this point the deformation of the specimen is elastic, i.e. when the stress is removed the specimen returns to its original length. Beyond the point *L* there is permanent deformation when the stress is removed, i.e. the material has ceased to be *elastic* and has become *plastic*. In the plastic stages individual materials vary somewhat; in general, however, at a point *B* there is a sudden increase in strain with further increases of stress – this is the *yield point*. Beyond the point *C*, the *breaking stress*, the wire will snap (which occurs at point *D*).

elastic modulus The ratio of the *stress applied to a body to the *strain produced. The *Young modulus of elasticity*, named after the British physicist Thomas Young (1773–1829), refers to longitudinal stress and strain. The *bulk modulus* is the ratio of the pressure on a body to its fractional decrease in volume. The *shear* (or *rigidity*) *modulus* is the tangential force per unit area divided by the angular deformation in radians.

electret A permanently electrified substance or body that has opposite charges at its extremities. They resemble permanent magnets in many ways. An electret can be made by cooling certain waxes in a strong electric field.

electrical energy A form of energy related to the position of an electric charge in an electric field. For a body with charge Q and an electric potential V, its electrical energy is QV. If V is a potential difference, the same expression gives the energy transformed when the charge moves through the p.d.

electric arc A luminous discharge between two electrodes. The discharge raises the electrodes to incandescence, the resulting thermal ionization largely providing the carriers to maintain the high current between the electrodes.

electric-arc furnace A furnace used in melting metals to make alloys, especially in steel manufacture, in which the heat source is an electric arc. In the direct-arc furnace, such as the Héroult furnace, an arc is formed between the metal and an electrode.

In the indirect-arc furnace, such as the Stassano furnace, the arc is formed between two electrodes and the heat is radiated onto the metal.

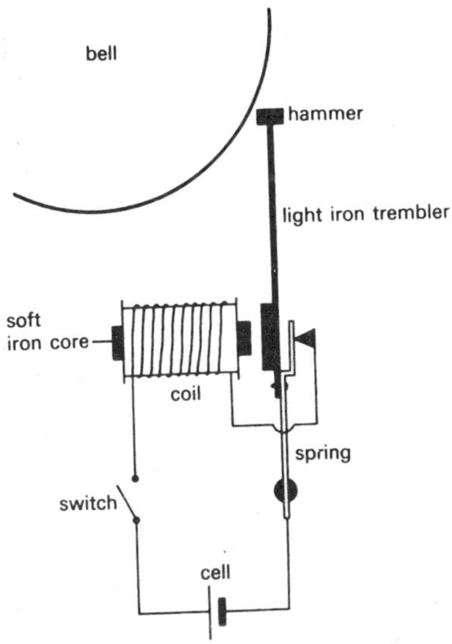

Electric bell

electric bell A device in which an electromagnetically operated hammer strikes a bell. Pressing the bell-push closes a circuit, causing current to flow from a battery or mains step-down transformer through an electromagnet. The electromagnet attracts a piece of soft iron attached to the hammer, which strikes the bell and at the same time breaks the circuit. The hammer springs back into its original position again, closing the circuit and causing the magnet to attract the soft iron. This process continues until the bell-push is released.

electric charge *See* charge.

electric constant *See* permittivity.

electric current *See* current.

electric displacement (electric flux density) Symbol *D*. The charge per unit area that would be displaced across a layer of conductor placed across an *electric field. This describes also the charge density on an extended surface that could be causing the field.

electric field A region in which an electric charge experiences a force usually because of a distribution of other charges. The *electric field strength* or *electric intensity* (*E*) at any point in an electric field is defined as the force per unit charge experienced by a small charge placed at that point. This is equivalent to a potential gradient along the field and is measured in volts per metre. The strength of the field can alternatively be described by its *electric displacement *D*. The ratio D/E for measurements in a vacuum is the electric constant ε_0. In a substance the observed potential gradient is reduced by electron movement so that D/E appears to increase: the new ratio (ε) is called the *permittivity of the substance. An electric field can be created by an isolated electric charge, in which case the field strength at a distance *r* from a point charge *Q* is given by $E = Q/4\pi r^2\varepsilon$, where ε is the permittivity of the intervening medium (*see* Coulomb's law). An electric field can also be created by a changing magnetic field.

electric flux Symbol Ψ. In an *electric field, the product of the electric flux density and the relevant area. *See* electric displacement.

electric flux density *See* electric displacement.

electricity Any effect resulting from the existence of stationary or moving electric charges.

electric lighting Illumination provided by electric currents. The devices used are the *arc lamp*, the *light bulb* (incandescent filament lamp), and the *fluorescent tube*. In the arc lamp, which is no longer used as a general means of illumination, an electric current flows through a gap between two carbon electrodes, between which a high potential difference is maintained. The current is carried by electrons and ions in the vapour produced by the electrodes and a mechanism is required to bring the electrodes closer together as they are vaporized. The device produces a strong white light but has many practical disadvantages. However, arcs enclosed in an inert gas (usually xenon) are increasingly used for such purposes as cinema projectors. The common light bulb is a glass bulb containing a tungsten filament and usually an inert gas. The passage of an electric current through the filament heats it to a white heat. Inert gas is used in the bulb to minimize blackening of the glass by evaporation of tungsten. In the fluorescent tube a glass tube containing mercury vapour (or some other gas) at a low pressure has its inner surface coated with a fluorescent substance. A discharge is created within the tube between two electrodes. Electrons emitted by the cathode collide with gas atoms or molecules and raise them to an excited state (*see* excitation). When they fall back to the *ground state they emit photons of ultraviolet radiation, which is converted to visible light by the coating of phosphor on the inner walls of the tube. In some lamps, such as the *sodium-vapour and *mercury-vapour lamps used in street lighting, no fluorescent substance is used, the light being emitted directly by the excited atoms of sodium or mercury. Vapour lights are more efficient than filament lights as less of the energy is converted into heat.

electric motor A machine for converting electrical energy into mechanical energy. They are quiet, clean, and have a high efficiency (75–95%). They work on the principle that a current passing through a coil within a magnetic field will experience forces that can be used to rotate the coil. In the *induction motor*, alternating current is fed to a stationary coil (the *stator*), which both creates the magnetic field and induces a current in the rotating coil (*rotor*), which it surrounds. The advantage of this kind of motor is that current does not have to be fed through a commutator to a moving part. In the *synchronous motor*, alternating current fed to the stator produces a magnetic field that rotates and locks with the field of the rotor, in this case an independent magnet, causing the rotor to rotate at the same speed as the stator field rotates. The rotor is either a permanent magnet or an electromagnet fed by a direct current through slip rings. In the *universal motor*, current is fed to the stator and, through a commutator, to the rotor. In the series-wound motor the two are in series; in the shunt-wound motor they are in parallel. These motors can be used with either a.c. or d.c. but some small motors use a permanent magnet as the stator and require d.c. for the rotor (via the commutator). *See also* linear motor.

electric polarization *See* dielectric.

electric potential Symbol *V*. The energy required to bring unit electric charge from infinity to the point in an electric field at which the potential is being specified. The unit of electric potential is the volt. The *potential difference* (*p.d.*) between two points in an electric field or circuit is the difference in the values of the electric potentials at the two points, i.e. it is the work done in moving unit charge from one point to the other.

electric power The rate of expending energy or doing work in an electrical system. For a direct-current circuit, it is given by the product of the current passing through a system and the potential difference across it. In alternating-current circuits, the power is given by $VI\cos\phi$, where V and I are the RMS values and ϕ is the *phase angle.

Cosϕ is called the *power factor* of the circuit.

electric spark The passage of an electric current through a gas between two points of high opposite potential, with the emission of light and sound. *Lightning consists of a spark between a cloud and earth or between two oppositely charged parts of the same cloud.

electric susceptibility *See* susceptibility.

electrocardiogram (ECG) A tracing or graph of the electrical activity of the heart. Recordings are made from electrodes fastened over the heart and usually on both arms and a leg. Changes in the normal pattern of an ECG may indicate heart irregularities or disease.

electrochemical cell *See* cell.

electrochemical equivalent Symbol z. The mass of a given element liberated from a solution of its ions in electrolysis by one coulomb of charge. *See* Faraday's laws (of electrolysis).

electrochemical series *See* electromotive series.

electrochemistry The study of chemical properties and reactions involving ions in solution, including electrolysis and electric cells.

electrochromatography *See* electrophoresis.

electrode A conductor that emits or collects electrons in a cell, thermionic valve, semiconductor device, etc. The *anode* is the positive electrode and the *cathode* is the negative electrode.

electrodeposition The process of depositing one metal on another by electrolysis, as in *electroforming and *electroplating.

electrode potential The potential difference produced between the electrode and the solution in a *half cell. It is not possible to measure this directly since any measurement involves completing the circuit with the electrolyte, thereby introducing another half cell. *Standard electrode potentials* $E^{\ominus}$ are defined by measuring the potential relative to a standard *hydrogen half cell using 1.0 molar solution at 25°C. The convention is to designate the cell so that the oxidized form is written first. For example,

$$Pt(s)|H_2(g)H^+(aq)|Zn^{2+}(aq)|Zn(s)$$

The e.m.f. of this cell is −0.76 volt (i.e. the zinc electrode is negative). Thus the standard electrode potential of the $Zn^{2+}|Zn$ half cell is −0.76 V. Electrode potentials are also called *reduction potentials. See also* electromotive series.

electrodialysis A method of obtaining pure water from water containing a salt, as in desalination. The water to be purified is fed into a cell containing two electrodes. Between the electrodes is placed an array of semipermeable membranes alternately semipermeable to positive ions and negative ions. The ions tend to segregate between alternate pairs of membranes, leaving pure water in the other gaps between membranes. In this way, the feed water is separated into two streams: one of pure water and the other of more concentrated solution.

electrodynamics The study of electric charges in motion, the forces created by electric and magnetic fields, and the relationship between them, especially with respect to the theory and design of generators and motors. *Compare* electrostatics.

electroencephalogram (EEG) A tracing or graph of the electrical activity of the brain. Electrodes taped to the scalp record electrical waves from different parts of the brain. The pattern of an EEG reflects an individual's level of consciousness and can be used to detect such disorders as epilepsy, tumours, or brain damage.

electroforming A method of forming intricate metal articles or parts by *electrodeposition of the metal on a removable conductive mould.

electroluminescence *See* luminescence.

electrolysis The production of a chemical reaction by passing an electric current through an electrolyte. In electrolysis, positive ions migrate to the cathode and negative ions to the anode.

electrolyte A liquid that conducts electricity as a result of the presence of positive or negative ions. Electrolytes are molten ionic compounds or solutions containing ions, i.e. solutions of ionic salts or of compounds that ionize in solution. Liquid metals, in which the conduction is by free electrons, are not usually regarded as electrolytes.

electrolytic capacitor *See* capacitor.

electrolytic cell A cell in which electrolysis occurs; i.e. one in which current is passed through the electrolyte from an external source.

electrolytic corrosion Corrosion that occurs through an electrochemical reaction.

electrolytic gas The highly explosive gas formed by the electrolysis of water. It consists of two parts hydrogen and one part oxygen by volume.

electrolytic rectifier A *rectifier consisting of two dissimilar electrodes immersed in an electrolyte. By suitable choice of electrodes and electrolyte the cell can be made to pass current easily in one direction but hardly at all in the other. Examples include a lead–aluminium cell with ammonium phosphate(V) electrolyte and a tantalum–lead cell with sulphuric acid as the electrolyte.

electrolytic refining The purification of metals by electrolysis. It is commonly applied to copper. A large piece of impure copper is used as the anode with a thin strip of pure copper as the cathode. Copper (II) sulphate solution is the electrolyte. Copper dissolves at the anode: $Cu \rightarrow Cu^{2+} + 2e$, and is deposited at the cathode. The net result is transfer of pure copper from anode to cathode. Gold and silver in the impure copper form a so-called *anode sludge* at the bottom of the cell, which is recovered.

electrolytic separation A method of separating isotopes by exploiting the different rates at which they are released in electrolysis. It was formerly used for separating deuterium and hydrogen. On electrolysis of water, hydrogen is formed at the cathode more readily than deuterium, thus the water becomes enriched with deuterium oxide.

electromagnet A magnet consisting of a soft ferromagnetic core with a coil of insulated wire wound round it. When a current flows through the wire the core becomes magnetized; when the current ceases to flow the core loses its magnetization. Electromagnets are used in switches, solenoids, electric bells, metal-lifting cranes, and many other applications.

electromagnetic induction The production of an electromotive force in a conductor when there is a change of magnetic flux linkage with the conductor or when there is relative motion of the conductor across a magnetic field. The magnitude of the e.m.f. is proportional (and in modern systems of units equal) to the rate of change of the flux linkage or the rate of cutting flux $d\Phi/dt$; the sense of the induced e.m.f. is such that any induced current opposes the change causing the induction, i.e. $E = -d\Phi/dt$. *See* Faraday's laws; Lenz's law; Neumann's law; inductance.

electromagnetic interaction *See* fundamental interactions.

electromagnetic pump A pump used for moving liquid metals, such as the liquid-sodium coolant in a fast nuclear reactor. The liquid is passed through a flattened pipe over two electrodes between which a direct current flows. A magnetic field at right angles to the current causes a force to be created directly on the liquid, along the axis of the tube. The pump has no moving parts and is therefore safe and trouble free.

electromagnetic radiation Energy resulting from the acceleration of electric charge and the associated electric fields and magnetic fields. The energy can be regarded as waves propagated through space (requiring no supporting medium) involving oscillating electric and magnetic fields at right angles to each other and to the direction of propagation. In a vacuum the waves travel with a constant speed (the speed of light) of 2.9979×10^8 metres per second; if material is present they are slower. Alternatively, the energy can be regarded as a stream of *photons travelling at the speed of light, each photon having an energy hc/λ, where h is the Planck constant, c is the speed of light, and λ is the wavelength of the associated wave. A fusion of these apparently conflicting concepts is possible using the methods of *quantum mechanics or *wave mechanics. The characteristics of the radiation depend on its wavelength. *See* electromagnetic spectrum.

electromagnetic spectrum The range of wavelengths over which *electromagnetic radiation extends. The longest waves ($10^5 - 10^{-3}$ metres) are radio waves, the next longest ($10^{-3} - 10^{-6}$ m) are infrared waves, then comes the narrow band ($4-7 \times 10^{-7}$ m) of visible light, followed by ultraviolet waves ($10^{-7} - 10^{-9}$ m), X-rays ($10^{-9} - 10^{-11}$ m), and gamma rays ($10^{-11} - 10^{-14}$ m).

electromagnetic units (EMU) A system of electrical units formerly used in the *c.g.s. system. The EMU of electric current is the *abampere* (all EMU have the prefix *ab-* attached to the names of practical units). The abampere is the current that, flowing in an arc of a circle (1 centimetre in diameter), exerts a force of 1 dyne on unit magnetic pole at the centre of the circle. In EMU the magnetic constant is of unit magnitude. The system has now been replaced by *SI units. *Compare* electrostatic units.

electromagnetic wave *See* electromagnetic radiation; wave.

electrometallurgy The uses of electrical processes in the separation of metals from their ores, the refining of metals, or the forming or plating of metals.

electrometer A measuring instrument for determining a voltage difference without drawing an appreciable current from the source. Originally electrostatic instruments based on the electroscope, they are now usually based on operational amplifiers, solid-state devices with high input impedances. Electrometers are also used to measure low currents (nanoamperes), by passing the current through a high resistance.

electromotive force (e.m.f.) The greatest potential difference that can be generated by a particular source of electric current. In practice this may be observable only when the source is not supplying current, because of its *internal resistance.

electromotive series (electrochemical series) A series of chemical elements arranged in order of their *electrode potentials. The hydrogen electrode ($H^+ + e \rightarrow$

½H_2) is taken as having zero electrode potential. Elements that have a greater tendency than hydrogen to lose electrons to their solution are taken as *electropositive*; those that gain electrons from their solution are below hydrogen in the series and are called *electronegative*. The series shows the order in which metals replace one another from their salts; electropositive metals will replace hydrogen from acids. The chief metals and hydrogen, placed in order in the series, are: potassium, calcium, sodium, magnesium, aluminium, zinc, cadmium, iron, nickel, tin, lead, hydrogen, copper, mercury, silver, platinum, gold.

electron An *elementary particle, classed as a *lepton, with a rest mass of 9.109 558 $\times 10^{-31}$ kg and a negative charge of 1.602 192 $\times 10^{-19}$ coulomb. Electrons are present in all atoms in groupings called shells around the nucleus; when they are detached from the atom they are called *free electrons*. The antiparticle of the electron is the *positron*.

electron affinity Symbol A. The energy change occurring when an atom or molecule gains an electron to form a negative ion. For an atom or molecule X, it is the energy released for the electron-attachment reaction

$$X(g) + e \rightarrow X^-(g)$$

Often this is measured in electronvolts. Alternatively, the molar enthalpy change, ΔH, can be used.

electron capture 1. The formation of a negative ion by an atom or molecule when it acquires an extra free electron. **2.** A radioactive transformation in which a nucleus acquires an electron from an inner orbit of the atom, thereby transforming, initially, into a nucleus with the same mass number but an atomic number one less than that of the original nucleus (capture of the electron transforms a proton into a neutron). This type of capture is accompanied by emission of an X-ray photon as the vacancy in the inner orbit is filled by an outer electron.

electron diffraction *Diffraction of a beam of electrons by atoms or molecules. The fact that electrons can be diffracted in a similar way to light and X-rays shows that particles can act as waves (*see* de Broglie wavelength). An electron (mass m, charge e) accelerated through a potential difference V acquires a kinetic energy $mv^2/2 = eV$, where v is the velocity of the electron. The momentum (p) of the electron is $\sqrt{(2eVm)}$. As the de Broglie wavelength (λ) of an electron is given by h/p, where h is the Planck constant, then $\lambda = h/\sqrt{(2eVm)}$. For an accelerating voltage of 3600 V, the wavelength of the electron beam is 0.02 nanometre, some 3×10^4 times shorter than visible radiation.

Electrons then, like X-rays, show diffraction effects with molecules and crystals in which the interatomic spacing is comparable to the wavelength of the beam. They have the advantage that their wavelength can be set by adjusting the voltage. Unlike X-rays they have very low penetrating power. The first observation of electron diffraction was by George Thomson (1892–1975) in 1927, in an experiment in which he passed a beam of electrons in a vacuum through a very thin gold foil onto a photographic plate. Concentric circles were produced by diffraction of electrons by the lattice. The same year Clinton J. Davisson (1881–1958) and Lester Germer (1896–1971) performed a classic experiment in which they obtained diffraction patterns by glancing an electron beam off the surface of a nickel crystal. Both experiments were important verifications of de Broglie's theory and the new quantum theory.

Electron diffraction, because of the low penetration, cannot easily be used to investigate crystal structure. It is, however, employed to measure bond lengths and angles of molecules in gases. Moreover, it is extensively used in the study of solid surfaces and absorption. The main techniques are low-energy electron diffraction (*LEED*) in which the electron beam is reflected onto a fluorescent screen, and high-energy electron diffraction (*HEED*) used either with reflection or transmission in investigating thin films.

electronegative Describing elements that tend to gain electrons and form negative ions. The halogens are typical electronegative elements. For example, in hydrogen chloride, the chlorine atom is more electronegative than the hydrogen and the molecule is polar, with negative charge on the chlorine atom. There are various ways of assigning values for the *electronegativity* of an element. *Mulliken electronegativities* are calculated from $E = (I + A)/2$, where I is ionization potential and A is electron affinity. More commonly, *Pauling electronegativities* are used. These are based on bond dissociation energies using a scale in which fluorine, the most electronegative element, has a value 4. Some other values on this scale are B 2, C 2.5, N 3.0, O 3.5, Si 1.8, P 2.1, S 2.5, Cl 3.0, Br 2.8.

electron gun A device used in *cathode-ray tubes (including television tubes), electron microscopes, etc., to produce a steady narrow beam of electrons. It usually consists of a heated cathode, control grid, and two or more annular anodes inserted in an evacuated tube. The electrons emitted by the cathode are attracted to the final anode, through which they pass. The intensity of the beam is regulated by the control grid and potential differences between the anodes create electric fields that focus the diverging electrons into a narrow beam.

electronics The study and design of control, communication, and computing devices that rely on the movement of electrons in circuits containing semiconductors, thermionic valves, resistors, capacitors, and inductors.

electron lens A device used to focus an electron beam. It is analogous to an optical lens but instead of using a refracting material, such as glass, it uses a coil or coils to produce a magnetic field or an arrangement of electrodes between which an electric field is created. Electron lenses are used in *electron microscopes and *cathode-ray tubes.

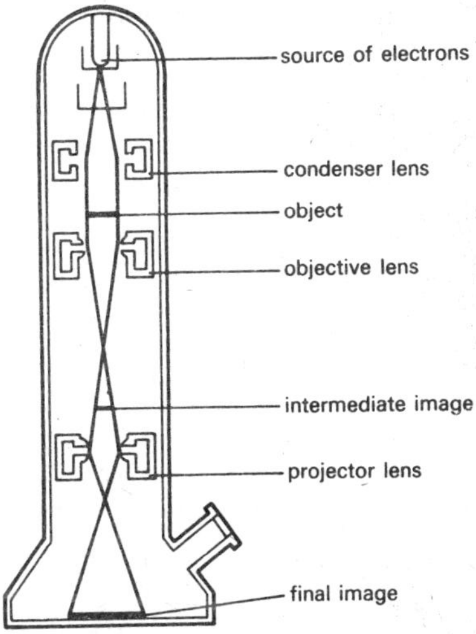

Principle of transmission electron microscope

electron microscope A form of microscope that uses a beam of electrons instead of a beam of light (as in the optical microscope) to form a large image of a very small object. In optical microscopes the resolution is limited by the wavelength of the light. High-energy electrons, however, can be associated with a considerably shorter wavelength than light; for example, electrons accelerated to an energy of 10^5 electronvolts have a wavelength of 0.04 nanometre (*see* de Broglie wavelength) enabling a resolution of 0.2–0.5 nm to be achieved. The *transmission electron microscope* has an electron beam, sharply focused by *electron lenses, passing through a very thin metallized specimen (less than 50 nanometres thick) onto a fluorescent screen, where a visual image is formed. This image can be photographed. The *scanning electron microscope* can be used with thicker specimens and forms a perspective image, although the resolution and magnification are lower. In this type of instrument a beam of primary electrons scans the specimen and those that are reflected, together with any secondary electrons emitted, are collected. This current is used to modulate a separate electron beam in a TV monitor, which scans the screen at the same frequency, consequently building up a picture of the specimen. The resolution is limited to about 10–20 nm.

electron optics The study of the use of *electron lenses in the *electron microscope, *cathode-ray tubes, and other similar devices. The focusing of beams of positive or negative ions also relies on these methods.

electron probe microanalysis (EPM) A method of analysing a very small quantity of a substance (as little as 10^{-13} gram). The method consists of directing a very finely focused beam of electrons on to the sample to produce the characteristic X-ray spectrum of the elements present. It can be used quantitatively for elements with atomic numbers in excess of 11.

electron-spin resonance (ESR) A spectroscopic method of locating electrons within the molecules of a paramagnetic substance (*see* magnetism) in order to provide information regarding its bonds and structure. The spin of an unpaired electron is associated with a *magnetic moment that is able to align itself in one of two ways with an applied external magnetic field. These two alignments correspond to different *energy levels, with a statistical probability, at normal temperatures, that there will be slightly more in the lower state than in the higher. By applying microwave radiation to the sample a transition to the higher state can be achieved. The precise energy difference between the two states of an electron depends on the surrounding electrons in the atom or molecule. In this way the position of unpaired electrons can be investigated. The technique is used particularly in studying free radicals and paramagnetic substances such as inorganic complexes. *See also* nuclear magnetic resonance.

electronvolt Symbol eV. A unit of energy equal to the work done on an electron in moving it through a potential difference of one volt. It is used as a measure of particle energies although it is not an *SI unit. 1 eV $= 1.602 \times 10^{-19}$ joule.

electrophoresis (cataphoresis) A technique for the analysis and separation of colloids, based on the movement of charged colloidal particles in an electric field. There are various experimental methods. In one the sample is placed in a U-tube and a buffer solution added to each arm, so that there are sharp boundaries between buffer and sample. An electrode is placed in each arm, a voltage applied, and the motion of the boundaries under the influence of the field is observed. The rate of migration of the particles depends on the field, the charge on the particles, and on other factors, such as the size and shape of the particles. More simply, electrophoresis can be carried out using an adsorbent, such as a strip of filter paper, soaked in a buffer with two electrodes making contact. The sample is placed between the electrodes and a voltage applied. Different components of the mixture migrate at different rates, so the sample separates into zones. The components can be identified by the rate at which they move. This technique has also been called *electrochromatography*.

Electrophoresis is used extensively in studying mixtures of proteins, nucleic acids, carbohydrates, enzymes, etc. In clinical medicine it is used for determining the protein content of body fluids.

electrophorus An early form of *electrostatic generator. It consists of a flat dielectric plate and a metal plate with an insulated handle. The dielectric plate is charged by friction and the metal plate is placed on it and momentarily earthed, which leaves the metal plate with an induced charge of opposite polarity to that of the dielectric plate. The process can be repeated until all of the original charge has leaked away.

electroplating A method of plating one metal with another by *electrodeposition. The articles to be plated are made the cathode of an electrolytic cell and a rod or bar of the plating metal is made the anode. Electroplating is used for covering metal with a decorative, more expensive, or corrosion-resistant layer of another metal.

electropositive Describing elements that tend to lose electrons and form positive ions. The alkali metals are typical electropositive elements.

electroscope A device for detecting electric charge and for identifying its polarity. In the *gold-leaf electroscope* two rectangular gold leaves are attached to the end of a conducting rod held in an insulated frame. When a charge is applied to a plate attached to the other end of the conducting rod, the leaves move apart owing to the mutual repulsion of the like charges they have received.

electrostatic field The *electric field that surrounds a stationary charged body.

electrostatic generator A device used to build up electric charge to an extreme potential usually for experimental purposes. The *electrophorus and the *Wimshurst machine were early examples; a more usual device now is the *Van de Graaff generator.

electrostatic precipitation A method of removing solid and liquid particles from suspension in a gas. The gas is exposed to an electric field so that the particles are attracted to and deposited on a suitably placed electrode. Electrostatic precipitation is widely used to remove dust and other pollutants from waste gases and from air. *See also* Cottrell precipitator.

electrostatics The study of electric charges at rest, the forces between them (*see* Coulomb's law), and the electric fields associated with them. *Compare* electrodynamics.

electrostatic units (ESU) A system of electrical units in the *c.g.s. system. The ESU of electric charge is the *statcoulomb* (all ESU have the prefix *stat-* attached to the names of practical units). The statcoulomb is the quantity of electric charge that will repel an equal quantity 1 centimetre distant with a force of 1 dyne. In ESU the electric constant is of unit magnitude. The system has now been replaced by *SI units. *Compare* electromagnetic units.

electrostriction A change in the dimensions of a body as a result of reorientation of its molecules when it is placed in an electric field. If the field is not homogeneous the body will tend to move; if its relative permittivity is higher than that of its surroundings it will tend to move into a region of higher field strength. *Compare* magnetostriction.

electrum **1.** An alloy of gold and silver containing 55–88% of gold. **2.** A *German silver alloy containing 52% copper, 26% nickel, and 22% zinc.

elementary particles The fundamental constituents of all the matter in the universe. Until the discovery of the electron by J. J. Thomson in 1898, it was assumed that atoms were the fundamental constituents of matter. This discovery, and Rutherford's discovery of the atomic nucleus and the proton in 1914, made it apparent that atoms were not themselves elementary, in the sense that they have an internal structure. Chadwick's discovery of the neutron in 1932 completed the atomic model based on an atomic nucleus consisting of protons and neutrons surrounded by sufficient electrons to balance the nuclear charge. It did not, however, explain the great stability of the nucleus, which clearly could not be held together by an electromagnetic interaction, as the neutron has no electric charge. In 1935 Yukawa suggested that the *exchange forces* that held them together involved short-lived particles, called *mesons*, which jumped from proton to neutron and back again. This concept led to the discovery of strong interactions, and weak interactions, bringing the total of *fundamental interactions to four. It also led to the discovery of some 200 short-lived 'elementary' particles, some of which were clearly more elementary than others. In the current classification two main classes of particles are recognized: *leptons (electron, muon, neutrinos, tau particle), which interact either by the electromagnetic interaction or the weak interaction and have no apparent internal structure; and *hadrons (nucleons, pions, etc.), which interact by the strong interaction and have a complex internal structure.

Hadron structure is currently based on Murray Gell-Mann's concept of the *quark*, introduced in 1963. In this model, hadrons are divided into *baryons* (which decay into protons) and *mesons* (which decay into leptons and photons). Baryons consist of three quarks and mesons consist of two quarks (a quark and antiquark). In the quark theory, therefore, the only truly elementary particles are leptons and quarks. Unlike electrons and protons, which have exactly equal but opposite charges, quarks have charges that are fractions of the electronic charge ($+2/3$ or $-1/3$ of the electronic charge). Quarks occur in six *flavours* (no connection with taste): up (u; $+2/3$ charge), down (d; $-1/3$), charmed (c; $+2/3$), strange (s; $-1/3$), top (t; $+2/3$), and bottom (b; $-1/3$). The evidence for the existence of top is not totally established. The proton, being a baryon, consists of three quarks uud ($2/3 + 2/3 - 1/3 = 1$) and the neutron consists of udd ($2/3 - 1/3 - 1/3 = 0$). For each flavour there are equivalent antiquarks ($\bar{u}$, $\bar{d}$, etc.), which have opposite values of electric charge ($\bar{u} = -2/3$, etc.).

In order to avoid conflict with the *Pauli exclusion principle it has proved necessary to add the concept of *colour charge* to the six flavours. Each flavour of quark occurs in the three primary colours red, green, and blue; each antiquark has the complementary anticolours cyan, magenta, and yellow. There are therefore a total of 18 quarks and 18 antiquarks. The rules of combination to form hadrons are that the combinations of colours should always give white, either by mixing three primary colours to make baryons or by mixing one primary colour with its complementary anticolour to make mesons. The use of the word 'colour' in this context implies an analogy to visual colours, not that the particles are coloured. The theory governing these combinations is modelled on quantum electrodynamics and is called *quantum chromodynamics*.

The strong interaction between quarks is visualized as occurring by the exchange of eight types of chargeless particles, with no rest mass, called *gluons* (because they glue the quarks together). Although gluons, like the photons that perform an analogous function between leptons, have no electrical charge, they do have a colour charge. Each gluon carries one colour and one anticolour. There are nine colour–anticolour pairs but one is excluded as it is equivalent to white. In an interaction a quark can change its colour, but all colour changes are accompanied by the emission of a gluon. This gluon is then absorbed by another quark, whose colour changes to compensate for the original change. For example, if a red quark changes to green it emits a gluon bearing the colours red + antigreen. When this is absorbed by a green quark the green of the quark and the antigreen of the gluon annihilate each other leaving the second quark with the colour red, acquired from the gluon. The net change is nil, because as a result of the interaction there is still one green quark and one red quark. All hadrons therefore remain white, even though quark colours move from point to point. The strong force can be envisaged as the system of interactions needed to maintain this condition.

The whole elaborate quark theory is now well established by circumstantial evidence, but as neither quarks nor gluons have ever been identified in experiments the theory does not claim to have been directly verified. Individual quarks would have the curious property of being much more massive than the hadrons they usually form (because of the enormous potential energy they would have when separated) and some theorists believe it is in consequence fundamentally impossible for them to exist in isolation. However some experimenters have reported results consistent with the presence of the fractional charges that unattached quarks would have.

elements of an orbit Six parameters that can be used to define the path of a celestial body. The shape of the orbit is defined by its eccentricity (*see* conic) and semimajor axis. The orientation of the orbit is specified by the *inclination of the orbital plane to the reference plane (usually the *ecliptic) and by the longitude of the ascending *node (the angular distance from the vernal equinox to the ascending node). The position of the body in its orbit is defined by its eccentric *anomaly and the position as a function of time is calculated from the time of periapsis passage (*see* apsides).

elevation of boiling point An increase in the boiling point of a liquid when a solid is dissolved in it. The elevation is proportional to the number of particles dissolved (molecules or ions) and is given by $\Delta t = k_B C$, where C is the molal concentration of solute. The constant k_B is the *ebullioscopic constant* of the solvent and if this is known, the molecular weight of the solute can be calculated from the measured value of Δt. The elevation is measured by a Beckmann thermometer. *See also* colligative property.

Elinvar Trade name for a nickel–chromium steel containing about 36% nickel, 12% chromium, and smaller proportions of tungsten and manganese. Its elasticity does not vary with temperature and it is therefore used to make hairsprings for watches.

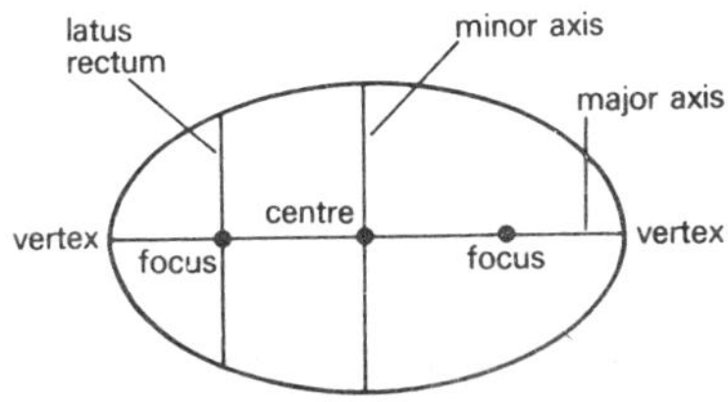

An ellipse

ellipse A *conic formed by the intersection of a plane with a right circular cone, so that the plane is inclined to the axis of the cone at an angle in excess of half the apex angle of the cone. The ellipse has two vertices, which are joined by a line called the *major axis*. The centre of the ellipse falls on this line, midway between the vertices. The *minor axis* is the line perpendicular to the major axis that passes through the centre and joins two points on the ellipse. The *foci* of an ellipse are two points on the major axis so placed that for any point on the ellipse the sum of the distances from that point to each focus is constant. The area of an ellipse is πab, where a and b are half the major and minor axes, respectively. For an ellipse centred at the origin, the equation in Cartesian coordinates is $x^2/a^2 + y^2/b^2 = 1$. The foci are at $(ea, 0)$ and $(-ea, 0)$, where e is the eccentricity. Each of the two chords of the ellipse passing through a focus and parallel to the minor axis is called a *latus rectum* and has a length equal to $2b^2/a$.

ellipsoid A solid body formed when an *ellipse is rotated about an axis. If it is rotated about its major axis it is a *prolate ellipsoid*; if it is rotated about its minor axis it is an *oblate ellipsoid*. For an ellipsoid centred at the origin the equation in Cartesian coordinates is:

$$x^2/a^2 + y^2/b^2 + z^2/c^2 = 1.$$

elliptical galaxy *See* galaxy.

elliptical polarization *See* polarization of light.

eluate *See* elution.

eluent *See* elution.

elution The process of removing an adsorbed material (*adsorbate*) from an adsorbent by washing it in a liquid (*eluent*). The solution consisting of the adsorbate dissolved in the eluent is the *eluate*.

elutriation The process of suspending finely divided particles in an upward flowing stream of air or water to wash and separate them into sized fractions.

emanation The former name for the gas radon, of which there are three isotopes: Rn–222 (radium emanation), Rn–220 (thoron emanation), and Rn–219 (actinium emanation).

e.m.f. *See* electromotive force.

emission spectrum *See* spectrum.

emissivity Symbol ε. The ratio of the power per unit area radiated by a surface to that radiated by a *black body at the same temperature. A black body therefore has an emissivity of 1 and a perfect reflector has an emissivity of 0. The emissivity of a surface is equal to its *absorptance.

emittance *See* exitance.

emitter *See* transistor.

empirical Denoting a result that is obtained by experiment or observation rather than from theory.

empirical formula *See* formula.

emulsion A *colloid in which small particles of one liquid are dispersed in another liquid. Usually emulsions involve a dispersion of water in an oil or a dispersion of oil in water, and are stabilized by an *emulsifier*. Commonly emulsifiers are substances, such as detergents, that have lyophobic and lyophilic parts in their molecules.

enantiomers *See* optical activity.

enantiomorphism *See* optical activity.

endoergic Denoting a nuclear process that absorbs energy. *Compare* exoergic.

endothermic Denoting a chemical reaction that takes heat from its surroundings. *Compare* exothermic.

energy A measure of a system's ability to do work. Like work itself, it is measured in joules. Energy is conveniently classified into two forms: *potential energy* is the energy stored in a body or system as a consequence of its position, shape, or state (this includes gravitational energy, electrical energy, nuclear energy, and chemical energy); *kinetic energy* is energy of motion and is usually defined as the work that will be done by the body possessing the energy when it is brought to rest. For a body of mass m having a speed v, the kinetic energy is $mv^2/2$. The rotational kinetic energy of a body having an angular velocity ω is $I\omega^2/2$, where I is its moment of inertia.

The *internal energy of a body is the sum of the potential energy and the kinetic energy of its component atoms and molecules.

energy band A range of energies that electrons can have in a solid. In a single atom, electrons exist in discrete *energy levels. In a crystal, in which large numbers of atoms are held closely together in a lattice, electrons are influenced by a number of adjacent nuclei and the sharply defined levels of the atoms become bands of allowed energy; this approach to energy levels in solids is often known as the *band theory*. Each band represents a large number of allowed quantum states. Between the bands are *forbidden bands*. The outermost electrons of the atoms (i.e. the ones responsible for chemical bonding) form the *valence band* of the solid. This is the band, of those occupied, that has the highest energy.

The band structure of solids accounts for their electrical properties. In order to move through the solid, the electrons have to change from one quantum state to another. This can only occur if there are empty quantum states close to the electrons. In general, if the valence band is full, electrons cannot change to new quantum states in the same band. For conduction to occur, the electrons have to be in an unfilled band – the *conduction band*. Metals are good conductors either because the valence band and the conduction band are only half-filled or because the conduction band overlaps with the valence band; in either case vacant states are available. In insulators the conduction band and valence band are separated by a wide forbidden band and electrons do not have enough energy to 'jump' from one to the other.

In intrinsic *semiconductors the forbidden gap is narrow and, at normal temperatures, electrons at the top of the valence band can move by thermal agitation into the conduction band (at absolute zero, a semiconductor would act as an insulator). Doped semiconductors have extra bands in the forbidden gap.

energy level A definite fixed energy that a molecule, atom, electron, or nucleus can have. In an atom, for example, the atom has a fixed energy corresponding to the *orbitals in which its electrons move around the nucleus. The atom can accept a quantum of energy to become an excited atom (*see* excitation) if that extra energy will raise an electron to a permitted orbital. Between the *ground state*, which is the lowest possible energy level for a particular system, and the first excited state there are no permissi-

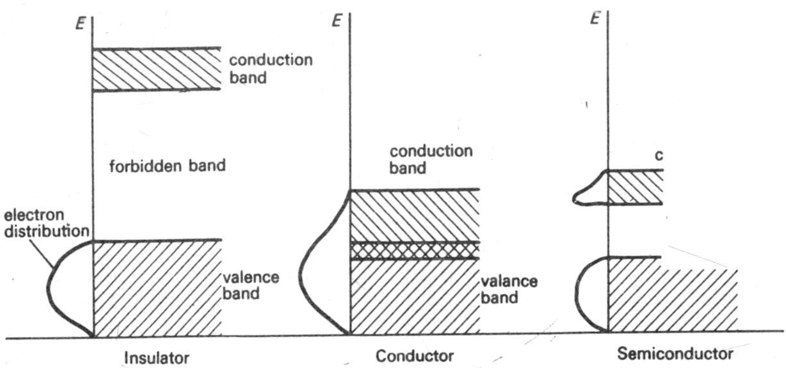

Energy bands

ble energy levels. According to the *quantum theory, only certain energy levels are possible. An atom passes from one energy level to the next without passing through fractions of that energy transition. These levels are usually described by the energies associated with the individual electrons in the atoms, which are always lower than an arbitrary level for a free electron. The energy levels of molecules also involve quantized vibrational and rotational motion.

engine Any device for converting some other form of energy into mechanical work. *See* heat engine; Carnot cycle; internal-combustion engine; steam engine.

enrichment The process of increasing the abundance of a specified isotope in a mixture of isotopes. It is usually applied to an increase in the proportion of U–235, or the addition of Pu–239 to natural uranium for use in a nuclear reactor or weapon.

enthalpy Symbol H. A thermodynamic property of a system defined by $H = U + pV$, where H is the enthalpy, U is the internal energy of the system, p its pressure, and V its volume. In a chemical reaction carried out in the atmosphere the pressure remains constant and the enthalpy of reaction, ΔH, is equal to $\Delta U + p\Delta V$. For an exothermic reaction ΔH is taken to be negative.

entropy Symbol S. A measure of the unavailability of a system's energy to do work; an increase in entropy is accompanied by a decrease in energy availability. When a system undergoes a reversible change the entropy (S) changes by an amount equal to the energy (Q) absorbed by the system divided by the thermodynamic temperature (T) at which the energy is absorbed, i.e. $\Delta S = \Delta Q/T$. However, all real processes are to a certain extent irreversible changes and in any closed system an irreversible change is always accompanied by an increase in entropy.

In a wider sense entropy can be interpreted as a measure of a system's disorder; the higher the entropy the greater the disorder. As any real change to a closed system tends towards higher entropy, and therefore higher disorder, it follows that the entropy of the universe (if it can be considered a closed system) is increasing and its available energy is decreasing (*see* heat death of the universe). This increase in the entropy of the universe is one way of stating the second law of *thermodynamics.

ephemeris A tabulation showing the calculated future positions of the sun, moon, and planets, together with other useful information for astronomers and navigators. It is published at regular intervals.

ephemeris time (ET) A time system that has a constant uniform rate as opposed to other systems that depend on the earth's rate of rotation, which has inherent irregularities. It is reckoned from an instant in 1900 (Jan 0d 12h) when the sun's mean longitude was 279.696 677 8°. The unit of ephemeris time is the tropical year, which contains 31 556 925.9747 *ephemeris seconds.* This fundamental definition of the *second was replaced in 1964 by the caesium second of atomic time.

epicentre The point on the surface of the earth directly above the focus of an earthquake or directly above or below a nuclear explosion.

epicycle A small circle whose centre rolls around the circumference of a larger fixed circle. The curve traced out by a point on the epicycle is called an *epicycloid.*

epidiascope An optical instrument used by lecturers, etc., for projecting an enlarged image of either a translucent object (such as a slide or transparency) or an opaque object (such as a diagram or printed page) onto a screen.

epitaxy (epitaxial growth) Growth of a layer of one substance on a single crystal of another, such that the crystal structure in the layer is the same as that in the substrate. It is used in making semiconductor devices.

epithermal neutron A neutron with an energy in excess of that associated with a thermal neutron (*see* moderator) but less than that of a *fast neutron, i.e. a neutron having an energy in the range 0.1 to 100 eV.

EPM *See* electron probe microanalysis.

equation of motion (kinematic equation) Any of four equations that apply to bodies moving linearly with uniform acceleration (a). The equations, which relate distance covered (s) to the time taken (t), are:

$$v = u + at$$
$$s = (u + v)t/2$$
$$s = ut + at^2/2$$
$$v^2 = u^2 + 2as$$

where u is the initial velocity of the body and v is its final velocity.

equation of state An equation that relates the pressure p, volume V, and thermodynamic temperature T of an amount of substance n. The simplest is the ideal *gas law:

$$pV = nRT,$$

where R is the universal gas constant. Applying only to ideal gases, this equation takes no account of the volume occupied by the gas molecules (according to this law if the pressure is infinitely great the volume becomes zero), nor does it take into account any forces between molecules. A more accurate equation of state would therefore be

$$(p + k)(V - nb) = nRT,$$

where k is a factor that reflects the decreased pressure on the walls of the container as a result of the attractive forces between particles, and nb is the volume occupied by the particles themselves when the pressure is infinitely high. In the *van der Waals equation of state*, proposed by the Dutch physicist J. D. van der Waals (1837–1923),

$$k = n^2a/V^2,$$

where a is a constant. This equation more accurately reflects the behaviour of real gases; several others have done better but are more complicated.

equation of time The length of time that must be added to the mean solar time, as shown on a clock, to give the apparent solar time, as shown by a sundial. The amount varies during the year, being a minimum of −14.2 minutes in February and a maximum of +16.4 minutes in October. It is zero on four days (April 15/16, June 14/15, Sept. 1/2, Dec. 25/26). The difference arises as a result of two factors: the eccentricity of the earth's orbit and the inclination of the ecliptic to the celestial equator.

equator 1. The great circle around the earth that lies in a plane perpendicular to the earth's axis. It is equidistant from the two geographical poles. **2.** The *magnetic equator* is a line of zero magnetic dip (*see* geomagnetism) that is close to the geographical equator but lies north of it in Africa and south of it in America. **3.** The *celestial equator* is the circle formed on the *celestial sphere by the extension of the earth's equatorial plane.

equilibrium A state in which a system has its energy distributed in the statistically most probable manner; a state of a system in which forces, influences, reactions, etc., balance each other out so that there is no net change.

A body is in *static equilibrium* if the resultants of all forces and all couples acting on it are both zero; it may be at rest and will certainly not be accelerated. Such a body at rest is in *stable equilibrium* if after a slight displacement it returns to its original position – for a body whose weight is the only downward force this will be the case if the vertical line through its centre of gravity always passes through its base. If at any time this vertical line does not pass through the base, the body will be in *unstable equilibrium* and any displacement will tend to increase.

A body is said to be in *thermal equilibrium* if no net heat exchange is taking place within it or between it and its surroundings. A system is in chemical equilibrium when a reaction and its reverse are proceeding at equal rates. These are examples of *dynamic equilibrium*, in which activity in one sense or direction is in aggregate balanced by comparable reverse activity.

equinox 1. Either of the two points on the *celestial sphere at which the *ecliptic intersects the celestial equator. The sun appears to cross the celestial equator from south to north at the *vernal equinox* and from north to south at the *autumnal equinox*. **2.** Either of the two instants at which the centre of the sun appears to cross the celestial equator. In the northern hemisphere the vernal equinox occurs on or about March 21 and the autumnal equinox on or about Sept. 23. In the southern hemisphere the dates are reversed. *See* precession of the equinoxes.

equipartition of energy The theory, proposed by Ludwig Boltzmann (1844–1906) and given some theoretical support by James Clerk Maxwell (1831–79), that the energy of gas molecules in a large sample under thermal *equilibrium is equally divided among their available *degrees of freedom, the average energy for each degree of freedom being $kT/2$, where k is the *Boltzmann constant and T is the thermodynamic temperature. The proposition is not generally true if *quantum considerations are important, but is frequently a good approximation.

erecting prism A glass prism used in optical instruments to convert an inverted image into an erect image, as in prismatic binoculars.

erg A unit of work or energy used in the c.g.s. system and defined as the work done by a force of 1 dyne when it acts through a distance of 1 centimetre. 1 erg = 10^{-7} joule.

ergonomics The study of the engineering aspects of the relationship between workers and their working environment.

Esaki diode *See* tunnel diode.

ESCA *See* photoelectron spectroscopy.

escapement A device in a clock or watch that controls the transmission of power from the spring or falling weight to the hands. It is usually based on a balance wheel or pendulum. It thus allows energy to enter the mechanism in order to move the hands round the face, overcome friction in the gear trains, and maintain the balance wheel or pendulum in continuous motion.

escape velocity The minimum speed needed by a space vehicle, rocket, etc., to escape from the gravitational field of the earth, moon, or other celestial body. The gravitational force between a rocket of mass m and a celestial body of mass M and radius r is MmG/r^2 (*see* Newton's law of gravitation). Therefore the gravitational potential energy of the rocket with respect to its possible position very far from the celestial body on which it is resting can be shown to be $-GmM/r$, assuming (by convention) that the potential energy is zero at an infinite distance from the celestial body. If the rocket is to escape from the gravitational field it must have a kinetic energy that exceeds this potential energy, i.e. the kinetic energy $mv^2/2$ must be greater than MmG/r, or $v > \sqrt{(2MG/r)}$. This is the value of the escape velocity. Inserting numerical values for the earth and moon into this relationship gives an escape velocity from the earth of 11 200 m s^{-1} and from the moon of 2370 m s^{-1}.

ESR *See* electron-spin resonance.

ether (aether) A hypothetical medium once believed to be necessary to support the propagation of electromagnetic radiation. It is now regarded as unnecessary and in modern theory electromagnetic radiation can be propagated through empty space. The existence of the ether was first called into question as a result of the *Michelson–Morley experiment.

eudiometer An apparatus for measuring changes in volume of gases during chemical reactions. A simple example is a graduated glass tube sealed at one end and inverted in mercury. Wires passing into the tube allow the gas mixture to be sparked to initiate the reaction between gases in the tube.

eutectic mixture A solid solution consisting of two or more substances and having the lowest freezing point of any possible mixture of these components. The minimum freezing point for a set of components is called the *eutectic point*. Low melting-point alloys are usually eutectic mixtures.

evaporation The change of state of a liquid into a vapour at a temperature below the boiling point of the liquid. Evaporation occurs at the surface of a liquid, some of those molecules with the highest kinetic energies escaping into the gas phase. The result is a fall in the average kinetic energy of the molecules of the liquid and consequently a fall in its temperature.

even–even nucleus An atomic nucleus containing an even number of protons and an even number of neutrons.

even–odd nucleus An atomic nucleus containing an even number of protons and an odd number of neutrons.

event horizon *See* black hole.

evolute The locus of the centres of curvature of all the points on a given curve (called the *involute*).

exa- Symbol E. A prefix used in the metric system to denote 10^{18} times. For example, 10^{18} metres = 1 exametre (Em).

excess electron An electron in a *semiconductor that is not required in the bonding system of the crystal lattice and has been donated by an impurity atom. It is available for conduction (*excess conduction*).

exchange force **1.** A force resulting from the continued interchange of particles in a manner that bonds their hosts together. Examples are the covalent bond involving electrons, and the strong interaction (*see* fundamental interactions) in which mesons are exchanged between nucleons or gluons are exchanged between quarks (*see* elementary particles). **2.** *See* magnetism.

excitation **1.** A process in which a nucleus, electron, atom, ion, or molecule acquires energy that raises it to a quantum state (*excited state*) higher than that of its *ground state. The difference between the energy in the ground state and that in the excited state is called the *excitation energy*. *See* energy level. **2.** The process of applying current to the winding of an electromagnet, as in an electric motor. **3.** The process of applying a signal to the base of a transistor or the control electrode of a thermionic valve.

exciton An electron–hole pair in a crystal that is bound in a manner analogous to the electron and proton of a hydrogen atom. It behaves like an atomic excitation that passes from one atom to another and may be long-lived. Exciton behaviour in *semiconductors is important.

exclusion principle *See* Pauli exclusion principle.

exitance Symbol M. The radiant or luminous flux emitted per unit area of a surface. The *radiant exitance* (M_e) is measured in watts per square metre ($W\,m^{-2}$), while the *luminous exitance* (M_v) is measured in lumens per square metre ($lm\,m^{-2}$). Exitance was formerly called *emittance*.

exoergic Denoting a nuclear process that gives out energy. *Compare* endoergic.

exosphere *See* earth's atmosphere.

exothermic Denoting a chemical reaction that releases heat into its surroundings. *Compare* endothermic.

expansion The writing of a function or quantity as a *series of terms. The series may be finite or infinite. *See* binomial theorem; Taylor series.

expansion of the universe The hypothesis, based on the evidence of the *redshift, that the distance between the galaxies is continuously increasing. The original theory, which was proposed in 1929 by Edwin Hubble (1889–1953), assumes that the galaxies are flying apart like fragments from a bomb as a consequence of the big bang with which the universe originated. Several variants have since been proposed. *See also* big-bang theory; Hubble constant.

expansivity (thermal expansion) 1. *Linear expansivity* is the fractional increase in length of a specimen of a solid, per unit rise in temperature. If a specimen increases in length from l_1 to l_2 when its temperature is raised $\theta°$, then the expansivity (α) is given by:

$$l_2 = l_1(1 + \alpha\theta).$$

This relationship assumes that α is independent of temperature. This is not, in general, the case and a more accurate relationship is:

$$l_2 = l_1(1 + a\theta + b\theta^2 + c\theta^3 \ldots),$$

where a, b, and c are constants. **2.** *Superficial expansivity* is the fractional increase in area of a solid surface caused by unit rise in temperature, i.e.

$$A_2 = A_1(1 + \beta\theta),$$

where β is the superficial expansivity. To a good approximation $\beta = 2\alpha$. **3.** *Volume expansivity* is the fractional increase in volume of a solid, liquid, or gas per unit rise in temperature, i.e.

$$V_2 = V_1(1 + \gamma\theta),$$

where γ is the cubic expansivity and $\gamma = 3\alpha$. For liquids, the expansivity observed directly is called the *apparent expansivity* as the container will also have expanded with the rise in temperature. The *absolute expansivity* is the apparent expansivity plus the volume expansivity of the container. For the expansion of gases, *see* Charles' Law.

exponent A number or symbol that indicates the power to which another number or expression is raised. For example, $(x + y)^n$ indicates that the expression $(x + y)$ is raised to the nth power; n is the exponent. Any number or expression in which the exponent is zero is equal to 1, i.e. $x^0 = 1$.

exponential A function that varies as the power of another quantity. If $y = a^x$, y varies exponentially with x. The function e^x is called the *exponential function* (*see* e). It is equal to the sum of the *exponential series*, i.e.

$$e^x = 1 + x + x^2/2! + x^3/3! + \ldots + x^n/n!$$

exponential growth A form of population growth in which the rate of growth is related to the number of individuals present. Increase is slow when numbers are low but rises sharply as numbers increase. If population number is plotted against time on a graph a characteristic J-shaped curve results. In animal and plant populations, such factors as overcrowding, lack of nutrients, and disease limit population increase beyond a certain point and the J-shaped exponential

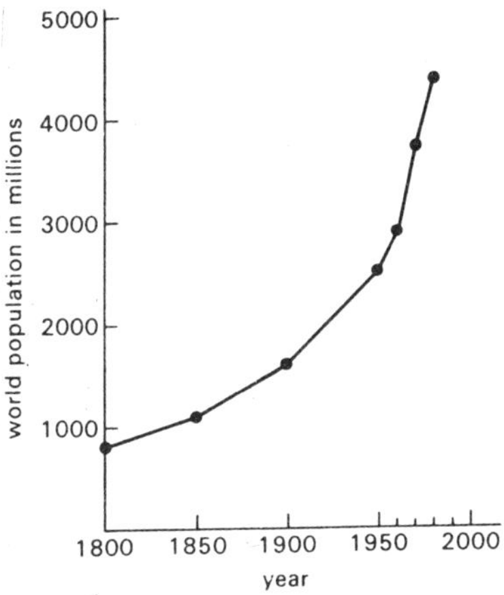

Graph showing exponential growth of the human population

curve tails off giving an S-shaped (sigmoid) curve.

exposure meter A photocell that operates a meter to indicate the correct exposure for a specified film in photography. It enables the correct shutter speed and aperture to be chosen for any photographic circumstances. Some cameras have a built-in exposure meter that automatically sets the aperture according to the amount of light available and the chosen shutter speed.

extensometer Any device for measuring the extension of a specimen of a material under longitudinal stress. A common method is to make the specimen form part of a capacitor, the capacitance of which will change with a change in the specimen's dimensions.

extinction coefficient A measure of the extent by which the intensity of a beam of light is reduced by passing through a distance d of a solution having a molar concentration c of the dissolved substance. If the intensity of the light is reduced from I_1 to I_2, the extinction coefficient is $[\log(I_1/I_2)]/cd$.

extraction 1. The process of obtaining a metal from its ore. **2.** The separation of a component from a mixture by selective solubility.

extraordinary ray *See* double refraction.

extremely high frequency (EHF) A radio frequency between 30 000 megahertz and 300 gigahertz.

extrinsic semiconductor *See* semiconductor.

eye The organ of sight. They normally occur in pairs, are nearly spherical, and filled with fluid. Light is refracted by the *cornea through the pupil in the *iris and onto the *lens, which focuses images onto the retina. These images are received by light-sensitive cells in the retina (*see* cone; rod), which transmit impulses to the brain via the optic nerve.

eyepiece (ocular) The lens or system of lenses in an optical instrument that is nearest to the eye. It usually produces a magnified image of the previous image formed by the instrument.

F

Fabry–Pérot interferometer A type of *interferometer in which monochromatic light is passed through a pair of parallel half-silvered glass plates producing circular interference fringes. One of the glass plates is adjustable, enabling the separation of the plates to be varied. The wavelength of the light can be determined by observing the fringes while adjusting the separation. This type of instrument is used in spectroscopy.

face-centred cubic (f.c.c.) *See* cubic crystal.

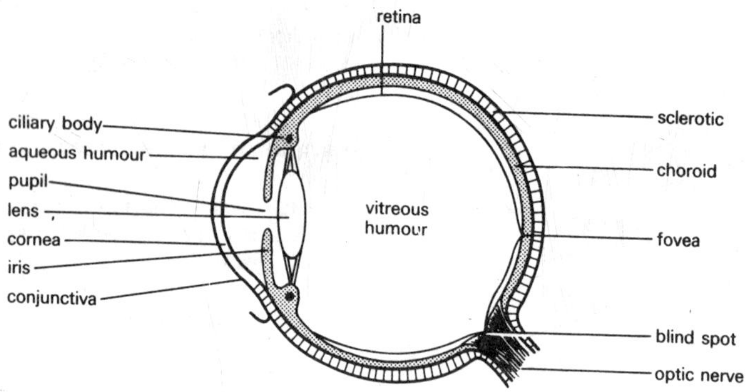

Structure of the vertebrate eye

factorial The product of a given number and all the whole numbers below it. It is usually writen $n!$, e.g. factorial $4 = 4! = 4 \times 3 \times 2 \times 1 = 24$. Factorial 0 is defined as 1.

Fahrenheit scale A temperature scale in which (by modern definition) the temperature of boiling water is taken as 212 degrees and the temperature of melting ice as 32 degrees. It was invented in 1714 by the German scientist G. D. Fahrenheit (1686–1736), who set the zero at the lowest temperature he knew how to obtain in the laboratory (by mixing ice and common salt) and took his own body temperature as 96°F. The scale is no longer in scientific use. To convert to the *Celsius scale the formula is $C = 5(F - 32)/9$.

fall-out (*or* **radioactive fall-out)** Radioactive particles deposited from the atmosphere either from a nuclear explosion or from a nuclear accident. *Local fall-out*, within 250 km of an explosion, falls within a few hours of the explosion. *Tropospheric fall-out* consists of fine particles deposited all round the earth in the approximate latitude of the explosion within about one week. *Stratospheric fall-out* may fall anywhere on earth over a period of years. The most dangerous radioactive isotopes in fall-out are the fission fragments iodine–131 and strontium–90. Both can be taken up by grazing animals and passed on to human populations in milk, milk products, and meat. Iodine–131 accumulates in the thyroid gland and strontium–90 accumulates in bones.

farad Symbol F. The SI unit of capacitance, being the capacitance of a capacitor that, if charged with one coulomb, has a potential difference of one volt between its plates. $1\ F = 1\ C V^{-1}$. The farad itself is too large for most applications; the practical unit is the microfarad (10^{-6} F). The unit is named after Michael Faraday (1791–1867).

Faraday cage An earthed screen made of metal wire that surrounds an electric device in order to shield it from external electrical fields.

Faraday constant Symbol *F*. The electric charge carried by one mole of electrons or singly ionized ions, i.e. the product of the *Avogadro constant and the charge on an electron (disregarding sign). It has the value $9.648\,670 \times 10^4$ coulombs per mole. This number of coulombs is sometimes treated as a unit of electric charge called the *faraday*.

Faraday effect The rotation of the plane of polarization of electromagnetic radiation on passing through an isotropic medium exposed to a magnetic field. The angle of rotation is proportional to Bl, where l is the length of the path of the radiation in the medium and B is the magnetic flux density.

Faraday's laws Two laws describing electrolysis:
(1) The amount of cnemical change during electrolysis is proportional to the charge passed.
(2) The charge required to deposit or liberate a mass m is given by $Q = Fmz/M$, where F is the Faraday constant, z the charge of the ion, and M the relative ionic mass.

These are the modern forms of the laws. Originally, they were stated by Faraday in a different form:
(1) The amount of chemical change produced is proportional to the quantity of electricity passed.
(2) The amount of chemical change produced in different substances by a fixed quantity of electricity is proportional to the electrochemical equivalent of the substance.

Faraday's laws of electromagnetic induction (1) An e.m.f. is induced in a conductor when the magnetic field surrounding it changes. (2) The magnitude of the e.m.f. is proportional to the rate of change of the field. (3) The direction of the induced e.m.f. depends on the orientation of the field.

fast neutron A neutron resulting from nuclear fission that has an energy in excess of 0.1 MeV (1.6×10^{-14} J), having lost little of its energy by collision. In some contexts *fast fission* is defined as fission brought about by fast neutrons, i.e. neutrons having energies in excess of 1.5 MeV (2.4×10^{-13} J), the fission threshold of uranium–238. *See also* nuclear reactor; slow neutron.

fast reactor *See* nuclear reactor.

fatigue *See* metal fatigue.

f.c.c. Face-centred cubic. *See* cubic crystal.

feedback The use of part of the output of a system to control its performance. In *positive feedback*, the output is used to enhance the input; an example is an electronic oscillator, or the howl produced by a loudspeaker that is placed too close to a microphone in the same circuit. A small random noise picked up by the microphone is amplified and reproduced by the loudspeaker. The microphone now picks it up again; it is further amplified, and fed from the speaker to microphone once again. This continues until the system is overloaded. In *negative feedback*, the output is used to reduce the input. In electronic amplifiers, stability is achieved, and distortion reduced, by using a system in which the input is decremented in proportion as the output increases. A similar negative feedback is used in *governors that reduce the fuel supply to an engine as its speed increases.

femto- Symbol f. A prefix used in the metric system to denote 10^{-15}. For example, 10^{-15} second = 1 femtosecond (fs).

Fermat's principle The path taken by a ray of light between any two points in a system is always the path that takes the least time. This principle leads to the law of the rectilinear propagation of light and the laws of reflection and refraction. It was discovered by the French mathematician, Pierre de Fermat (1601–65).

fermi A unit of length formerly used in nuclear physics. It is equal to 10^{-15} metre. In SI units this is equal to 1 femtometre (fm). It was named after the Italian-born US physicist Enrico Fermi (1901–54).

Fermi–Dirac statistics *See* quantum statistics.

Fermi level The energy level in a solid at which the probability of finding an electron is 1/2. The Fermi level in conductors lies in the conduction band (*see* energy bands), in insulators it lies in the valence band, and in semiconductors it falls in the gap between the conduction band and the valence band. At absolute zero all the electrons would occupy energy levels up to the Fermi level and no higher levels would be occupied.

fermion An *elementary particle with half-integral spin; i.e. a particle that conforms to Fermi–Dirac statistics (*see* quantum statistics). *Compare* boson.

fermium Symbol Fm. A radioactive metallic transuranic element belonging to the actinoids; a.n. 100; mass number of the most stable isotope 257 (half-life 10 days). Ten isotopes are known. The element was first identified by A. Ghiorso and associates in debris from the first hydrogen-bomb explosion in 1952.

ferrimagnetism *See* magnetism.

ferrite 1. A member of a class of mixed oxides $MO.Fe_2O_3$, where M is a metal such as cobalt, manganese, nickel, or zinc. The ferrites are ceramic materials that show either ferrimagnetism or ferromagnetism, but are not electrical conductors. For this reason they are used in high-frequency circuits as magnetic cores. **2.** *See* steel.

ferroalloys Alloys of iron with other elements made by smelting mixtures of iron ore and the metal ore; e.g. ferrochromium, ferrovanadium, ferromanganese, ferrosilicon, etc. They are used in making alloy *steels.

ferroelectric materials Ceramic dielectrics, such as Rochelle salt and barium titanate, that have a domain structure making them analogous to ferromagnetic materials (*see* magnetism). They exhibit hysteresis and usually the *piezoelectric effect.

ferromagnetism *See* magnetism.

fertile material An isotope of a radioactive element that can absorb a neutron to form a *fissile material. Uranium–238, for example, absorbs a neutron to form uranium–239, which decays to plutonium–239. This is the conversion that occurs in a breeder reactor (*see* nuclear reactor).

FET *See* transistor.

Feynman diagram *See* quantum electrodynamics.

fibre optics *See* optical fibres.

field A region in which a body experiences a *force as the result of the presence of some other body or bodies. A field is thus a method of representing the way in which bodies not in contact can influence each other. For example, a body that has mass is surrounded by a region in which another body that has mass experiences a force tending to draw the two bodies together. This is the gravitational field (*see* Newton's law of gravitation). The other three *fundamental interactions can also be represented by means of fields of force. However in the case of the *magnetic field and *electric field that together create the electromagnetic interaction, the force can vary in direction according to the character of the field. For example, in the field surrounding a negatively charged body, a positively charged body will experience a force of attraction, while another negatively charged body is repelled.

The strength of any field can be described as the ratio of the force experienced by a small appropriate specimen to the relevant property of that specimen, e.g. force/mass for the gravitational field.

field coil The coil in an electrical machine that produces the magnetic field.

field-effect transistor (FET) *See* transistor.

field-emission microscope A type of electron microscope in which a high negative voltage is applied to a metal tip placed in an evacuated vessel some distance from a glass screen with a fluorescent coating. The tip produces electrons by *field emission*, i.e. the emission of electrons from an unheated sharp metal part as a result of a high electric field. The emitted electrons form an enlarged pattern on the fluorescent screen, related to the individual exposed planes of atoms. As the resolution of the instrument is limited by the vibrations of the metal atoms, it is helpful to cool the tip in liquid helium. Although the individual atoms forming the point are not displayed, individual adsorbed atoms of other substances can be, and their activity is observable.

field-ionization microscope (field-ion microscope) A type of electron microscope that is similar in principle to the *field-emission microscope, except that a high positive voltage is applied to the metal tip, which is surrounded by low pressure gas (usually helium) rather than a vacuum. The image is formed in this case by *field ionization*: ionization at the surface of an unheated solid as a result of a strong electric field creating positive ions by electron transfer from surrounding atoms or molecules. The image is formed by ions striking the fluorescent screen. Individual atoms on the surface of the tip can be resolved and, in certain cases, adsorbed atoms may be detected.

field lens The lens in the compound eyepiece of an optical instrument that is furthest from the eye. Its function is to increase the field of view by refracting towards the main eye lens rays that would otherwise miss it.

field magnet The magnet that provides the magnetic field in an electrical machine. In some small dynamos and motors it is a permanent magnet but in most machines it is an electromagnet.

file A collection of data stored in a computer. It may consist of program instructions or numerical, textual, or graphical information. It usually consists of a set of similar or related records.

film badge A lapel badge containing masked photographic film worn by personnel who could be exposed to ionizing radiation. The film is developed to indicate the extent that the wearer has been exposed to harmful radiation.

filter 1. (in physics) A device placed in the path of a beam of radiation to alter its frequency distribution. For example, a plane pigmented piece of glass may be placed over a camera lens to alter the relative intensity of the component wavelengths of the beam entering the camera. **2.** A device for separating solid particles from a fluid.

filter pump A simple laboratory vacuum pump in which air is removed from a system by a jet of water forced through a narrow nozzle. The lowest pressure possible is the vapour pressure of water.

filtrate The clear liquid obtained by filtration.

filtration The process of separating solid particles using a filter. In vacuum filtration, the liquid is drawn through the filter by a vacuum pump.

finder A small low-powered astronomical telescope, with a wide field of view, that is fixed to a large astronomical telescope so that the large telescope can be pointed in the correct direction to observe a particular celestial body.

fine structure Closely spaced optical spectral lines arising from *transitions between energy levels that are split by the vibrational or rotational motion of a molecule or by electron spin. They are visible only at high resolution. *Hyperfine structure*, visible only at very high resolution, results from the influence of the atomic nucleus on the allowed energy levels of the atom.

finite series *See* series.

fissile material An isotope of a radioactive element that undergoes nuclear fission, either spontaneously or when irradiated by neutrons. Fissile nuclides, such as uranium–235 and plutonium–239, are used in *nuclear reactors and nuclear weapons. *Compare* fertile material.

fission *See* nuclear fission.

fission products *See* nuclear fission.

fission-track dating A method of estimating the age of glass and other mineral objects by observing the tracks made in them by the fission fragments of the uranium nuclei that they contain. By irradiating the objects with neutrons to induce fission and comparing the density and number of the tracks before and after irradiation it is possible to estimate the time that has elapsed since the object solidified.

Fitzgerald contraction *See* Lorentz–Fitzgerald contraction.

fixed point A temperature that can be accurately reproduced to enable it to be used as the basis of a *temperature scale.

fixed star One of very many heavenly bodies that does not appear to alter its position on the *celestial sphere. They were so called to distinguish them from the planets, which were once known as *wandering stars*. The discovery of the *proper motion of stars in the 18th century established that stars are not fixed in the sky although, because of their immense distances from the solar system, they may appear to be so.

Fizeau's method A method of measuring the speed of light, invented by the French physicist Armand Fizeau (1819–96) in 1849. A cogwheel rotating at high speed enables a series of flashes to be transmitted to a distant mirror. The light reflected back to the cogwheel is observed and the speed of light calculated from the rates of rotation of the wheel required to produce an eclipse of the returning light.

flame A hot luminous mixture of gases undergoing combustion. The chemical reactions in a flame are mainly free-radical chain reactions and the light comes from fluorescence of excited molecules or ions or from incandescence of small solid particles (e.g. carbon).

flash photolysis A technique for studying free-radical reactions in gases. The apparatus used typically consists of a long glass or quartz tube holding the gas, with a lamp outside the tube suitable for producing an intense flash of light. This dissociates molecules in the sample creating free radicals, which can be detected spectroscopically by a beam of light passed down the axis of the tube. It is possible to focus the spectrometer on an absorption line for a particular product and measure its change in intensity with time using an oscilloscope. In this way the kinetics of very fast free-radical gas reactions can be studied.

flash point The temperature at which the vapour above a volatile liquid forms a combustible mixture with air. At the flash point the application of a naked flame gives a momentary flash rather than sustained combustion, for which the temperature is too low.

flavour *See* elementary particles.

Fleming's rules Rules to assist in remembering the relative directions of the field, current, and motion in electrical machines. The left hand refers to motors, the right hand to generators. If the forefinger, second finger, and thumb of the left hand are extended at right angles to each other, the forefinger indicates the direction of the field, the second finger the direction of the current, and the thumb the direction of motion. If the right hand is used the digits indicate these directions in a generator. The mnemonic was invented by Sir John Ambrose Fleming (1849–1945).

flip-flop (bistable circuit) An electronic circuit that has two stable states. It is switched from one stable state to the other by means of a triggering pulse. They are extensively used as *logic circuits in computers.

floppy disk A flexible plastic disk with a magnetic coating encased in a stiff envelope. It is used to store information in a small computer system. *See* magnetic disk.

fluidics The use of jets of fluid in pipes to perform many of the control functions usually performed by electronic devices. Being about one million times slower than electronic devices, fluidic systems are useful where delay lines are required. They are also less sensitive to high temperatures, strong magnetic fields, and ionizing radiation than electronic devices.

fluidization A technique used in some industrial processes in which solid particles suspended in a stream of gas are treated as if they were in the liquid state. Fluidization is useful for transporting powders, such as coal dust. *Fluidized beds*, in which solid particles are suspended in an upward stream, are extensively used in the chemical industry, particularly in catalytic reactions where the powdered catalyst has a high surface area.

fluid mechanics The study of fluids at rest and in motion. *Fluid statics* is concerned with the pressures and forces exerted on liquids and gases at rest. *Hydrostatics is specifically concerned with the behaviour of liquids at rest. In *fluid dynamics* the forces exerted on fluids, and the motion that results from these forces, are examined. It can be divided into *hydrodynamics: the motion of liquids (not only water); and aerodynamics: the motion of gases.

Fluid dynamics is an important science used to solve many of the problems arising in aeronautical, chemical, mechanical, and civil engineering. It also enables many natural phenomena, such as the flight of birds, the swimming of fish, and the development of weather conditions, to be studied scientifically.

fluorescence *See* luminescence.

fluorescent light *See* electric lighting.

flux 1. *See* luminous flux. **2.** *See* magnetic flux. **3.** *See* electric flux. **4.** The product of the number of particles flowing per unit area of beam cross section.

flux density 1. *See* magnetic flux density. **2.** *See* electric flux density.

fluxmeter An instrument used to measure *magnetic flux. It is used in conjunction with a coil (the *search coil*) and resembles a moving-coil galvanometer except that there are no restoring springs. A change in the magnetic flux induces a momentary current in the search coil and in the coil of the meter, which turns in proportion and stays in the deflected position. This type of instrument has been largely superseded by a type using the Hall probe (*see* Hall effect).

FM (frequency modulation) *See* modulation.

f-number *See* aperture.

focal length The distance between the *optical centre of a lens or pole of a spherical mirror and its *principal focus.

focal point *See* focus.

focal ratio *See* aperture.

focus 1. (in optics) Any point in an optical system through or towards which rays of light are converged. It is sometimes called the *focal point* and sometimes loosely used to mean *principal focus or (particularly by photographers) *focal length. **2.** (in mathematics) *See* conic; ellipse.

foot The unit of length in *f.p.s units. It is equal to one-third of a yard and is now therefore defined as 0.3048 metre. Several units based on the foot were formerly used in science, including the units of work, the *foot-pound-force* and the *foot-poundal*, and the illumination units, the *foot-candle* and the *foot-lambert*. These have all been replaced by SI units.

forbidden band *See* energy bands.

force Symbol *F*. The agency that tends to change the momentum of a massive body, defined as being proportional to the rate of increase of momentum. For a body of mass *m* travelling at a velocity *v*, the momentum is *mv*. In any coherent system of units the force is therefore given by $F = d(mv)/dt$. If the mass is constant $F = m dv/dt = ma$, where *a* is the acceleration (*see* Newton's laws of motion). The SI unit of force is the newton. Forces occur always in equal and opposite action–reaction pairs between bodies, though it is often convenient to think of one body being in a force *field.

forced convection *See* convection.

force ratio (mechanical advantage) The ratio of the output force (load) of a machine to the input force (effort).

formula 1. (in chemistry) A way of representing a chemical compound using symbols for the atoms present. Subscripts are used for the numbers of atoms. The *molecular formula* simply gives the types and numbers of atoms present. For example, the molecular formula of ethanoic acid is $C_2H_4O_2$. The *empirical formula* gives the atoms in their simplest ratio; for ethanoic acid it is CH_2O. The *structural formula* gives an indication of the way the atoms are arranged. Commonly, this is done by dividing the formula into groups; ethanoic acid can be written $CH_3.CO.OH$ (or more usually simply CH_3COOH). Structural formulae can also show the arrangement of atoms or groups in space. **2.** (in mathematics and physics) A rule or law expressed in algebraic symbols.

Fortin barometer *See* barometer.

fossil fuel Coal, oil, and natural gas, the fuels used by man as a source of energy. They are formed from the remains of living organisms and all have a high carbon or hydrogen content. Their value as fuels relies on the exothermic oxidation of carbon to form carbon dioxide ($C + O_2 \rightarrow CO_2$) and the oxidation of hydrogen to form water ($H_2 + \frac{1}{2}O_2 \rightarrow H_2O$).

Foucault pendulum A simple pendulum in which a heavy bob attached to a long wire is free to swing in any direction. As a result of the earth's rotation, the plane of the pendulum's swing slowly turns (at the poles of the earth it makes one complete revolution in 24 hours). It was devised by the French physicist Jean Bernard Léon Foucault (1819–68) in 1851, when it provided the first direct evidence of the earth's rotation.

Fourier series An expansion of a periodic function as a series of trigonometric functions. Thus,

$$f(x) = a_0 + (a_1 \cos x + b_1 \sin x) + (a_2 \cos 2x + b_2 \sin 2x) + \ldots,$$

where a_0, a_1, b_1, b_2, etc., are constants, called *Fourier coefficients*. The series was first formulated by J. B. J. Fourier (1768–1830) and is used in harmonic analysis (*Fourier analysis*) to determine the harmonic components of a complex wave.

fourth dimension *See* space–time.

f.p.s. units The British system of units based on the foot, pound, and second. It has now been replaced for all scientific purposes by SI units.

frame of reference A set of axes, taken as being for practical purposes at rest, that enables the position of a point (or body) in space to be defined at any instant of time. In a four-dimensional continuum (*see* space–time) a frame of reference consists of a set of four coordinate axes, three spatial and one of time.

Frasch process A method of obtaining sulphur from underground deposits using a tube consisting of three concentric pipes. Superheated steam is passed down the outer pipe to melt the sulphur, which is forced up through the middle pipe by compressed air fed through the inner tube. The steam in the outer casing keeps the sulphur molten in the pipe.

Fraunhofer diffraction A form of *diffraction in which the light source and the receiving screen are in effect at infinite distances from the diffracting object, so that the wave fronts can be treated as planar rather than spherical. In practice it involves parallel beams of light. It can be regarded as an extreme case of *Fresnel diffraction but is of more practical use in explaining single and multiple slit patterns. It was studied by the German optician Joseph von Fraunhofer (1787–1826).

Fraunhofer lines Dark lines in the solar spectrum that result from the absorption by elements in the solar chromosphere of some of the wavelengths of the visible radiation emitted by the hot interior of the sun.

free electron *See* electron.

free energy A measure of a system's ability to do work. The *Gibbs free energy* (or *Gibbs function*), G, is defined by $G = H - TS$, where G is the energy liberated or absorbed in a reversible process at constant pressure and constant temperature (T), H is the *enthalpy and S the *entropy of the system. Changes in Gibbs free energy, ΔG, are useful in indicating the conditions under which a chemical reaction will occur. If ΔG is positive the reaction will only occur if energy is supplied to force it away from the equilibrium position (i.e. when $\Delta G = 0$). If ΔG is negative the reaction will proceed spontaneously to equilibrium.

The *Helmholtz free energy* (or *Helmholtz function*), F, is defined by $F = U - TS$, where U is the *internal energy. For a reversible isothermal process, ΔF represents the useful work available.

free fall Motion resulting from a gravitational field that is unimpeded by a medium that would provide a viscous retarding force or bouyancy. In the earth's gravitational field, free fall is a downward motion at a constant acceleration, known as the *acceleration of free fall.

free space A region in which there is no matter and no electromagnetic or gravitational fields. It has a temperature of absolute zero, unit refractive index, and the speed of light is its maximum value. The electric constant (*see* permittivity) and the magnetic constant (*see* permeability) are defined for free space.

freeze drying A process used in dehydrating food, blood plasma, and other heat-sensitive substances. The product is deep-frozen and the ice trapped in it is removed by reducing the pressure and causing it to sublime. The water vapour is then removed, leaving an undamaged dry product.

freezing mixture A mixture of components that produces a low temperature. For example, a mixture of ice and sodium chloride gives a temperature of −20°C.

Frenkel defect *See* defect.

frequency Symbol *f*. The rate of repetition of a regular event. The number of cycles of a wave, or some other oscillation or vibration, per second is expressed in *hertz (cycles per second). The frequency (f) of a wave motion is given by $f = c/\lambda$, where c is the velocity of propagation and λ is the wavelength. The frequency associated with a quantum of electromagnetic energy is given by $f = E/h$, where E is the quantum's energy and h is the Planck constant.

frequency modulation (FM) *See* modulation; radio.

fresnel A unit of frequency equal to 10^{12} hertz. In SI units this is equal to 1 terahertz (THz). It was named after the French physicist A. J. Fresnel (1788–1827).

Fresnel diffraction A form of *diffraction in which the light source or the receiving screen, or both, are at finite distances from the diffracting object, so that the wavefronts are not plane, as in *Fraunhofer diffraction. It was studied by A. J. Fresnel (1788–1827).

Fresnel lens A lens with one face cut into a series of steps. It enables a relatively light and robust lens of short focal length though poor optical quality to be used in projectors (as condenser lenses), searchlights, spotlights, car headlights, and lighthouses. Such lenses have been made fine enough to serve as magnifiers for reading small print.

friction The force that resists the motion of one surface relative to another with which it

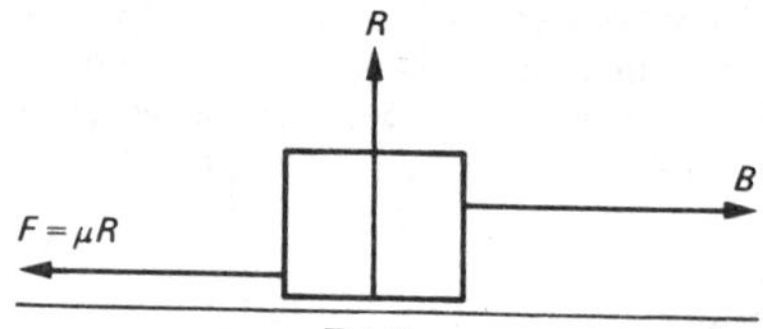

Friction

is in contact. For a body resting on a horizontal surface there is a normal contact force, R, between the body and surface, acting perpendicularly to the surface. If a horizontal force B is applied to the body with the intention of moving it to the right, there will be an equal horizontal friction force, F, to the left, resisting the motion. If B is increased until the body just moves, the value of F will also increase until it reaches the *limiting frictional force* (F_L), which is the maximum value of F. F_L is then equal to $\mu_s R$, where μ_s is the *coefficient of static friction*, the value of which depends on the nature of the surfaces. Once the body is moving with constant velocity, the value of F falls to a value F_k, which is equal to $\mu_k R$, where μ_k is the *coefficient of kinetic friction*. Both μ_s and μ_k are independent of the surface area of the body unless this is very small and μ_k is almost independent of the relative velocity of the body and surface.

The cause of friction is that surfaces, however smooth they may look to the eye, on the microscopic scale have many humps and crests. Therefore the actual area of contact is very small indeed, and the consequent very high pressure leads to local pressure welding of the surfaces. During motion the welds are broken and remade continually. *See also* rolling friction.

froth flotation A method of separating mixtures of solids, used industrially for separating ores from the unwanted gangue. The mixture is ground to a powder and water and a frothing agent added. Air is blown through the water. With a suitable frothing agent, the bubbles adhere only to particles of ore and carry them to the surface, leaving the gangue particles at the bottom.

frustum A solid figure produced when two parallel planes cut a larger solid or when one plane parallel to the base cuts it.

fuel A substance that is oxidized or otherwise changed in a furnace or heat engine to release useful heat or energy. For this purpose wood, vegetable oil, and animal products have largely been replaced by *fossil fuels since the 18th century.

The limited supply of fossil fuels and the expense of extracting them from the earth has encouraged the development of nuclear fuels to produce electricity (*see* nuclear energy).

fuel cell A cell in which the chemical energy of a fuel is converted directly into electrical energy. The simplest fuel cell is one in which hydrogen is oxidized to form water over porous sintered nickel electrodes. A supply of gaseous hydrogen is fed to a compartment containing the porous cathode and a supply of oxygen is fed to a compartment containing the porous anode; the electrodes are separated by a third compartment containing a hot alkaline electrolyte, such as potassium hydroxide. The electrodes are porous to enable the gases to react with the electrolyte, with the nickel in the electrodes acting as a catalyst. At the cathode the hydrogen reacts with the hydroxide ions in the electrolyte to form water, with the release of two electrons per hydrogen molecule:

$$H_2 + 2OH^- \rightarrow 2H_2O + 2e^-$$

At the anode, the oxygen reacts with the water, taking up electrons, to form hydroxide ions:

$$\tfrac{1}{2}O_2 + H_2O + 2e^- \rightarrow 2OH^-$$

The electrons flow from the cathode to the anode through an external circuit as an electric current. The device is a more efficient converter of electric energy than a heat engine, but it is bulky and requires a continuous supply of gaseous fuels. Their use to power electric vehicles is being actively explored.

fuel element *See* nuclear reactor.

fugacity Symbol *f*. A thermodynamic function used in place of partial pressure in reactions involving real gases and mixtures. For a component of a mixture, it is defined by $d\mu = RTd(\ln f)$, where μ is the chemical potential. It has the same units as pressure and the fugacity of a gas is equal to the pressure if the gas is ideal. The fugacity of a liquid or solid is the fugacity of the vapour with which it is in equilibrium. The ratio of the fugacity to the fugacity in some standard state is the *activity. For a gas, the standard state is chosen to be the state at which the fugacity is 1. The activity then equals the fugacity.

full-wave rectifier *See* rectifier.

function Any operation or procedure that relates one variable to one or more other variables. If y is a function of x, written $y = f(x)$, a change in x produces a change in y, and if x is known, y can be determined. x is known as the *independent variable* and y is the *dependent variable*.

fundamental *See* harmonic.

fundamental constants (universal constants) Those parameters that do not change throughout the universe. The charge on an electron, the speed of light in free space, the Planck constant, the gravitational constant, the electric constant, and the magnetic constant are all thought to be examples.

fundamental interactions The four different types of interaction that can occur between bodies. These interactions can take place even when the bodies are not in physical contact and together they account for all the observed forces that occur in the universe. While the unification of these four types of interaction into one model, theory, or set of equations has long been the aim of physicists, this has not yet been achieved, although progress has been made in the unification of the electromagnetic and weak interactions (*see* unified-field theory).

The *gravitational interaction*, some 10^{40} times weaker than the electromagnetic interaction, is the weakest of all. The force that it generates acts between all bodies that have mass and the force is always attractive. The interaction can be visualized in terms of a classical *field of force in which the strength of the force falls off with the square of the distance between the interacting bodies (*see* Newton's law of gravitation). The hypothetical gravitational quantum, the *graviton*, is also a useful concept in some contexts. On the atomic scale the gravitational force is negligibly weak, but on the cosmological scale, where masses are enormous, it is immensely important in holding the components of the universe together.

The *weak interaction*, some 10^{10} times weaker than the electromagnetic interaction, functions only between *leptons and is responsible for the *beta decay of particles and nuclei. In the current model, the weak interaction is visualized as a force mediated by the exchange of virtual particles, called intermediate vector bosons, between interacting leptons.

The *electromagnetic interaction* is responsible for the forces that control atomic structure, chemical reactions, and all electromagnetic phenomena. It functions only between charged particles, but unlike the gravitational interaction, can be either attractive or repulsive. The interaction is either visualized as a classical field of force (*see* Coulomb's law), analogous to the gravitational field, or as an exchange of virtual *photons, a more convenient form for quantum theory considerations.

The *strong interaction*, some 10^{2} times stronger than the electromagnetic interaction, functions only between *hadrons and is responsible for the force between nucleons that gives the atomic nucleus its great stability. It operates at very short range inside the nucleus (as little as 10^{-15}metre) and is visualized as an exchange of virtual mesons.

fundamental units A set of independently defined *units of measurement that forms the basis of a system of units. Such a set requires three mechanical units (usually of length, mass, and time) and one electrical unit; it has also been found convenient to treat certain other quantities as fundamental, even though they are not strictly independent. In the metric system the centimetre–gram–second (c.g.s.) system was replaced by the metre–kilogram–second (m.k.s.) system; the latter has now been adapted to provide the basis for *SI units. In British Imperial units the foot–pound–second (f.p.s.) system was formerly used.

fuse A length of wire made of a metal alloy of low melting point that is designed to melt at a specified current loading in order to protect an electrical device or circuit from overloading. The wire is often enclosed in a small glass or ceramic cartridge with metal ends.

fusible alloys Alloys that melt at low temperature (around 100°C). They have a number of uses, including constant-temperature baths, pipe bending, and automatic sprinklers to provide a spray of water to prevent fires from spreading. Fusible alloys are usually *eutectic mixtures of bismuth, lead, tin, and cadmium. Wood's metal and Lipowitz's alloy are examples of alloys that melt at about 70°C.

fusion 1. Melting. **2.** *See* nuclear fusion.

fusion reactor *See* thermonuclear reactor.

G

gain *See* amplifier.

galaxy A vast collection of stars, dust, and gas held together by the gravitational attraction between its components. Galaxies are usually classified as elliptical, spiral, or irregular in shape. The *elliptical galaxies* have no internal structure but appear on photographs of the sky as fuzzy elliptical patches. The *spiral galaxies* are flat disc-shaped collections of stars with prominent spiral arms. *Irregular galaxies* have no apparent structure or shape.

The sun belongs to a spiral galaxy known as the *Galaxy* (with a capital G) or the *Milky Way System*. There are some 10^{11} stars in the system, which is about 30 000 parsecs across with a maximum thickness at the centre of about 4000 parsecs. The sun is about 10 000 parsecs from the centre of the Galaxy.

The galaxies are separated from each other by enormous distances, the nearest galaxy to our own being about 5×10^{15} parsecs away.

Galilean telescope *See* telescope.

Galilean transformations A set of equations for transforming the position and motion parameters from a frame of reference with origin at O and coordinates (x,y,z) to a frame with origin at O′ and coordinates at (x',y',z'). They are:

$$x' = x - vt$$
$$y' = y$$
$$z' = z$$
$$t' = t$$

The equations conform to Newtonian mechanics. *Compare* Lorentz transformations.

gallon 1. (*or* **Imperial gallon)** The volume occupied by exactly ten pounds of distilled water of density 0.998 859 gram per millilitre in air of density 0.001 217 gram per millilitre. 1 gallon = 4.546 09 litres (cubic decimetres). **2.** A unit of volume in the US Customary system equal to 0.832 68 Imperial gallon, i.e. 3.785 44 litres.

galvanic cell *See* voltaic cell.

galvanized iron Iron or steel that has been coated with a layer of zinc to protect it from corrosion. Corrugated mild-steel sheets for roofing and mild-steel sheets for dustbins, etc., are usually galvanized by dipping them in molten zinc. The formation of a brittle zinc–iron alloy is prevented by the addition of small quantities of aluminium or magnesium. Wire is often galvanized by a cold electrolytic process as no alloy forms in this process. Galvanizing is an effective method of protecting steel because even if the surface is scratched, the zinc still protects the underlying metal. *See* sacrificial protection.

galvanometer An instrument for detecting and measuring small electric currents. In the moving-coil instrument a pivoted coil of fine insulated copper wire surrounds a fixed soft-iron core between the poles of a permanent magnet. The interaction between the field of the permanent magnet and the sides of the coil, produced when a current flows through it, causes a torque on the coil. The moving coil carries either a pointer or a mirror that deflects a light beam when it moves; the extent of the deflection is a measure of the strength of the current. The galvanometer can be converted into an *ammeter or a *voltmeter. Digital electronic instruments are increasingly replacing the moving-coil type. *See also* ballistic galvanometer.

gamma-iron *See* iron.

gamma radiation Electromagnetic radiation emitted by excited atomic nuclei during the process of passing to a lower excitation state. Gamma radiation ranges in energy from about 10^{-15} to 10^{-11} joule (10 keV to 100 MeV) corresponding to a wavelength range of about 10^{-10} to 10^{-14} metre. A common source of gamma radiation is cobalt–60, the decay process of which is:

$$^{60}_{27}Co \xrightarrow{\beta} {}^{60}_{28}Ni \xrightarrow{\gamma} {}^{60}_{28}Ni$$

The de-excitation of nickel–60 is accompanied by the emission of a gamma-ray photon having an energy of 1.87×10^{-13} J (1.17 MeV).

gas A state of matter in which the matter concerned occupies the whole of its container irrespective of its quantity. In an *ideal gas, which obeys the *gas laws exactly, the molecules themselves would have a negligible volume and negligible forces between them, and collisions between molecules would be perfectly elastic. In practice, however, the behaviour of real gases deviates from the gas laws because their molecules occupy a finite volume, there are small forces between molecules, and in polyatomic gases collisions are to a certain extent inelastic (*see* equation of state).

gas constant (universal molar gas constant) Symbol R. The constant that appears in the *universal gas equation* (*see* gas laws). It has the value 8.314 34 J K^{-1} mol^{-1}.

gas-cooled reactor *See* nuclear reactor.

gas equation *See* gas laws.

gas laws Laws relating the temperature, pressure, and volume of an *ideal gas. *Boyle's law states that the pressure (p) of a specimen is inversely proportional to the volume (V) at constant temperature (pV = constant). The modern equivalent of *Charles' law states that the volume is directly proportional to the thermodynamic temperature (T) at constant pressure (V/T = constant); originally this law stated the constant expansivity of a gas kept at constant pressure. The pressure law states that the pressure is directly proportional to the thermodynamic temperature for a specimen kept at constant volume. The three laws can be combined in the *universal gas equation*, $pV = nRT$, where n is the amount of gas in the specimen and R is the *gas constant. The gas laws were first established experimentally for real gases, although they are obeyed by real gases to only a limited extent; they are obeyed best at high temperatures and low pressures. *See also* equation of state.

gas thermometer A device for measuring temperature in which the working fluid is a gas. It provides the most accurate method of measuring temperatures in the range 2.5 to 1337 K. Using a fixed mass of gas a *constant-volume thermometer* measures the pressure of a fixed volume of gas at relevant temperatures, usually by means of a mercury *manometer and a *barometer.

gas turbine An internal-combustion engine in which the products of combustion of a fuel burnt in compressed air are expanded through a turbine. Atmospheric air is compressed by a rotary compressor driven by the turbine, fed into a combustion chamber, and mixed with the fuel (kerosene, natural gas, etc.); the expanding gases drive the turbine and power is taken from the unit by means of rotation of the turbine shaft (as in locomotives) or thrust from a jet (as in aircraft).

gate 1. An electronic circuit with a single output and one or more inputs; the output is a function of the input or inputs. In the *transmission gate* the output waveform is a replica of a selected input during a specific interval. In the *switching gate* a constant output is obtained for a specified combination of inputs. These gates are the basic components of digital computers. *See* logic circuits. **2.** The electrode in a field-effect *transistor that controls the current through the channel.

gauss Symbol G. The c.g.s. unit of magnetic flux density. It is equal to 10^{-4} tesla.

Gauss' law The total electric flux normal to a closed surface in an electric field is proportional to the algebraic sum of the electric charges within the surface. A similar law applies to surfaces drawn in a magnetic field and the law can be generalized for any vector field through a closed surface. It was first stated by Karl Gauss (1777–1855).

gaussmeter A *magnetometer, especially one calibrated in gauss.

Gay Lussac's law 1. When gases combine chemically the volumes of the reactants and the volume of the product, if it is gaseous, bear simple relationships to each other when measured under the same conditions of temperature and pressure. The law was first stated in 1808 by J. L. Gay Lussac (1778–1850) and led to *Avogadro's law. **2.** *See* Charles' law.

Gegenschein (German: counterglow) A faint elliptical patch of light visible on a moonless night on the ecliptic at a point 180° from the position of the sun. It is caused by the reflection of sunlight by meteoric particles (*see also* zodiacal light).

Geiger counter (Geiger–Müller counter) A device used to detect and measure *ionizing radiation. It consists of a tube containing a low-pressure gas (usually argon or neon and methane) and a cylindrical hollow cathode through the centre of which runs a fine-wire anode. A potential difference of about 1000 volts is maintained between the electrodes. An ionizing particle or photon passing through a window into the tube will cause an ion to be produced and the high p.d. will accelerate it towards its appropriate electrode, causing an avalanche of further ionizations by collision. The consequent current pulses can be counted in electronic circuits or simply amplified to work a small loudspeaker in the instrument. It was first devised in 1908 by Hans Geiger (1882–1947). Geiger and W. Müller produced an improved design in 1928.

Geissler tube An early form of gas-discharge tube designed to demonstrate the luminous effects of an electric discharge passing through a low-pressure gas between two electrodes. Modified forms are used in spectroscopy as a source of light. It was invented in 1858 by Heinrich Geissler (1814–79).

gel A lyophilic *colloid that has coagulated to a rigid or jelly-like solid. In a gel, the disperse medium has formed a loosely-held network of linked molecules through the dispersion medium. Examples of gels are silica gel and gelatin.

general theory of relativity *See* relativity.

generation time The average time that elapses between the creation of a neutron by fission in a nuclear reactor and a fission produced by that neutron.

generator Any machine that converts mechanical power into electrical power. Electromagnetic generators are the main source of electricity and may be driven by steam turbines, water turbines, internal-combustion engines, windmills, or by some moving part of any other machine. In power stations, generators produce alternating current and are often called *alternators*.

geocentric universe A theological, rather than scientific, view of the universe in which the earth is regarded as being at its centre. Galileo finally established that the earth revolves round the sun (not the other way round, as the church believed); during the 20th century it has become clear from advances in observational astronomy that the earth is no more than one of nine planets orbiting the sun, which is one of countless millions of similar stars, many of which undoubtedly possess planetary bodies on which life could have evolved.

geodesic (geodesic line) The shortest distance between two points on a curved surface.

geodesy The science concerned with surveying and mapping the earth's surface to determine, for example, its exact size, shape, and gravitational field. The information supplied by geodesy in the form of locations, distances, directions, elevations, and gravity information is of use in civil engineering, navigation, geophysics, and geography.

geomagnetism The science concerned with the earth's magnetic field. If a bar magnet is suspended at any point on the earth's surface so that it can move freely in all planes, the north-seeking end of the magnet (N-pole) will point in a broadly northerly direction. The angle (D) between the horizontal direction in which it points and the geographic meridian at that point is called the *magnetic declination*. This is taken to be positive to the east of geographic north and negative to the west. The needle will not, however, be horizontal except on the *magnetic equator*. In all other positions it will make an angle (I) with the horizontal, called the *inclination* (or *magnetic dip*). At the *magnetic poles* $I = 90°$ ($+90°$ at the N-pole, $-90°$ at the S-pole) and the needle will be vertical. The positions of the poles, which vary with time, were in the 1970s approximately 76.1°N, 100°W (N) and 65.8°S, 139°E (S). The vector intensity F of the geomagnetic field is specified by I, D, and F, where F is the local magnetic intensity of the field measured in gauss or tesla (1 gauss $= 10^{-4}$ tesla). F, I, and D, together with the horizontal and vertical components of F, and its north and east components, are called the *magnetic elements*. The value of F varies from about 0.2 gauss to 0.6 gauss, in general being higher in the region of the poles than at the equator, but values vary irregularly over the earth's surface with no correlation with surface features. There is also a slow unpredictable change in the local values of the magnetic elements called the *secular magnetic variation*. For example, in London between 1576 and 1800 D changed from $+11°$ to $-24°$ and I varied between 74° and 67°. The study of *palaeomagnetism has extended knowledge of the secular magnetic variation into the geological past and it is clear that the direction of the geomagnetic field has reversed several times. The source of the field and the cause of the variations are not known with any certainty but the source is believed to be associated with electric currents in the earth's liquid core.

geometrical optics *See* optics.

geometric average *See* average.

geometric series A series of numbers or terms in which the ratio of any term to the subsequent term is constant. For example, 1, 4, 16, 64, 256, . . ., has a *common ratio* of 4. In general, a geometric series can be written $a + ar + ar^2 \ldots + ar^{n-1}$ and the sum of n terms is $a(r^n - 1)/(r - 1)$.

geophysics The geological science in which the principles of mathematics and physics are applied to the study of the earth's crust and interior. It includes the study of earthquake waves, geomagnetism, gravitational fields, and electrical conductivity using precise quantitative principles. In applied geophysics the techniques are applied to the discovery and location of economic minerals (e.g. petroleum).

geostationary orbit *See* synchronous orbit.

geosynchronous orbit *See* synchronous orbit.

geothermal energy Heat within the earth's interior that is a potential source of energy. Volcanoes, geysers, hot springs, and fumaroles are all sources of geothermal energy. The main areas of the world in which these energy sources are used to generate power include Larderello (Italy), Wairakei (New Zealand), Geysers (California, USA), and Reykjavik (Iceland). High-temperature porous rock also occurs in the top few kilometres of the earth's crust. Thermal energy from these reservoirs can be tapped by drilling into them and extracting their thermal energy by conduction to a fluid. The hot fluid can then be used for direct heating or to raise steam to drive a turbogenerator.

German silver (nickel silver) An alloy of copper, zinc, and nickel, often in the proportions 5:2:2. It resembles silver in appearance and is used in cheap jewellery and cutlery and as a base for silver-plated wire. *See also* electrum.

getter A substance with a strong affinity for specified other substances. A getter may be used to remove the last traces of a gas when achieving a high vacuum. It may also be used to remove impurities from semiconductors.

GeV Gigaelectronvolt, i.e. 10^9 eV. In the USA this is often written BeV, billion-electronvolt.

giant star A very large star that is highly luminous. Lying above the main sequence on a *Hertzsprung–Russell diagram, giant stars represent a late stage in *stellar evolution. *See also* red giant; supergiant.

gibbous *See* phases of the moon.

Gibbs free energy (Gibbs function) *See* free energy. It is named after the US chemist J. W. Gibbs (1839–1903).

giga- Symbol G. A prefix used in the metric system to denote one thousand million times. For example, 10^9 joules = 1 gigajoule (GJ).

gilbert Symbol Gb. The c.g.s. unit of *magnetomotive force equal to $10/4\pi$ (= 0.795 77) ampere-turn. It is named after the English physician William Gilbert (1544–1603).

gimbal A type of mount for an instrument (such as a *gyroscope or compass) in which the instrument is free to rotate about two perpendicular axes.

Giorgi units *See* m.k.s. units.

glass fibres Melted glass drawn into thin fibres some 0.005 mm–0.01 mm in diameter. The fibres may be spun into threads and woven into fabrics, which are then impregnated with resins to give a material that is both strong and corrosion resistant for use in car bodies and boat building.

globular cluster An approximately spherical cluster of stars, usually containing between 10^4 and 10^6 stars. Globular clusters appear to move in elliptical orbits around the galaxies. The Milky Way system has about 200 such clusters moving around it.

glove box A metal box that has gloves fitted to ports in its walls. It is used to manipulate mildly radioactive materials and in laboratory techniques in which an inert, sterile, dry, or dust-free atmosphere has to be maintained.

glow discharge An electrical discharge that passes through a gas at low pressure and causes the gas to become luminous. The glow is produced by the return to the ground state of ions excited by the discharge.

gluon *See* elementary particles.

governor A device that maintains a motor or engine at a constant speed despite variations in the load, using the principle of negative feedback. A common method uses a set of flying balls that reduce the fuel intake as the speed increases. The balls, attached by flexible steel strips to a collar capable of moving vertically up and down a rotating shaft, move outwards as the speed increases. The collar rises as the balls fly out and is coupled to a lever that controls the fuel intake.

grad (gradient) Symbol ∇. The operator

$$\boldsymbol{i}\partial/\partial x + \boldsymbol{j}\partial/\partial y + \boldsymbol{k}\partial/\partial z$$

where $\boldsymbol{i}$, $\boldsymbol{j}$, and $\boldsymbol{k}$ are unit vectors in the x, y, and z directions. In physics, $\nabla \boldsymbol{F}$ is a vector in the direction in which the rate of change of $\boldsymbol{F}$ is a maximum, if there is one.

gradient 1. The slope of a line. In Cartesian coordinates, a straight line $y = mx + c$, has a gradient m. For a curve, $y = \mathrm{f}(x)$, the gradient at a point is the derivative $\mathrm{d}y/\mathrm{d}x$ at that point, i.e. the slope of the tangent to the curve at that point. **2.** *See* grad.

Graham's law The rates at which gases diffuse is inversely proportional to the square roots of their densities. This principle is made use of in the diffusion method of separating isotopes. The law was formulated in 1829 by Thomas Graham (1805–69).

gram Symbol g. One thousandth of a kilogram. The gram is the fundamental unit of mass in *c.g.s. units and was formerly used in such units as the *gram-atom*, *gram-molecule*, and *gram-equivalent*, which have now been replaced by the *mole.

graph A diagram that illustrates the relationship between two variables. It usually consists of two perpendicular axes, calibrated in the units of the variables and crossing at a point called the *origin*. Points are plotted in the spaces between the axes and the points are joined to form a curve. *See also* Cartesian coordinates; polar coordinates.

graphite-modulated reactor *See* nuclear reactor.

graticule (in optics) A network of fine wires or a scale in the eyepiece of a telescope or microscope or on the stage of a microscope for measuring purposes.

grating *See* diffraction grating.

gravitation *See* Newton's law of gravitation.

gravitational constant Symbol G. The constant that appears in *Newton's law of gravitation; it has the value 6.664×10^{-11} $\mathrm{N\,m^2\,kg^{-2}}$. G is usually regarded as a universal constant although, in some models of the universe, it is proposed that it decreases with time as the universe expands.

gravitational field The region of space surrounding a body that has the property of *mass. In this region any other body that has mass will experience a force of attraction. The ratio of the force to the mass of the second body is the *gravitational field strength*.

gravitational interaction *See* fundamental interactions.

gravitational mass *See* mass.

gravitational shift *See* redshift.

gravitational waves Waves propagated through a *gravitational field. The prediction that an accelerating mass will radiate gravitational waves (and lose energy) comes from the general theory of *relativity. Many attempts have been made to detect waves from space directly using large metal detectors. The theory suggests that a pulse of gravitational radiation (as from a supernova explosion or *black hole) vibrates the detector and the disturbance is detected by a transducer. The interaction is very weak and extreme care is required to avoid external disturbances and the effects of thermal noise in the detecting system. So far, no accepted direct observations have been made. However, indirect evidence of gravitational waves has been produced by long-term observation of a pulsar.

graviton A hypothetical particle or quantum of energy exchanged in a gravitational interaction (*see* fundamental interactions). Such a particle has not been observed but is postulated to make the gravitational interaction consistent with quantum mechanics. It would be expected to travel at the speed of light and have zero rest mass and charge.

gravity The phenomenon associated with the gravitational force acting on any object that has mass and is situated within the earth's *gravitational field. The weight of a body (*see* mass) is equal to the force of gravity acting on the body. According to Newton's second law of motion $F = ma$, where F is the force producing an acceleration a on a body of mass m. The weight of a body is therefore equal to the product of its mass and the acceleration due to gravity (g), which is now called the *acceleration of free fall. By combining the second law of motion with *Newton's law of gravitation ($F = GM_1M_2/d^2$) it follows that: $g = GM/d^2$, where G is the *gravitational constant, M is the mass of the earth, and d is the distance of the body from the centre of the earth. For a body on the earth's surface $g = 9.806\,65$ m s^{-2}.

A force of gravity also exists on other planets, moons, etc., but because it depends on the mass of the planet and its diameter, the strength of the force is not the same as it is on earth. If F_e is the force acting on a given mass on earth, the force F_p acting on the same mass on another planet will be given by:

$$F_p = F_e d_e{}^2 M_p / M_e d_p{}^2,$$

where M_p and d_p are the mass and diameter of the planet, respectively. Substituting values of M_p and d_p for the moon shows that the force of gravity on the moon is only 1/6 of the value on earth.

gray Symbol Gy. The derived SI unit of absorbed *dose of ionizing radiation (*see* radiation units). It is named after the British radiobiologist L. H. Gray (1905–65).

great circle Any circle on a sphere formed by a plane that passes through the centre of the sphere. The equator and the meridians of longitude are all great circles on the earth's surface.

greenhouse effect 1. The effect within a greenhouse in which solar radiation mainly in the visible range of the spectrum passes through the glass roof and walls and is absorbed by the floor, earth, and contents, which re-emit the energy as infrared radiation. Because the infrared radiation cannot escape through the glass, the temperature inside the greenhouse rises. **2.** A similar effect in which the earth's atmosphere behaves like the greenhouse and the surface of the earth absorbs most of the solar radiation, re-emitting it as infrared radiation. This is absorbed by carbon dioxide, water, and ozone in the atmosphere as well as by clouds and reradiated back to earth. At night this absorption prevents the temperature falling rapidly after a hot day, especially in regions with a high atmospheric water content.

Gregorian telescope *See* telescope.

grid 1. (in electricity) The system of overhead wires or underground cables by which electrical power is distributed from power stations to users. The grid is at a high voltage, up to 750 kV in some countries. **2.** (in electronics) *See* control grid.

ground state The lowest stable energy state of a system, such as a molecule, atom, or nucleus. *See* energy level.

ground wave A radio wave that travels in approximately a straight line between points on the earth's surface. For transmission over longer distances sky waves have to be involved. *See* radio transmission.

gun metal A type of bronze usually containing 88–90% copper, 8–10% tin, and 2–4% zinc. Formerly used for cannons, it is still used for bearings and other parts that require high resistance to wear and corrosion.

gyrocompass A *gyroscope that is driven continuously so that it can be used as a nonmagnetic compass. When the earth rotates the gyroscope experiences no torque if its spin axis is parallel to the earth's axis; if these axes are not parallel, however, the gyroscope experiences a sequence of restoring torques that tend to make it align itself with the earth's axis. The gyrocompass is therefore an accurate north-seeking device that is uninfluenced by metallic or magnetic objects and it is also more consistent than the magnetic compass. It is therefore widely used on ships, aircraft, missiles, etc.

gyromagnetic ratio Symbol γ. The ratio of the angular momentum of an atomic system to its magnetic moment. The inverse of the gyromagnetic ratio is called the *magnetomechanical ratio.*

gyroscope A disc with a heavy rim mounted in a double *gimbal so that its axis can adopt any orientation in space. When the disc is set spinning the whole contrivance has two useful properties: (1) Gyroscopic inertia, i.e. the direction of the axis of spin resists change so that if the gimbals are turned the spinning disc maintains the same orientation in space. This property forms the basis of the *gyrocompass and other navigational devices. (2) Precession, i.e. when a gyroscope is subjected to a torque that tends to alter the direction of its axis, the gyroscope turns about an axis at right angles both to the axis about which the torque was applied and to its main axis of spin. This is a consequence of the need to conserve *angular momentum.

In the gyrostabilizer for stabilizing a ship, aircraft, or platform, three gyroscopes are kept spinning about mutually perpendicular axes so that any torque tending to alter the orientation of the whole device affects one of the gyroscopes and thereby activates a servomechanism that restores the original orientation.

H

habit *See* crystal habit.

hadron Any of a class of subatomic particles that interact by the strong interaction (*see* fundamental interactions). The class includes protons, neutrons, and pions. Hadrons are believed to have an internal structure and to consist of quarks; they are therefore not truly elementary. Hadrons are either *baryons, which decay into protons and are believed to consist of three quarks, or *mesons, which decay into *leptons and photons or into proton pairs and are believed to consist of a quark and an antiquark. *See* elementary particles.

half cell An electrode in contact with a solution of ions, forming part of a *cell. Various types of half cell exist, the simplest consisting of a metal electrode immersed in a solution of metal ions. Gas half cells have a gold or platinum plate in a solution with gas bubbled over the metal plate. The commonest is the *hydrogen half cell. Half cells can also be formed by a metal in contact with an insoluble salt or oxide and a solution. The calomel half cell is an example of this. Half cells are commonly referred to as *electrodes*.

half-life *See* decay.

half-thickness The thickness of a specified material that reduces the intensity of a beam of radiation to half its original value.

half-wave plate *See* retardation plate.

half-wave rectifier *See* rectifier.

half-width Half the width of a spectrum line (or in some cases the full width) measured at half its height.

Hall effect The production of an e.m.f. within a conductor or semiconductor through which a current is flowing when there is a strong transverse magnetic field. The potential difference develops at right angles to both the current and the field. It is caused by the deflection of charge carriers by the field and was first discovered by Edwin Hall (1855–1938). The strength of the electric field E_H produced is given by the relationship $E_H = R_H jB$, where j is the current density, B is the magnetic flux density, and R_H is a constant called the *Hall coefficient*. The value of R_H can be shown to be $1/ne$, where n is the number of charge carriers per unit volume and e is the electronic charge. The effect is used to investigate the nature of charge carriers in metals and semiconductors, in the *Hall probe* for the measurement of magnetic fields, and in magnetically operated switching devices.

Halley's comet A bright *comet with a period of 76 years. Last seen in 1910, it is due to be visible in the early spring of 1986. The comet moves around the sun in the opposite direction to the planets. Its orbit was first calculated in 1705 by Edmund Halley (1656–1742), after whom it is named.

halo A luminous ring that sometimes can be observed around the sun or the moon. It is caused by diffraction of their light by particles in the earth's atmosphere; the radius of the ring is inversely proportional to the predominant particle radius.

Hamiltonian Symbol H. A function used to express the energy of a system in terms of its momentum and positional coordinates. In simple cases this is the sum of its kinetic and potential energies. In *Hamiltonian equations*, the usual equations used in mechanics (based on forces) are replaced by equations expressed in terms of momenta. These concepts, often called *Hamiltonian mechanics*, were formulated by Sir William Rowan Hamilton (1805–65).

hard ferromagnetic materials *See* soft iron.

hard radiation Ionizing radiation of high penetrating power, usually gamma rays or short-wavelength X-rays. *Compare* soft radiation.

hardware *See* computer.

harmonic An oscillation having a frequency that is a simple multiple of a *fundamental* sinusoidal oscillation. The fundamental frequency of a sinusoidal oscillation is usually called the *first harmonic*. The *second harmonic* has a frequency twice that of the fundamental and so on. A taut string or column of air, as in a violin or organ, will sound upper harmonics at the same time as the fundamental sounds. This is because the string or column of air divides itself into

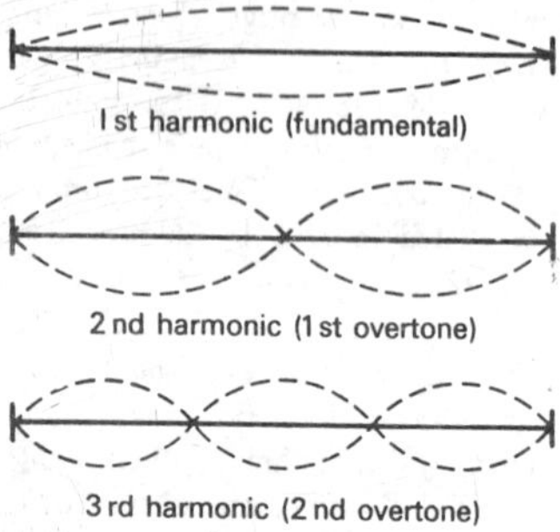

Harmonics

sections, each section then vibrating as if it were a whole. The upper harmonics are also called *overtones*, but the second harmonic is the first overtone, and so on. Musicians, however, often regard harmonic and overtone as synonymous, not counting the fundamental as a harmonic.

harmonic motion *See* simple harmonic motion.

harmonic series (harmonic progression) A series or progression in which the reciprocals of the terms have a constant difference between them, e.g. $1 + 1/2 + 1/3 + 1/4 \dots + 1/n$.

Harvard classification *See* spectral class.

health physics The branch of medical physics concerned with the protection of medical, scientific, and industrial workers from the hazards of ionizing radiation and other dangers associated with atomic physics. Establishing the maximum permissible *dose of radiation, the disposal of radioactive waste, and the shielding of dangerous equipment are the principal activities in this field.

heat Energy in the course of being transferred from one body or system to another as a result of a difference in temperature. The energy in the body or system before or after transfer is also sometimes called heat but this leads to confusion, especially in thermodynamics.

A body in equilibrium with its surroundings contains energy (the kinetic and potential energies of its atoms and molecules) but this is called *internal energy*, U, rather than heat. When such a body changes its temperature or phase there is a change in internal energy, ΔU, which (according to the first law of *thermodynamics) is given by $\Delta U = Q - W$, where Q is the heat absorbed by the body from the surroundings and W is the work done simultaneously on the surroundings. To use the word 'heat' for both U and Q is clearly confusing. Note also that certain physical quantities are described as *heat of atomization, *heat of combustion, etc. What is usually used is a standard molar *enthalpy change for the process under consideration. The units are kJ mol^{-1}; a negative value indicates that energy is liberated. *See also* heat capacity; heat transfer; latent heat.

heat balance 1. A balance sheet showing all the heat inputs to a system (such as a chemical process, furnace, etc.) and all the heat outputs. **2.** The equilibrium that exists on the average between the radiation received from the sun by the earth and its atmosphere and that reradiated or reflected by the earth and the atmosphere. In general, the regions of the earth nearer the equator than about 35°N or S receive more energy from the sun than they are able to reradiate, whereas those regions polewards of 35°N or S receive less energy than they lose. The excess of heat received by the low latitudes is carried to the higher latitudes by atmospheric and oceanic circulations.

heat capacity (thermal capacity) The ratio of the heat supplied to an object or specimen to its consequent rise in temperature. The *specific heat capacity* is the ratio of the heat supplied to unit mass of a substance to its consequent rise in temperature. The *molar heat capacity* is the ratio of the heat supplied to unit amount of a substance to its consequent rise in temperature. In practice, heat capacity (*C*) is measured in joules per kelvin, specific heat capacity (*c*) in $J K^{-1} kg^{-1}$, and molar heat capacity (C_m) in $J K^{-1} mol^{-1}$. For a gas, the values of *c* and C_m are commonly given either at *constant volume*, when only its *internal energy is increased, or at *constant pressure*, which requires a greater input of heat as the gas is allowed to expand and do work against the surroundings. The symbols for the specific and molar heat capacities at constant volume are c_v and C_v, respectively; those for the specific and molar heat capacities at constant pressure are c_p and C_p.

heat death of the universe The condition of the universe when *entropy is maximized and all large-scale samples of matter are at a uniform temperature. In this condition no energy is available for doing work and the universe is finally unwound. The condition was predicted by Rudolph Clausius (1822–88), the German physicist, who introduced the concept of entropy. Clausius's dictum that "the energy of the universe is constant, its entropy tends to a maximum" is a statement of the first two laws of thermodynamics. These laws apply in this sense only to closed systems and, for the predicted heat death to occur, the universe must be a closed system.

heat engine A device for converting heat into work. The heat is derived from the combustion of a fuel. In an *internal-combustion engine the fuel is burnt inside the engine, whereas in a *steam engine or steam turbine, examples of external-combustion engines, the fuel is used to raise steam outside the engine and then some of the steam's internal energy is used to do work inside the engine. Engines usually work on cycles of operation, the most efficient of which would be the *Carnot cycle. This cannot be realized in practice, but the *Rankine cycle is approximated by some engines.

heat exchanger A device for transferring heat from one fluid to another without permitting the two fluids to contact each other. A simple industrial heat exchanger consists of a bundle of parallel tubes, through which one fluid flows, enclosed in a container, through which the other fluid flows in the opposite direction (a *counter-current heat exchanger*).

heat of atomization The energy required to dissociate one mole of a given substance into atoms. *See* heat.

heat of combustion The energy liberated when one mole of a given substance is completely oxidized. *See* heat.

heat of formation The energy liberated or absorbed when one mole of a compound is formed from its constituent elements.

heat of neutralization The energy liberated in neutralizing one mole of an acid or base.

heat of reaction The energy liberated or absorbed as a result of the complete chemical reaction of molar amounts of the reactants.

heat of solution The energy liberated or absorbed when one mole of a given substance is completely dissolved in a large volume of solvent (strictly, to infinite dilution).

heat pump A device for transferring heat from a low temperature source to a high temperature region by doing work. It is essentially a refrigerator with a different emphasis. The working fluid is at one stage a vapour, which is compressed by a pump adiabatically so that its temperature rises. It is then passed to the radiator, where heat is given out to the surroundings (the space to be heated) and the fluid condenses to a liquid. It is then expanded into an evaporator where it takes up heat from its surroundings and becomes a vapour again. The cycle is completed by returning the vapour to the compressor. Heat pumps are sometimes adapted to be used as dual-purpose space-heating-in-winter and air-conditioning-in-summer devices.

heat radiation (radiant heat) Energy in the form of electromagnetic waves emitted by a solid, liquid, or gas as a result of its temperature. It can be transmitted through space; if there is a material medium this is not warmed by the radiation except to the extent that it is absorbed. Although it covers the whole electromagnetic spectrum, the highest proportion of this radiation lies in the infrared portion of the spectrum. *See* black body; Planck's radiation law; Stefan's law; Wien's displacement law.

heat shield A specially prepared surface that prevents a spacecraft or capsule from overheating as it re-enters the earth's atmosphere. The surface is coated with a plastic impregnated with quartz fibres, which is heated by friction with air molecules as the craft enters the atmosphere, causing the outer layer to vaporize. In this way about 80% of the energy is reradiated and the craft is safeguarded from an excessive rise in temperature.

heat transfer The transfer of energy from one body or system to another as a result of a difference in temperature. The heat is transferred by *conduction (*see also* conductivity), *convection, and radiation (*see* heat radiation).

Heaviside–Kennelly layer *See* earth's atmosphere.

heavy hydrogen *See* deuterium.

heavy water (deuterium oxide) Water in which hydrogen atoms, ^1H, are replaced by the heavier isotope deuterium, ^2H. It is a colourless liquid, which forms hexagonal crystals on freezing. Its physical properties differ from those of 'normal' water; r.d. 1.105; m.p. 3.8°C; b.p. 101.4°C. Deuterium oxide occurs to a small extent (about 0.003% by weight) in natural water, from which it can be separated by fractional distillation or by electrolysis. It is particularly useful in the nuclear industry because of its ability to reduce the energies of fast neutrons to thermal energies (*see* moderator) and because its quenching cross-section is lower than that of hydrogen and consequently it does not appreciably reduce the neutron flux. In the laboratory it is used for *labelling other molecules for studies of reaction mechanisms.

hecto- Symbol h. A prefix used in the metric system to denote 100 times. For example, 100 coulombs = 1 hectocoulomb (hC).

Heisenberg uncertainty principle *See* uncertainty principle.

heliocentric universe A view of the universe in which the sun is taken to be at its centre. The model was first proposed by Aristarchus of Samos (310–230 BC) but dropped in favour of the *geocentric system proposed by Ptolemy (*c.* 90–168 AD). Copernicus (1473–1543) revived an essentially heliocentric view, which was upheld by Galileo (1564–1642) against strong opposition from the church on the grounds that if the earth was not at the centre of the universe man's position in it was diminished. In the modern view the sun is at the centre of the *solar system, but the solar system is one of an enormous number of stars in the Galaxy, which is itself one of an enormous number of *galaxies.

Helmholtz free energy *See* free energy.

henry Symbol H. The *SI unit of inductance equal to the inductance of a closed circuit in which an e.m.f. of one volt is produced when the electric current in the circuit varies uniformly at a rate of one ampere per second. It is named after Joseph Henry (1797–1878), a US physicist.

Henry's law At a constant temperature the mass of gas dissolved in a liquid at equilibrium is proportional to the partial pressure of the gas. The law, discovered in 1801 by the British chemist and physician William Henry (1775–1836), is a special case of the partition law. It applies only to gases that do not react with the solvent.

hertz Symbol Hz. The *SI unit of frequency equal to one cycle per second. It is named after Heinrich Hertz (1857–94), a German physicist.

Hertzsprung–Russell diagram (H–R diagram) A graphical representation of the absolute magnitude of stars (usually along the *y*-axis) plotted against the spectral class or colour index (*x*-axis). The *y*-axis then represents the energy output of the star and the *x*-axis its surface temperature. The majority of stars on such a diagram fall on a band running from the top left to the bottom right of the graph. These are called *main-sequence stars* (the sun falls into this class). The few stars falling in the lower left portion are called *white dwarfs. The *giants fall in a cluster above the main sequence and the *supergiants are above them. The diagram, which was first devised in 1911 by Ejnar Hertzsprung (1873–1969) and in 1913 by H. N. Russell (1897–1957), forms the basis of the theory of *stellar evolution.

Hess's law If reactants can be converted into products by a series of reactions, the sum of the heats of these reactions (with due regard to their sign) is equal to the heat of reaction for direct conversion from reactants to products. More generally, the overall energy change in going from reactants to products does not depend on the route taken. The law can be used to obtain thermodynamic data that cannot be measured directly. For example, the heat of formation

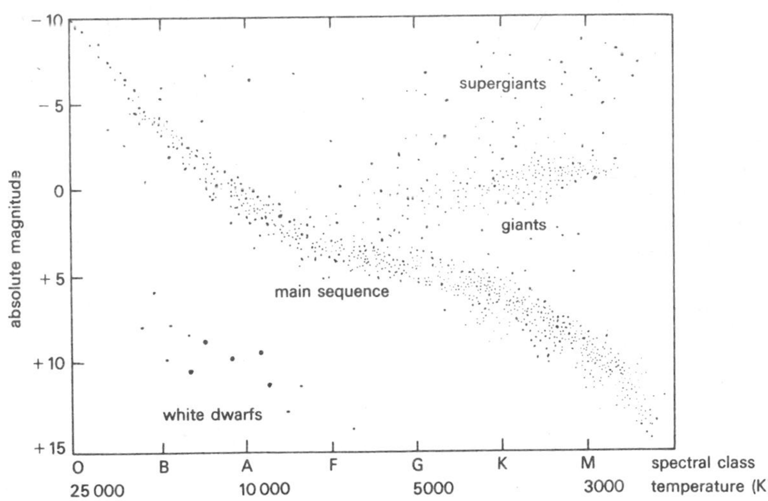

Hertzsprung–Russell diagram

of ethane can be found by considering the reactions:

$$2C(s) + 3H_2(g) + 3\tfrac{1}{2}O_2(g) \rightarrow 2CO_2(g) + 3H_2O(l)$$

The heat of this reaction is $2\Delta H_C + 3\Delta H_H$, where ΔH_C and ΔH_H are the heats of combustion of carbon and hydrogen respectively, which can be measured. By Hess' law, this is equal to the sum of the energies for two stages:

$$2C(s) + 3H_2(g) \rightarrow C_2H_6(g)$$

(the heat of formation of ethane, ΔH_f) and

$$C_2H_6(g) + 3\tfrac{1}{2}O_2 \rightarrow 2CO_2(g) + 3H_2O(l)$$

(the heat of combustion of ethane, ΔH_E). As ΔH_E can be measured and as

$$\Delta H_f + \Delta H_E = 2\Delta H_c + 3\Delta H_H$$

ΔH_f can be found. Another example is the use of the *Born–Haber cycle to obtain lattice energies. The law was first put forward in 1840 by the Russian chemist Germain Henri Hess (1802–50). It is sometimes called the *law of constant heat summation* and is a consequence of the law of conservation of energy.

heterodyne Denoting a device or method of radio reception in which *beats are produced by superimposing a locally generated radio wave on an incoming wave. In the *superheterodyne receiver the intermediate frequency is amplified and demodulated. In the *heterodyne wavemeter*, a variable-frequency local oscillator is adjusted to give a predetermined beat frequency with the incoming wave, enabling the frequency of the incoming wave to be determined.

heuristic Denoting a method of solving a problem for which no *algorithm exists. It involves trial and error, as in *iteration.

Heusler alloys Ferromagnetic alloys containing no ferromagnetic elements. The original alloys contained copper, manganese, and tin and were first made by Conrad Heusler (19th century mining engineer).

high frequency (HF) A radio frequency in the range 3–30 megahertz; i.e. having a wavelength in the range 10–100 metres.

high-speed steel A steel that will remain hard at dull red heat and can therefore be used in cutting tools for high-speed lathes. It usually contains 12–22% tungsten, up to 5% chromium, and 0.4–0.7% carbon. It may also contain small amounts of vanadium, molybdenum, and other metals.

high tension (HT) A high potential difference, usually one of several hundred volts or more. However, batteries supplying the anode circuits of radio devices using valves are usually called *high-tension batteries*, even when they supply a lower potential difference.

hole A vacant electron position in the lattice structure of a solid that behaves like a mobile positive *charge carrier with a negative *rest energy. *See* semiconductor.

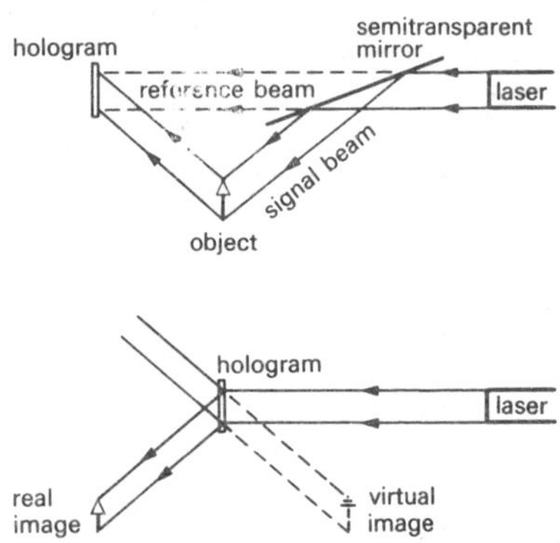

Holography

holography A method of recording and displaying a three-dimensional image of an object, usually using *coherent radiation from a *laser and photographic plates. The light from a laser is divided so that some of it (the reference beam) falls directly on a photographic plate. The other part illuminates the object, which reflects it back onto the photographic plate. The two beams form interference patterns on the plate, which when developed is called the *hologram.* To reproduce the image of the object, the hologram is illuminated by coherent light, ideally the original reference beam. The hologram produces two sets of diffracted waves; one set forms a virtual image coinciding with the original object position and the other forms a real image on the other side of the plate. Both are three-dimensional. The method was invented by Dennis Gabor (1900–79) in 1948. More recent techniques can produce holograms visible in white light.

Hooke's law The *stress applied to any solid is proportional to the *strain it produces within the elastic limit for that solid. The ratio of longitudinal stress to strain is equal to the Young modulus of elasticity (*see* elastic modulus). The law was first stated by the English scientist Robert Hooke (1635–1703) in the form "Ut tensio, sic vis."

horsepower (hp) An imperial unit of power originally defined as 550 foot-pound force per second; it is equal to 745.7 watts.

hot-wire instrument An electrical measuring instrument (basically an ammeter) in which the current to be measured is passed through a thin wire and causes its temperature to rise. The temperature rise, which is proportional to the square of the current, is measured by the expansion of the wire. Such instruments can be used for either direct current or alternating current.

Hubble constant The rate at which the velocity of recession of the galaxies increases with distance as determined by the *redshift. The value is not agreed upon but current measurements indicate that it lies between 49 and 95 $km\ s^{-1}$ per megaparsec. The reciprocal of the Hubble constant, the *Hubble time,* is a measure of the age of the universe, assuming that the expansion rate has remained constant. If, as seems likely, the expansion rate has slowed down with time as a result of gravitational attraction between the galaxies, the Hubble time gives an upper limit to the age of the universe of some 2×10^{10} years. The constant is named after the US astronomer Edwin Hubble (1889–1953).

hue *See* colour.

humidity The concentration of water vapour in the atmosphere. The *absolute humidity* is the mass of water vapour per unit volume of air, usually expressed in $kg\ m^{-3}$. A useful measure is the *relative humidity,* the ratio, expressed as a percentage, of the moisture in the air to the moisture it would contain if it were saturated at the same temperature and pressure. The *specific humidity* is also sometimes used: this is the mass of water vapour in the atmosphere per unit mass of air.

Huygen's construction (Huygen's principle) Every point on a wavefront may itself be regarded as a source of secondary waves. Thus, if the position of a wavefront at any instant is known, a simple construction enables its position to be drawn at any subsequent time. The construction was first used by Christian Huygens (1629–95).

hydraulic press A device in which a force applied to a small piston creates a pressure, which is transmitted through a fluid to a larger piston, where it gives rise to a larger force. This depends on Pascal's principle that the pressure applied anywhere in an enclosed fluid is transmitted equally in all directions. The principle of the hydraulic press is widely used in jacks, vehicle brakes, presses, and earth-moving machinery, usually with oil as the working fluid.

hydraulics The study of water or other fluids at rest or in motion, particularly with

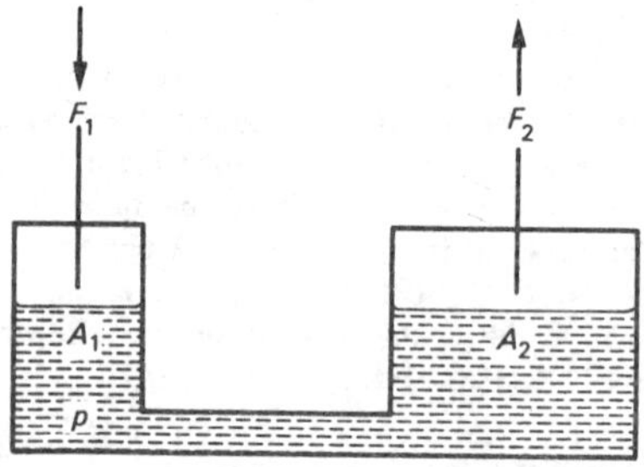

Hydraulic press

respect to their engineering uses. The study is based on the principles of *hydrostatics and *hydrodynamics.

hydrodynamics The study of the motion of incompressible fluids and the interaction of such fluids with their boundaries.

hydroelectric power Electric power generated by a flow of water. A natural waterfall provides a source of energy, in the form of falling water, which can be used to drive a water *turbine. This turbine can be coupled to a generator to provide electrical energy. Hydroelectric generators can be arranged to work in reverse so that during periods of low power demand current can be fed to the generator, which acts as a motor. This motor drives the turbine, which then acts as a pump. The pump then raises water to an elevated reservoir so that it can be used to provide extra power at peak-load periods.

hydrogen bomb *See* nuclear weapons.

hydrogen electrode *See* hydrogen half cell.

hydrogen half cell (hydrogen electrode) A type of half cell in which a metal foil is immersed in a solution of hydrogen ions and hydrogen gas is bubbled over the foil. The standard hydrogen electrode, used in measuring standard *electrode potentials, uses a platinum foil with a 1.0 M solution of hydrogen ions, the gas at 1 atmosphere pressure, and a temperature of 25°C. It is written $Pt(s)|H_2(g)$, $H^+(aq)$, the effective reaction being $H_2 \rightarrow 2H^+ + 2e$.

hydrogen ion *See* pH.

hydrogen spectrum The atomic spectrum of hydrogen is characterized by lines corresponding to radiation quanta of sharply defined energy. A graph of the frequencies at which these lines occur against the ordinal number that characterizes their position in the series of lines, produces a smooth curve indicating that they obey a formal law. In 1885 J. J. Balmer (1825–98) discovered the law having the form:

$$1/\lambda = R(1/n_1^2 + 1/n_2^2)$$

This law gives the so-called *Balmer series* of lines in the visible spectrum in which $n_1 = 2$ and $n_2 = 3,4,5\ldots$, λ is the wavelength associated with the lines, and R is the *Rydberg constant.

In the *Lyman series*, discovered by Theodore Lyman (1874–1954), $n_1 = 1$ and the lines fall in the ultraviolet. The Lyman series is the strongest feature of the solar spectrum as observed by rockets and satellites above the earth's atmosphere. In the *Paschen series*, discovered by F. Paschen (1865–1947), $n_1 = 3$ and the lines occur in the far infrared. The *Brackett series* also occurs in the far infrared, with $n_1 = 4$.

hydrometer An instrument for measuring the density or relative density of liquids. It usually consists of a glass tube with a long bulb at one end. The bulb is weighted so that the device floats vertically in the liquid, the relative density being read off its calibrated stem by the depth of immersion.

hydrostatics The study of liquids at rest, with special reference to storage tanks, dams, bulkheads, and hydraulic machinery.

hygrometer An instrument for measuring *humidity in the atmosphere. The mechanical type uses an organic material, such as human hair, which expands and contracts with changes in atmospheric humidity. The expansion and contraction is used to operate a needle. In the electric type, the change in resistance of a hygroscopic substance is used as an indication of humidity. In *dew-point hygrometers* a polished surface is reduced in temperature until water vapour from the atmosphere forms on it. The temperature of this dew point enables the relative humidity of the atmosphere to be calculated. In the *wet-and-dry bulb hygrometer*, two thermometers are mounted side by side, the bulb of one being surrounded by moistened muslin. The thermometer with the wet bulb will register a lower temperature than that with a dry bulb owing to the cooling effect of the evaporating water. The temperature difference enables the relative humidity to be calculated. Only the dew-point hygrometer can be operated as an absolute instrument; all the others must ultimately be calibrated against this.

hyper- A prefix denoting over, above, high; e.g. hypersonic.

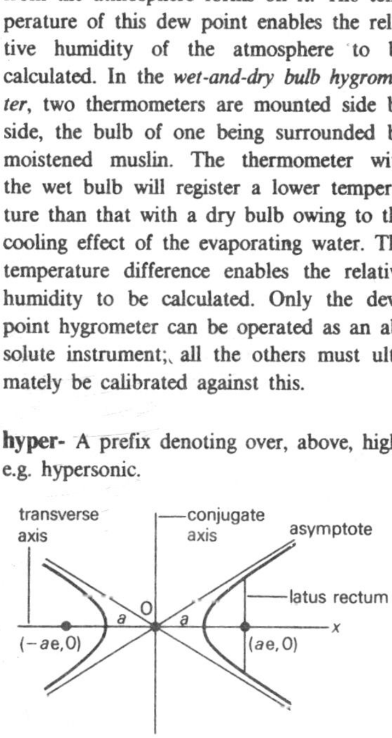

A hyperbola

hyperbola A *conic with eccentricity $e > 1$. It has two branches. For a hyperbola centred at the origin, the *transverse axis* runs along the x-axis between the vertices and has length $2a$. The *conjugate axis* runs along the y-axis and has length $2b$. There are two *foci* on the x-axis at $(ae, 0)$ and $(-ae, 0)$. The *latus rectum*, the chords through the foci perpendicular to the transverse axis, have length $2b^2/a$. The equation of the hyperbola is:

$$x^2/a^2 - y^2/b^2 = 1,$$

and the asymptotes are $y = \pm bx/a$.

hyperbolic functions A set of functions; *sinh*, *cosh*, and *tanh*, that have similar properties to *trigonometric functions but are related to the hyperbola in the manner that trigonometric functions are related to the circle. The hyperbolic sine (sinh) of the angle x is defined by:

$$\sinh x = \tfrac{1}{2}(e^x - e^{-x}).$$

Similarly,

$$\cosh x = \tfrac{1}{2}(e^x + e^{-x})$$

$$\tanh x = (e^x - e^{-x})/(e^x + e^{-x})$$

Hyperbolic secant (sech), cosecant (cosech), and cotangent (coth) are the reciprocals of sinh, cosh, and tanh, respectively.

hypercharge A quantized property of *baryons (*see* elementary particles) that provides a formal method of accounting for the nonoccurrence of certain expected decays by means of the strong interaction (*see* fundamental interactions). Hypercharge is in some respects analogous to electric charge but it is not conserved in weak interactions. Nucleons have a hypercharge of +1, and the *pion has a value of 0. Quarks would be expected to have fractional hypercharges.

hyperfine structure *See* fine structure.

hypermetropia (hyperopia) Long-sightedness. A vision defect in which the lens of the eye is unable to accommodate sufficiently to throw the image of near objects onto the retina. It is caused usually by shortness of the eyeball rather than any fault in the lens system. The subject requires spectacles with converging lenses to bring the image from behind the retina back on to its surface.

hyperon A shortlived *elementary particle; it is classified as a *baryon and has a nonzero *strangeness.

hypersonic Denoting a velocity in excess of Mach 5 (*see* Mach number). *Hypersonic* *flight* is flight at hypersonic speeds in the earth's atmosphere.

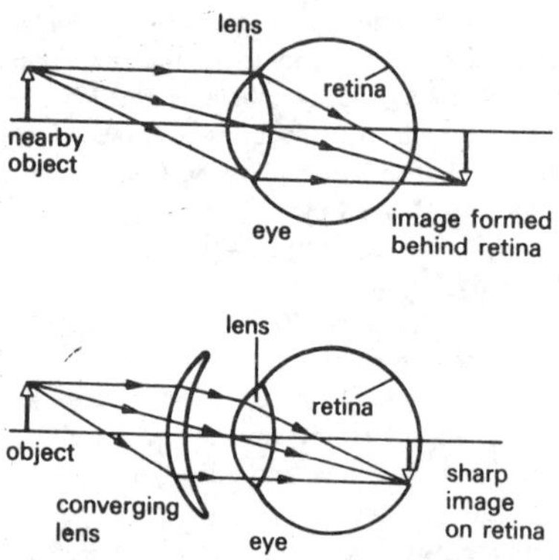

Hypermetropia

hypertonic solution A solution that has a higher osmotic pressure than some other solution.

hypo- A prefix denoting under, below, low; e.g. hypotonic.

hypotonic solution A solution that has a lower osmotic pressure than some other solution.

hypsometer A device for calibrating thermometers at the boiling point of water. As the boiling point depends on the atmospheric pressure, which in turn depends on the height above sea level, the apparatus can be used to measure height above sea level.

hysteresis A phenomenon in which two physical quantities are related in a manner that depends on whether one is increasing or decreasing in relation to the other.

The repeated measurement of *stress against *strain, with the stress first increasing and then decreasing, will produce for some specimens a graph that has the shape of a closed loop. This is known as a *hysteresis cycle*. The most familiar hysteresis cycle, however, is produced by plotting the magnetic flux density (B) within a ferromagnetic material against the applied magnetic field strength (H).

If the material is initially unmagnetized at O it will reach saturation at P as H is increased. As the field is reduced and again increased the loop PQRSTP is formed. The area of this loop is proportional to the energy loss (*hysteresis loss*) occurring during the cycle. The value of B equal to OQ is called the *remanance* (or retentivity) and is the magnetic flux density remaining in the material after the saturating field has been reduced to zero. This is a measure of the tendency of the magnetic domain patterns (*see* magnetism) to remain distorted even after the distorting field has been removed. The value of H equal to OR is called the *coercive force* (or coercivity) and is the field strength required to reduce the remaining flux density to zero. It is a measure of the difficulty of restoring the symmetry of the domain patterns.

I

ice point The temperature at which there is equilibrium between ice and water at standard atmospheric pressure (i.e. the freezing or melting point under standard conditions). It was used as a fixed point (0°) on the Celsius scale, but the kelvin and the International Practical Temperature Scale are based on the *triple point of water.

iconoscope A form of television camera tube (*see* camera) in which the beam of light from the scene is focused on to a thin mica plate. One side of the plate is faced with a thin metallic electrode, the other side being coated with a mosaic of small globules of a photoemissive material. The light beam falling on the mosaic causes photoemission of electrons, creating a pattern of positive charges in what is effectively an array of tiny capacitors. A high-velocity electron beam scans the mosaic, discharging each capacitor in turn through the metallic electrode. The resulting current is fed to amplification circuits, the current from a particular section of the mosaic depending on the illumination it has received. In this way the optical information in the light beam is converted into an electrical signal.

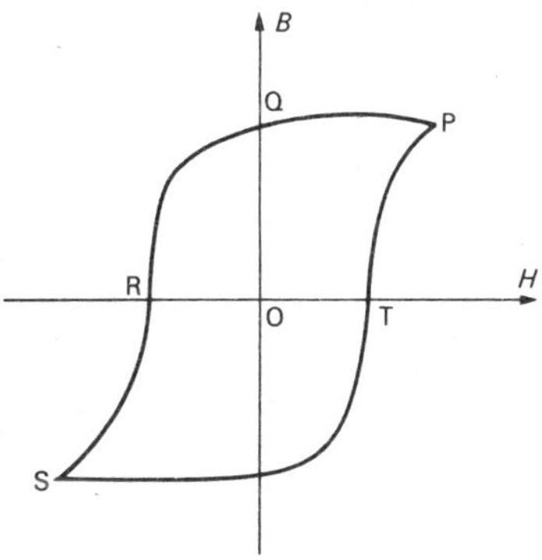

Hysteresis

ideal crystal A single crystal with a perfectly regular lattice that contains no impurities, imperfections, or other defects.

ideal gas (perfect gas) A hypothetical gas that obeys the *gas laws exactly. An ideal gas would consist of molecules that occupy negligible space and have negligible forces between them. All collisions made between molecules and the walls of the container or between molecules and other molecules would be perfectly elastic, because the molecules would have no means of storing energy except as translational kinetic energy.

ideal solution *See* Raoult's law.

identity Symbol $\equiv$. A statement of equality that applies for all values of the unknown quantity. For example, $5y \equiv 2y + 3y$.

ignition temperature **1.** The temperature to which a substance must be heated before it will burn in air. **2.** The temperature to which a *plasma has to be raised in order that nuclear fusion will occur.

illuminance (illumination) Symbol E. The energy in the form of visible radiation reaching a surface per unit area in unit time; i.e. the luminous flux per unit time. It is measured in *lux (lumens per square metre).

image A representation of a physical object formed by a lens, mirror, or other optical instrument. If the rays of light actually pass through the image, it is called a *real image*. If a screen is placed in the plane of a real image it will generally become visible. If the image is seen at a point from which the rays appear to come to the observer, but do not actually do so, the image is called a *virtual image*. No image will be formed on a screen placed at this point. Images may be *upright* or *inverted* and they may be *magnified* or *diminished*.

image converter An electronic device in which an image formed by invisible radiation (usually gamma rays, X-rays, ultraviolet, or infrared) is converted into a visible image. Commonly the invisible radiation is focused on to a photocathode, which emits electrons when it is illuminated. These electrons fall on a fluorescent anode screen, after acceleration and focusing by a system of electron lenses. The fluorescent screen produces a visible image. The device is used in fluoroscopes, infrared telescopes, ultraviolet microscopes, and other devices.

imaginary number A number that is a multiple of $\sqrt{-1}$, which is denoted by i; for example $\sqrt{-3} = i\sqrt{3}$. *See also* complex number.

immersion objective An optical microscope objective in which the front surface of the lens is immersed in a liquid on the cover glass of the microscope specimen slide. Cedar-wood oil or sugar solution is frequently used. It has the same refractive index as the glass of the slide, so that the object is effectively immersed in it. The presence of the liquid increases the effective aperture of the objective, thus increasing the resolution.

impedance Symbol *Z*. The quantity that measures the opposition of a circuit to the passage of a current and therefore determines the amplitude of the current. In a d.c. circuit this is the resistance (*R*) alone. In an a.c. circuit, however, the *reactance (*X*) also has to be taken into account, according to the equation: $Z^2 = R^2 + X^2$, where *Z* is the impedance. The *complex impedance* is given by $Z = R + iX$, where $i = \sqrt{-1}$. The real part of the complex impedance, the resistance, represents the loss of power according to *Joule's law. The ratio of the imaginary part, the reactance, to the real part is an indication of the difference in phase between the voltage and the current.

Imperial units The British system of units based on the pound and the yard. The former f.p.s. system was used in engineering and was loosely based on Imperial units; for all scientific purposes *SI units are now used. Imperial units are also being replaced for general purposes by metric units.

implosion An inward collapse of a vessel, especially as a result of evacuation.

impulse Symbol *J*. The product of a force *F* and the time *t* for which it acts. If the force is variable, the impulse is the integral of Fdt from t_0 to t_1. The impulse of a force acting for a given time interval is equal to the change in momentum produced over that interval, i.e. $J = m(v_1 - v_0)$, assuming that the mass (*m*) remains constant while the velocity changes from v_0 to v_1.

incandescence The emission of light by a substance as a result of raising it to a high temperature. An *incandescent lamp* is one in which light is emitted by an electrically heated filament. *See* electric lighting.

inclination 1. *See* geomagnetism. **2.** The angle between the orbital plane of a planet, satellite, or comet and the plane of the earth's *ecliptic.

indefinite integral *See* integration.

indeterminacy *See* uncertainty principle.

inductance The property of an electric circuit or component that causes an e.m.f. to be generated in it as a result of a change in the current flowing through the circuit (*self inductance*) or of a change in the current flowing through a neighbouring circuit with which it is magnetically linked (*mutual inductance*). In both cases the changing current is associated with a changing magnetic field, the linkage with which in turn induces the e.m.f. In the case of self inductance, *L*, the e.m.f., *E*, generated is given by $E = -L.dI/dt$, where *I* is the instantaneous current and the minus sign indicates that the e.m.f. induced is in opposition to the change of current. In the case of mutual inductance, *M*, the e.m.f., E_1, induced in one circuit is given by $E_1 = -M.dI_2/dt$, where I_2 is the instantaneous current in the other circuit.

induction A change in the state of a body produced by a field. *See* electromagnetic induction; inductance.

induction coil A type of *transformer used to produce a high-voltage alternating current or pulses of high-voltage current from a low-voltage direct-current source. The induction coil is widely used in spark-ignition *internal-combustion engines to produce the spark in the sparking plugs. In such an engine the battery is connected to the primary winding of the coil through a circuit-breaking device driven by the engine and the e.m.f. generated in the secondary winding of the coil is led to the sparking plugs through the distributor. The primary coil consists of relatively few turns, whereas the secondary consists of many turns of fine wire.

induction heating The heating of an electrically conducting material by *eddy currents induced by a varying electromagnetic field. Induction heating may be an undesirable effect leading to power loss in transformers and other electrical devices. It is, however, useful for melting, forging, rolling, and heat-treating metals, as well as for welding, brazing, and soldering. The material to be heated is inserted into a coil through which an alternating current flows and acts as the short-circuited secondary of a *transformer. Eddy currents induced in the material within the coil cause the temperature of the material to rise.

induction motor *See* electric motor.

inelastic collision A collision in which some of the kinetic energy of the colliding bodies is converted into internal energy in one body so that kinetic energy is not conserved. In collisions of macroscopic bodies some kinetic energy is turned into vibrational energy of the atoms, causing a heating effect. Collisions between molecules of a gas or liquid may also be inelastic as they cause changes in vibrational and rotational *energy levels. In nuclear physics, an inelastic collision is one in which the incoming particle causes the nucleus it strikes to become excited or to break up.

inertia The property of matter that causes it to resist any change in its motion. Thus, a body at rest remains at rest unless it is acted upon by an external force and a body in motion continues to move at constant speed in a straight line unless acted upon by an external force. This is a statement of Newton's first law of motion. The *mass of a body is a measure of its inertia. *See* Mach's principle; inertial frame.

inertial frame A *frame of reference in which bodies move in straight lines with constant speeds unless acted upon by external forces, i.e. a frame of reference in which bodies are not accelerated. Newton's laws of motion are valid in an inertial system but not in a system that is itself accelerated with respect to such a frame.

inertial mass *See* mass.

infinite series *See* series.

infinitesimal Vanishingly small but not zero. Infinitesimal changes are notionally made in the *calculus, which is sometimes called the *infinitesimal calculus*.

infinity Symbol ∞. A quantity having a value that is greater than any assignable value. In some contexts, $-\infty$ is also a useful concept – a quantity having a value that is less than any assignable value.

inflection A point on a curve at which the tangent changes from rotation in one direction to rotation in the opposite direction. If the curve $y = f(x)$ has a stationary point $dy/dx = 0$, there is either a maximum, minimum, or inflection at this point. If $d^2y/dx^2 = 0$, the stationary point is a point of inflection.

infrared astronomy The study of radiation from space in the infrared region of the spectrum (*see* infrared radiation). Some infrared radiation is absorbed by water and carbon dioxide molecules in the atmosphere but there are several narrow atmospheric *windows in the near-infrared (1.15–1.3 μm, 1.5–1.75 μm, 2–2.4 μm, 3.4–4.2 μm, 4.6–4.8 μm, 8–13 μm, and 16–18 μm). Longer wavelength observations must be made from balloons, rockets, or satellites. Infrared sources are either thermal, i.e. emitted by gas or dust atoms or molecules in the temperature range 100–3000 K, or electronic, i.e. emitted by high-energy electrons interacting with magnetic fields as in *synchrotron radiation. Detectors are either modified reflecting *telescopes or solid-state photon detectors, usually incorporating *photovoltaic devices.

infrared radiation (IR) Electromagnetic radiation with wavelengths longer than that of red light but shorter than radiowaves, i.e. radiation in the wavelength range 0.7 micrometre to 1 millimetre. It was discovered in 1800 by William Herschel (1738–1822) in the sun's spectrum. The natural vibrational frequencies of atoms and molecules and the rotational frequencies of some gaseous molecules fall in the infrared region of the electromagnetic spectrum. The infrared absorption spectrum of a molecule is highly characteristic of it and the spectrum can therefore be used for molecular identification. Glass is opaque to infrared radiation of wavelength greater than 2 micrometres and other materials, such as germanium, quartz, and polyethylene, have to be used to make lenses and prisms. Photographic film can be made sensitive to infrared up to about 1.2 μm.

infrasound Soundlike waves with frequencies below the audible limit of about 20 hertz.

insolation (from *in*coming *sol*ar radi*ation*) The solar radiation that is received at the earth's surface per unit area. It is related to the *solar constant, the duration of daylight, the altitude of the sun, and the latitude of the receiving surface. It is measured in $MJ\ m^{-2}$.

instantaneous value The value of any varying quantity at a specified instant.

insulator A substance that is a poor conductor of heat and electricity. Both properties usually occur as a consequence of a lack of mobile electrons. *See* energy bands.

integral calculus *See* calculus.

integrand *See* integration.

integrated circuit A miniature electronic circuit produced within a single crystal of a *semiconductor, such as silicon. They range from simple logic circuits, little more than 1 mm square, to large-scale circuits measuring up to 8 mm square and containing a million or so transistors (active components) and resistors or capacitors (passive components). They are widely used in memory circuits, microcomputers, pocket calculators, and electronic watches on account of their low cost and bulk, reliability, and high speed. They are made by introducing impurities into specific regions of the semiconductor crystal by a variety of techniques.

integration The process of continuously summing changes in a function f(x). It is the basis of the integral *calculus and the opposite process to *differentiation. The function to be integrated is called the *integrand* and the result of integration on the integrand is called the *integral*. For example, the integration of f(x) is written

$$\int f(x)dx,$$

the differential dx being added to indicate that f(x) must be integrated with respect to x. To complete the integration, a *constant of integration, C*, must be added where no interval over which the integration takes place is given. This is called an *indefinite integral*. If the interval is specified, e.g.

$$\int_{-r}^{r} f(x)dx$$

no constant of integration is required and the result is a *definite integral*. This means that f(x) is to be integrated between the values $x = +r$ and $x = -r$.

intensity 1. The rate at which radiant energy is transferred per unit area. *See* radiant intensity. **2.** The rate at which sound energy is transferred as measured relative to some reference value. *See* decibel. **3.** Magnetic intensity. *See* magnetic field. **4.** Electric intensity. *See* electric field. **5.** *See* luminous intensity.

interaction An effect involving a number of bodies, particles, or systems as a result of which some physical or chemical change takes place to one or more of them. *See also* fundamental interaction.

interference The interaction of two or more wave motions affecting the same part of a medium so that the instantaneous disturbances in the resultant wave are the algebraic sum of the instantaneous disturbances in the interfering waves.

The phenomenon was first described by Thomas Young (1773–1829) in 1801 in light waves; it provided strong evidence for the wave theory of light. In the apparatus known as *Young's slits*, light is passed from a small source through a slit in a screen and the light emerging from this slit is used to illuminate two adjacent slits on a second screen. By allowing the light from these two slits to fall on a third screen, a series of parallel interference fringes is formed. Where the maximum values of the two waves from the slits coincide a bright fringe occurs (*constructive interference*) and where the maxima of one wave coincide with the minima of the other dark fringes are produced (*destructive interference*). *Newton's rings are also an interference effect. Because *lasers produce *coherent radiation they are also used to produce interference effects, one application of their use being *holography. *See also* interferometer.

interferometer An instrument designed to produce optical *interference fringes for measuring wavelengths, testing flat surfaces, measuring small distances, etc. *See also* echelon; Fabry–Pérot interferometer; Michelson–Morley experiment. In astronomy, radio interferometers are one of the two basic types of *radio telescopes.

intermediate frequency *See* heterodyne; superheterodyne.

intermediate neutron A *neutron with kinetic energy in the range 10^2–10^5 electronvolts (1.6×10^{-17} – 1.6×10^{-14} joule).

intermediate vector boson *See* W-particle.

intermetallic compound A compound consisting of two or more metallic elements present in definite proportions in an alloy.

intermolecular forces Weak forces occurring between molecules. *See* van der Waals' forces.

internal-combustion engine A *heat engine in which fuel is burned in combustion chambers within the engine rather than in a separate furnace (as with the steam engine). The first working engine was the four-stroke *Otto engine* produced in 1876 by Nikolaus Otto (1832–91). In this type of engine a piston descends in a cylinder, drawing in a charge of fuel and air through an inlet valve; after reaching the bottom of its stroke the piston rises in the cylinder with the valves closed and compresses the charge; at or near the top of its stroke the charge is ignited by a spark and the resulting increase in pressure from the explosion forces the piston down again; on the subsequent upstroke the exhaust valve opens and the burnt gases are pushed out of the combustion chamber. The cycle is then repeated. Otto's engine used gas as a fuel; however, the invention of the carburettor and the development of the oil industry at the end of the 19th century enabled the Otto engine to become the source of power for the emerging motor car. A variation of the Otto four-stroke engine is the two-stroke engine that has no complicated valve system, the explosive charge entering and leaving the cylinder through ports in the cylinder that are covered and uncovered by the moving piston.

An alternative to the Otto engine, especially for heavy vehicles where weight is not a problem, is the compression-ignition *Diesel engine* invented by Rudolf Diesel (1858–1913) in about 1896. In this type of engine there are no sparking plugs; instead air is compressed in the cylinder, causing its temperature to rise to about 550°C. Oil is then sprayed into the combustion chamber and ignites on contact with the hot air. While the spark-ignition petrol engine typically works on a *compression ratio of 8 or 9 to 1, the Diesel engine has to have a compression ratio of 15 to 20 to 1. This requires a much heavier, and therefore more expensive, engine. *See also* Wankel engine; gas turbine.

internal conversion A process in which an excited atomic nucleus (*see* excitation) decays to the *ground state and the energy released is transferred by electromagnetic coupling to one of the bound electrons of that atom rather than being released as a photon. The coupling is usually with an electron in the K-, L-, or M-shell of the atom, and this *conversion electron* is ejected from the atom with a kinetic energy equal to the difference between the nuclear transition energy and the binding energy of the electron. The resulting ion is itself in an excited state and usually subsequently emits an *Auger electron or an X-ray photon.

internal energy Symbol U. The total of the kinetic energies of the atoms and molecules of which a system consists and the potential energies associated with their mutual interactions. It does not include the kinetic and potential energies of the system as a whole nor their nuclear energies or other intra-atomic energies. The value of the absolute internal energy of a system in any particular state cannot be measured; the significant quantity is the change in internal energy, ΔU. For a closed system (i.e. one that is not being replenished from outside its boundaries) the change in internal energy is equal to the heat absorbed by the system (Q) from its surroundings, less the work done (W) by the system on its surroundings, i.e. $\Delta U = Q - W$. *See also* energy; heat.

internal resistance The resistance within a source of electric current, such as a cell or generator. It can be calculated as the difference between the e.m.f. (E) and the potential difference (V) between the terminals divided by the current being supplied (I), i.e. $r = (E - V)/I$, where r is the internal resistance.

international candle A former unit of *luminous intensity. It has now been replaced by the *candela, to which it is approximately equal.

international date line An imaginary line on the earth's surface that joins the north and south poles and approximately follows the 180° meridian through the Pacific Ocean. This line has been agreed internationally to mark the beginning and end of a day. A traveller moving towards the east, against the sun's apparent movement, gains 1 hour for every 15° of longitude; westward he loses time at the same rate. In crossing the dateline therefore he is deemed to compensate for this by losing or gaining (respectively) one day. The 180° meridian was chosen as the date line by the International Meridian Conference in 1884.

International Practical Temperature Scale *See* temperature scales.

interplanetary space The space between the sun and the planets within the *solar system. The *interplanetary matter* that occupies this region of space mostly originates in the sun, from which it flows as the *solar wind. The solar wind consists primarily of protons emerging from the sun at a rate of about 10^9 kilograms per second. At the earth's distance from the sun the particle density has fallen to a few particles per cm^3. Apart from this very tenuous gas, there are also dust particles in interplanetary space, largely believed to originate in the belt of asteroids. Particles weighing about 1 g produce visible meteors in the earth's atmosphere; micrometeorites as small as 1 nanogram can be detected by their impact on spacecraft.

interstellar space The space between the stars. The *interstellar matter* that occupies this space constitutes several percent of the Galaxy's total mass and it is from this matter that new stars are formed. The matter is primarily hydrogen, ih which a number of other molecules and radicals have been detected, with some small solid dust grains floating in it. On average the density of matter in interstellar space is about one hydrogen atom per cm^3.

interstitial *See* defect.

interstitial compound A compound in which ions or atoms of a nonmetal occupy interstitial positions in a metal lattice. Such compounds often have metallic properties. Examples are found in the carbides, borides, and silicides.

intrinsic semiconductor *See* semiconductor.

Invar A tradename for an alloy of iron (63.8%), nickel (36%), and carbon (0.2%) that has a very low *expansivity over a a restricted temperature range. It is used in watches and other instruments to reduce their sensitivity to changes in temperature.

inverse functions If $y = f(x)$ and a function can be found so that $x = g(y)$, then g (y) is said to be the inverse function of $f(x)$. If y is a trigonometrical function of the angle x, say $y = \sin x$, then x is the *inverse trigonometrical function* of y, written $x = \arcsin y$ or $\sin^{-1}y$. Similarly, the other trigonometrical functions form the inverse trigonometrical functions $\cos^{-1}y$, $\tan^{-1}y$, $\cot^{-1}y$, $\sec^{-1}y$, and $\operatorname{cosec}^{-1}y$. *Inverse hyperbolic functions* are also formed in this way, e.g. $\sinh^{-1}y$, $\cosh^{-1}y$, and $\tanh^{-1}y$.

inverse-square law A law in which the magnitude of a physical quantity is proportional to the reciprocal of the square of the distance from the source of that property. *Newton's law of gravitation and *Coulomb's law are both examples.

inversion layer *See* transistor.

inversion temperature *See* Joule–Kelvin effect.

involute *See* evolute.

ion An atom or group of atoms that has either lost one or more electrons, making it positively charged (an anion), or gained one or more electrons, making it negatively charged (a cation). *See also* ionization.

ion engine A type of jet-propulsion engine that may become important for propelling or controlling spacecraft. It consists of a unit producing a beam of ions, which are accelerated by an electric or electromagnetic field. Reaction forces from the high-speed ions causes propulsion in much the same way as that caused by exhaust gas of a rocket. However, a separate beam of electrons or ions of opposite polarity to the propelling beam must also be ejected from the engine to enable recombination to take place behind the vehicle (to avoid the vehicle becoming charged). Ion engines provide high *specific impulse and therefore low propellant consumption. The three main components of an ion engine are the power generator, the propellant feed, and the thruster. The power generator may be a nuclear reactor or a solar-energy collector. If it is the former, a gas turbine is coupled to the reactor and the turbine drives an electric generator. A solar-energy unit provides electricity direct. The propellant chosen needs to have an ion of medium mass (low mass for high specific impulse, high mass for high thrust) and a low first *ionization potential. Caesium and mercury are materials currently envisaged as suitable propellants. The thruster consists of an ionizer to produce the ions, an accelerator to provide and shape the accelerating field, and a neutralizer (usually an electron emitter) to neutralize the fast-moving ion beam after ejection.

ion exchange The exchange of ions of the same charge between a solution (usually aqueous) and a solid in contact with it. The process occurs widely in nature, especially in the absorption and retention of water-soluble fertilizers by soils. For example, if a potassium salt is dissolved in water and applied to soil, potassium ions are absorbed by the soil and sodium and calcium ions are released from it.

The soil, in this case, is acting as an ion exchanger. Synthetic *ion-exchange resins* consist of various copolymers having a cross-linked three-dimensional structure to which ionic groups have been attached. An *anionic resin* has negative ions built into its structure and therefore exchanges positive ions. A *cationic resin* has positive ions built in and exchanges negative ions. Ion-exchange resins, which are used in sugar refining to remove salts, are synthetic organic polymers containing side groups that can be ionized. In anion exchange, the side groups are ionized basic groups, such as $-NH_3^+$ to which anions X^- are attached. The exchange reaction is one in which different anions in the solution displace the X^- from the solid. Similarly, cation exchange occurs with resins that have ionized acidic side groups such as $-COO^-$ or $-SO_2O^-$, with positive ions M^+ attached.

Ion exchange also occurs with inorganic polymers such as *zeolites, in which positive ions are held at sites in the silicate lattice. These are used for water-softening, in which Ca^{2+} ions in solution displace Na^+ ions in the zeolite. The zeolite can be regenerated with sodium chloride solution. *Ion-exchange membranes* are used as separators in electrolytic cells to remove salts from sea water (*see also* desalination) and in producing deionized water.

ionic radius A value assigned to the radius of an ion in a crystalline solid, based on the assumption that the ions are spherical with a definite size. X-ray diffraction can be used to measure the internuclear distance in crystalline solids. For example, in NaF the Na – F distance is 23.1 nm, and this is assumed to be the sum of the Na^+ and F^- radii. By making certain assumptions about the shielding effect that the inner electrons have on the outer electrons, it is possible to assign individual values to the ionic radii – Na^+ 9.6 nm; F^- 13.5 nm. In general, negative ions have larger ionic radii than positive ions. The larger the negative charge, the larger the ion; the larger the positive charge, the smaller the ion.

ionic strength Symbol I. A function expressing the effect of the charge of the ions in a solution, equal to the sum of the molality of each type of ion present multiplied by the square of its charge. $I = \Sigma m_i z_i^2$.

ion implantation The technique of implanting ions in the lattice of a semiconductor crystal in order to modify its electronic properties. It is used as an alternative to diffusion, or in conjunction with it, in the manufacture of integrated circuits and solid-state components.

ionization The process of producing *ions. Certain molecules (*see* electrolytes) ionize in solution; for example, acids ionize when dissolved in water:

$$HCl \rightarrow H^+ + Cl^-$$

Electron transfer also causes ionization in certain reactions; for example, sodium and chlorine react by the transfer of a valence electron from the sodium atom to the chlorine atom to form the ions that constitute a sodium chloride crystal:

$$Na + Cl \rightarrow Na^+Cl^-$$

Ions may also be formed when an atom or molecule loses one or more electrons as a result of energy gained in a collision with another particle or a quantum of radiation (*see* photoionization). This may occur as a result of the impact of *ionizing radiation or of *thermal ionization and the reaction takes the form

$$A \rightarrow A^+ + e$$

Alternatively, ions can be formed by electron capture, i.e.

$$A + e \rightarrow A^-$$

ionization chamber An instrument for detecting *ionizing radiation. It consists of two electrodes contained in a gas-filled chamber with a potential difference maintained between them. Ionizing radiation entering the chamber ionizes gas atoms, creating electrons and positive ions. The electric field between the electrodes drives the electrons to the anode and the positive ions to the cathode. This current is, in suitable conditions, proportional to the intensity of the radiation. *See also* Geiger counter.

ionization gauge A vacuum gauge consisting of a three-electrode system inserted into the container in which the pressure is to be measured. Electrons from the cathode are attracted to the grid, which is positively biased. Some pass through the grid but do not reach the anode, as it is maintained at a negative potential. Some of these electrons do, however, collide with gas molecules, ionizing them and converting them to positive ions. These ions are attracted to the anode; the resulting anode current can be used as a measure of the number of gas molecules present. Pressure as low as 10^{-6} pascal can be measured in this way.

ionization potential (IP) Symbol I. The minimum energy required to remove an electron from a specified atom or molecule to such a distance that there is no electrostatic interaction between ion and electron. Originally defined as the minimum potential through which an electron would have to fall to ionize an atom, the ionization potential was measured in volts. It is now, however, defined as the energy to effect an ionization and is conveniently measured in electronvolts (although this is not an SI unit).

The energy to remove the least strongly bound electrons is the *first ionization potential.* Second, third, and higher ionization potentials can also be measured, although there is some ambiguity in terminology. Thus, in chemistry the second ionization potential is often taken to be the minimum energy required to remove an electron from the singly charged ion; the second IP of lithium would be the energy for the process

$$Li^+ \rightarrow Li^{2+} + e$$

In physics, the second ionization potential is the energy required to remove an electron from the next to highest energy level in the neutral atom or molecule; e.g.

$$Li \rightarrow Li^{*+} + e,$$

where Li^{*+} is an excited singly charged ion produced by removing an electron from the K-shell.

ionizing radiation Radiation of sufficiently high energy to cause *ionization in the medium through which it passes. It may consist of a stream of high-energy particles (e.g. electrons, protons, alpha-particles) or short-wavelength electromagnetic radiation (ultraviolet, X-rays, gamma-rays). This type of radiation can cause extensive damage to the molecular structure of a substance either as a result of the direct transfer of energy to its atoms or molecules or as a result of the *secondary electrons released by ionization. In biological tissue the effect of ionizing radiation can be very serious, usually as a consequence of the ejection of an electron from a water molecule and the oxidizing or reducing effects of the resulting highly reactive species:

$$H_2O \rightarrow e^- + H_2O^* + H_2O^+ \rightarrow \cdot OH + H_3O^+ + \cdot H,$$

where the dot before a radical indicates an unpaired electron and an * denotes an excited species.

ion-microprobe analysis A technique for analysing the surface composition of solids. The sample is bombarded with a narrow beam (as small as 2 μm diameter) of high-energy ions. Ions ejected from the surface by sputtering are detected by mass spectrometry. The technique allows quantitative analysis of both chemical and isotopic composition for concentrations as low as a few parts per million.

ionosphere *See* earth's atmosphere; radio transmission.

ionospheric wave *See* radio transmission.

ion pair A pair of oppositely charged ions produced as a result of a single ionization; e.g.

$$HCl \rightarrow H^+ + Cl^-.$$

Sometimes a positive ion and an electron are referred to as an ion pair, as in

$$A \rightarrow A^+ + e^-.$$

ion pump A type of *vacuum pump that can reduce the pressure in a container to about 1 nanopascal by passing a beam of electrons through the residual gas. The gas is ionized and the positive ions formed are attracted to a cathode within the container where they remain trapped. The pump is only useful at very low pressures, i.e. below about 1 micropascal. The pump has a limited capacity because the absorbed ions eventually saturate the surface of the cathode. A more effective pump can be made by simultaneously *sputtering a film of metal, so that fresh surface is continuously produced. The device is then known as a *sputter-ion pump*.

IP *See* ionization potential.

IR *See* infrared radiation.

iris *See* diaphragm.

irradiance Symbol *E*. The *radiant flux per unit area reaching a surface; in SI units it is measured in watts per square metre ($W\,m^{-2}$). Irradiance refers to electromagnetic radiation of all kinds, whereas *illuminance refers only to visible radiation.

irradiation Exposure to any form of radiation, often exposure to *ionizing radiation is implied.

irrational number A number that cannot be expressed as the ratio of two integers. An irrational number may be a *surd, such as $\sqrt{2}$ or $\sqrt{3}$, which can be expressed to any desired degree of accuracy but cannot be assigned an exact value. Alternatively, it may be a *transcendental, such as π or e. *Compare* rational number.

irreversible process *See* reversible process.

isentropic process Any process that takes place without a change of *entropy. The quantity of heat transferred, δQ, in a reversible process is proportional to the change in entropy, δS, i.e. $\delta Q = T\delta S$, where T is the thermodynamic temperature. Therefore, a reversible *adiabatic process is isentropic, i.e. when $\delta Q = 0$, δS also equals 0.

isobar **1.** A line on a map or chart that joints points or places that have the same atmospheric pressure. **2.** A curve on a graph representing readings taken at constant pressure. **3.** One of two or more nuclides that have the same number of nucleons but different *atomic numbers. Radium–88, actinium–89, and thorium–90 are isobars as each has a *nucleon number of 228.

isobaric spin *See* isotopic spin.

isocline A line on a map or chart joining points or places of equal magnetic dip (*see* geomagnetism).

isodiaphere One of two or more nuclides in which the difference between the number of neutrons and the number of protons is the same. A nuclide and its product after losing an *alpha particle are isodiapheres.

isodynamic line A line on a map or chart joining points or places at which the total strengths of the earth's magnetic field are equal (*see* geomagnetism).

isoelectronic Describing compounds that have the same numbers of valence electrons. For example, nitrogen (N_2) and carbon monoxide (CO) are isoelectronic molecules.

isogonal line A line on a map or chart joining points or places of equal magnetic declination (*see* geomagnetism).

isomerism The existence of atomic nuclei that have the same atomic number and the same mass number but different energy states.

isomers *See* isomerism.

isometric **1.** (in technical drawing) Denoting a projection in which the three axes are equally inclined to the surface of the drawing and lines are drawn to scale. **2.** (in crystallography) Denoting a system in which the axes are perpendicular to each other, as in cubic crystals. **3.** (in physics) Denoting a line on a graph illustrating the way in which temperature and pressure are interrelated at constant volume.

isomorphism The existence of two or more substances (*isomorphs*) that have the same crystal structure, so that they are able to form *solid solutions.

isospin *See* isotopic spin.

isotherm **1.** A line on a map or chart joining points or places of equal temperature. **2.** A curve on a graph representing readings taken at constant temperature (e.g. the relationship between the pressure and volume of a gas at constant temperature).

isothermal process Any process that takes place at constant temperature. In such a process heat is, if necessary, supplied or removed from the system at just the right rate to maintain constant temperature. *Compare* adiabatic process.

isotone One of two or more nuclides that contain the same number of neutrons but different numbers of protons. The naturally occurring isotones, for example, strontium–88 and yttrium–89 (both with 50 neutrons), give an indication of the stability of certain nuclear configurations.

isotonic Describing solutions that have the same osmotic pressure.

isotope One of two or more atoms of the same element that have the same number of protons in their nucleus but different numbers of neutrons. Hydrogen (1 proton, no neutrons), deuterium (1 proton, 1 neutron), and tritium (1 proton, 2 neutrons) are isotopes of hydrogen. Most elements in nature consist of a mixture of isotopes. *See* isotope separation.

isotope separation The separation of the *isotopes of an element from each other on the basis of slight differences in their physical properties. For laboratory quantities the most suitable device is often the mass spectrometer. On a larger scale the methods used include gaseous diffusion (widely used for separating isotopes of uranium in the form of the gas uranium hexafluoride), distillation (formerly used to produce heavy water), electrolysis (requiring cheap electrical power), thermal diffusion (formerly used to separate uranium isotopes, but now considered uneconomic), centrifuging (a method in which there is renewed interest), and laser methods (involving the excitation of one isotope and its subsequent separation by electromagnetic means).

isotopic number (neutron excess) The difference between the number of neutrons in an isotope and the number of protons.

isotopic spin (isospin; isobaric spin) A quantum number applied to hadrons (*see* elementary particles) to distinguish between members of a set of particles that differ in their electromagnetic properties but are otherwise apparently identical. For example if electromagnetic interactions and weak interactions are ignored, the proton cannot be distinguished from the neutron in their strong interactions: isotopic spin was introduced to make a distinction between them. The use of the word 'spin' implies only an analogy to angular momentum, to which isotopic spin has a formal resemblance.

isotropic Denoting a medium whose physical properties are independent of direction. *Compare* anisotropic.

iteration The process of successive approximations used as a technique for solving a mathematical problem. The technique can be used manually but is widely used by computers.

J

jet propulsion (reaction propulsion) The propulsion of a body by means of a force produced by discharging a fluid in the form of a jet. The backward-moving jet of fluid reacts on the body in which it was produced, in accordance with Newton's third law of motion, to create a reactive force that drives the body forward. Jet propulsion occurs in nature, the squid using a form of it to propel itself through water. Although jet-propelled boats and cars have been developed, the main use of jet propulsion is in aircraft and spacecraft. Jet propulsion is the only known method of propulsion in space. In the atmosphere, jet propulsion becomes more efficient at higher altitudes, as efficiency is inversely proportional to the density of the medium through which a body is flying. The three principal means of providing jet propulsion are the turbojet, the ramjet, and the rocket. The *turbojet* is an air-breathing *heat engine based on the *gas turbine, used to power jet aircraft. The *ramjet* is also an air-breathing engine, but compression of the oxidant is achieved by the forward motion of the device through the atmosphere. This enables the compressor and turbine of the gas turbine to be dispensed with and the remaining system consists simply of an inlet diffuser, a combustion chamber in which fuel is burnt, and a jet nozzle through which the products of combustion are discharged. Used in guided missiles, the ramjet must be accelerated to its operating velocity before it can fly (*see also* pulse jet). These two forms of jet propulsion, being air-breathing engines, can only be used in the earth's atmosphere. The *rocket, however, carries its own oxidant and can therefore be used in space. *See also* ion engine.

Joly's steam calorimeter An apparatus invented by John Joly (1857–1933) to measure the specific heat capacity of a gas at constant volume. Two equal spherical containers are suspended from the opposite ends of a balance arm. One sphere is evacuated and the other contains the sample gas. The whole apparatus is enclosed in a steam bath, the specific heat capacity of the sample gas being calculated from the difference between the masses of the water that condenses on each sphere.

Josephson effects Electrical effects observed when two superconducting materials (at low temperature) are separated by a thin layer of insulating material (typically a layer of oxide less than 10^{-8} m thick). If normal metallic conductors are separated by such a barrier it is possible for a small current to flow between the conductors by the *tunnel effect. If the materials are superconductors (*see* superconductivity), several unusual phenomena occur:

(1) A supercurrent can flow through the barrier; i.e. it has zero resistance.

(2) If this current exceeds a critical value, this conductivity is lost; the barrier then only passes the 'normal' low tunnelling current and a voltage develops across the junction.

(3) If a magnetic field is applied below the critical current value, the current density changes regularly with distance across the junction. The net current through the barrier depends on the magnetic field applied. As the field is increased the net current increases from zero to a maximum, decreases to zero, increases again to a (lower) maximum, decreases, and so on. If the field exceeds a critical value the superconductivity in the barrier vanishes and a potential difference develops across the junction.

(4) If a potential difference is applied across the junction, a high-frequency alternating current flows through the junction. The frequency of this current depends on the size of the potential difference.

A junction of this type is called a *Josephson junction*; two or more junctions joined by superconducting paths form a *Josephson interferometer*. Such junctions can be used in measuring fundamental constants, in defining a voltage standard, and in the highly accurate measurement of magnetic fields. An important potential use is in logic components in high-speed computers. Josephson junctions can switch states very quickly (as low as 6 picoseconds). Moreover they have very low power consumption and can be packed closely without generating too much heat. It is possible that computers based on such devices could operate 50 times faster than the best existing machines. The effects are named after B. D. Josephson (1940–), who predicted them theoretically in 1962.

joule Symbol J. The *SI unit of work and energy equal to the work done when the point of application of a force of one newton moves, in the direction of the force, a distance of one metre. 1 joule = 10^7 ergs = 0.2388 calorie. It is named after James Prescott Joule (1818–89).

Joule heating The production of heat in a conductor as a result of the passage of an electric current through the conductor. The quantity of heat produced is given by *Joule's law.

Joule's laws **1.** The heat (Q) produced when an electric current (I) flows through a resistance (R) for a time (t) is given by $Q = I^2Rt$. **2.** The *internal energy of a given mass of gas is independent of its volume and pressure, being a function of temperature alone. This law applies only to *ideal gases (for which it provides a definition of thermodynamic temperature) as in a real gas intermolecular forces would cause changes in the internal energy should a change of volume occur. *See also* Joule–Thomson effect.

Joule–Thomson effect (Joule–Kelvin effect) The change in temperature that occurs when a gas expands through a porous plug into a region of lower pressure. For most real gases the temperature falls under these circumstances as the gas has to do internal work in overcoming the intermolecular forces to enable the expansion to take place. This is a deviation from *Joule's law. There is usually also a deviation from *Boyle's law, which can cause either a rise or a fall in temperature since any increase in the product of pressure and volume is a measure of external work done. At a given pressure, there is a particular temperature, called the *inversion temperature* of the gas, at which the rise in temperature from the Boyle's law deviation is balanced by the fall from the Joule's law deviation. There is then no temperature change. Above the inversion temperature the gas is heated by expansion, below it, it is cooled. The effect was discovered by James Joule working in collaboration with William Thomson (later Lord Kelvin; 1824–1907).

JUGFET *See* transistor.

junction detector (solid-state detector) A sensitive detector of *ionizing radiation in which the output is a current pulse proportional to the energy falling in or near the depletion region of a reverse-biased *semiconductor junction. The first types were made by evaporating a thin layer of gold on to a polished wafer of *n*-type germanium; however, gold–silicon devices can be operated at room temperature and these have superseded the germanium type, which have to be operated at the temperature of liquid nitrogen to reduce noise. When the gold–silicon junction is reverse-biased a depletion region, devoid of charge carriers (electrons and holes), forms in the silicon. Incoming ionizing radiation falling in this depletion region creates pairs of electrons and holes, which both have to be collected in order to give an output pulse proportional to the energy of the detected particle.

Junction detectors are used in medicine and biology as well as in space systems.

junction transistor *See* transistor.

K

kaon A K-meson. *See* meson.

Kater's pendulum A complex *pendulum designed by Henry Kater (1777–1835) to measure the acceleration of free fall. It consists of a metal bar with knife edges attached near the ends and two weights that can slide between the knife edges. The bar is pivoted from each knife edge in turn and the positions of the weights are adjusted so that the period of the pendulum is the same with both pivots. The period is then given by the formula for a simple pendulum, which enables g to be calculated.

katharometer An instrument for comparing the thermal conductivities of two gases by comparing the rate of loss of heat from two heating coils surrounded by the gases. The instrument can be used to detect the presence of a small amount of an impurity in air and is also used as a detector in gas chromatography.

keeper A piece of soft iron used to bridge the poles of a permanent magnet when it is not in use. It reduces the leakage field and thus preserves the magnetization.

kelvin Symbol K. The *SI unit of thermodynamic *temperature equal to the fraction 1/273.16 of the thermodynamic temperature of the *triple point of water. The magnitude of the kelvin is equal to that of the degree celsius (centigrade), but a temperature expressed in degrees celsius is numerically equal to the temperature in kelvins less 273.15 (i.e. °C = K − 273.15). The *absolute zero of temperature has a temperature of 0 K (−273.15°C). The former name *degree kelvin* (symbol °K) became obsolete by international agreement in 1967. The unit is named after Lord Kelvin (1824–1907).

Kelvin effect *See* Thomson effect.

Kepler's laws Three laws of planetary motion formulated by Johannes Kepler (1571–1630) in about 1610 on the basis of observations made by Tycho Brahe (1546–1601). They state that: (1) the orbits of the planets are elliptical with the sun at one *focus of the ellipse; (2) each planet revolves around the sun so that an imaginary line (the *radius vector*) connecting the planet to the sun sweeps out equal areas in equal times; (3) the ratio of the square of each planet's *sidereal period to the cube of its distance from the sun is a constant for all the planets.

Kerr effect The ability of certain substances to refract differently light waves whose vibrations are in two directions (*see* double refraction) when the substance is placed in an electric field. The effect, discovered in 1875 by John Kerr (1824–1907), is caused by the fact that certain molecules have electric *dipoles, which tend to be orientated by the applied field; the normal random motions of the molecules tends to destroy this orientation and the balance is struck by the relative magnitudes of the field strength, the temperature, and the magnitudes of the dipole moments.

The Kerr effect is observed in a *Kerr cell*, which consists of a glass cell containing the liquid or gaseous substance; two capacitor plates are inserted into the cell and light is passed through it at right angles to the electric field. There are two principal indexes of refraction: n_o (the ordinary index) and n_e (the extraordinary index). The difference in the velocity of propagation in the cell causes a phase difference, δ, between the two waves formed from a beam of monochromatic light, wavelength λ, such that

$$\delta = (n_o - n_e)x/\lambda,$$

where x is the length of the light path in the cell. Kerr also showed empirically that the ratio

$$(n_o - n_e)\lambda = BE^2,$$

where E is the field strength and B is a constant, called the *Kerr constant*, which is characteristic of the substance and approximately inversely proportional to the thermodynamic temperature.

The *Kerr shutter* consists of a Kerr cell filled with a liquid, such as nitrobenzene, placed between two crossed polarizers; the electric field is arranged to be perpendicular to the axis of the light beam and at 45° to the axis of the polarizers. In the absence of a field there is no optical path through the device. When the field is switched on the nitrobenzene becomes doubly refracting and a path opens between the crossed polarizers.

kilo- Symbol k. A prefix used in the metric system to denote 1000 times. For example, 1000 volts = 1 kilovolt (kV).

kilogram Symbol kg. The *SI unit of mass defined as a mass equal to that of the international platinum–iridium prototype kept by the International Bureau of Weights and Measures at Sèvres, near Paris.

kiloton weapon A nuclear weapon with an explosive power equivalent to one thousand tons of TNT. *Compare* megaton weapon.

kilowatt-hour Symbol kWh. The commercial unit of electrical energy. It is equivalent to a power consumption of 1000 watts for 1 hour.

kinematic equation *See* equation of motion.

kinematics The branch of mechanics concerned with the motions of objects without being concerned with the forces that cause the motion. In this latter respect it differs from *dynamics, which is concerned with the forces that cause motion. *See also* equation of motion.

kinematic viscosity Symbol ν. The ratio of the *viscosity of a liquid to its density. The SI unit is $m^2 s^{-1}$.

kinetic effect A chemical effect that depends on reaction rate rather than on thermodynamics. For example, diamond is thermodynamically less stable than graphite; its apparent stability depends on the vanishingly slow rate at which it is converted. *Overvoltage in electrolytic cells is another example of a kinetic effect. *Kinetic isotope effects* are changes in reaction rates produced by isotope substitution. For example, if the slow step in a chemical reaction is the breaking of a C–H bond, the rate for the deuterated compound would be slightly lower because of the lower vibrational frequency of the C–D bond.

kinetic energy *See* energy.

kinetics The branch of physical chemistry concerned with measuring and studying the rates of chemical reactions. The main aim of chemical kinetics is to determine the mechanism of reactions by studying the rate under different conditions (temperature, pressure, etc.).

kinetic theory A theory, largely the work of Count Rumford (1753–1814), James Joule (1818–89), and James Clerk Maxwell (1831–79), that explains the physical properties of matter in terms of the motions of its constituent particles. In a gas, for example, the pressure is due to the incessant impacts of the gas molecules on the walls of the container. If it is assumed that the molecules occupy negligible space, exert negligible forces on each other except during collisions, are perfectly elastic, and make only brief collisions with each other, it can be shown that the pressure p exerted by one mole of gas containing n molecules each of mass m in a container of volume V, will be given by:

$$p = nm\bar{c}^2/3V,$$

where $\bar{c}^2$ is the mean square speed of the molecules. As according to the *gas laws for one mole of gas: $pV = RT$, where T is the thermodynamic temperature, and R is the molar *gas constant, it follows that:

$$RT = nm\bar{c}^2/3$$

Thus, the thermodynamic temperature of a gas is proportional to the mean square speed of its molecules. As the average kinetic *energy of translation of the molecules is $m\bar{c}^2/2$, the temperature is given by:

$$T = (m\bar{c}^2/2)(2n/3R)$$

The number of molecules in one mole of any gas is the *Avogadro constant, N_A; therefore in this equation $n = N_A$. The ratio R/N_A is a constant called the *Boltzmann constant (k). The average kinetic energy of translation of the molecules of one mole of any gas is therefore $3kT/2$. For monatomic gases this is proportional to the *internal energy (U) of the gas, i.e.

$$U = N_A 3kT/2$$

and as $k = R/N_A$

$$U = 3RT/2$$

For diatomic and polyatomic gases the rotational and vibrational energies also have to be taken into account (*see* degrees of freedom).

In liquids, according to the kinetic theory, the atoms and molecules still move around at random, the temperature being proportional to their average kinetic energy. However, they are sufficiently close to each other for the attractive forces between molecules to be important. A molecule that approaches the surface will experience a resultant force tending to keep it within the liquid. It is, therefore, only some of the fastest moving molecules that escape; as a result the average kinetic energy of those that fail to escape is reduced. In this way evaporation from the surface of a liquid causes its temperature to fall.

In a crystalline solid the atoms, ions, and molecules are able only to vibrate about the fixed positions of a *crystal lattice; the attractive forces are so strong at this range that no free movement is possible.

Kirchhoff's law of radiation A law stating that the emissivity of a body is equal to its absorptance at the same temperature.

Kirchhoff's laws Two laws relating to electric circuits, first formulated by G. R. Kirchhoff (1824–87). (a) The current law states that the algebraic sum of the currents flowing through all the wires in a network that meet at a point is zero. (b) The voltage law states that the algebraic sum of the e.m.f.s within any closed circuit is equal to the sum of the products of the currents and the resistances in the various portions of the circuit.

klystron An electron tube that generates or amplifies microwaves by *velocity modulation*. Several types are used; in the simple two-cavity klystron a beam of high-energy electrons from an electron gun is passed through a *resonant cavity, where it interacts with high-frequency radio waves. This microwave energy modulates the velocities of the electrons in the beam, which then enters a drift space where the faster electrons overtake the slower ones to form bunches. The bunched beam now has an alternating component, which is transferred to an output cavity and thence to an output waveguide.

knocking The metallic sound produced by a spark-ignition petrol engine under certain conditions. It is caused by rapid combustion of the unburnt explosive mixture in the combustion chambers ahead of the flame front. As the flame travels from the sparking plug towards the piston it compresses and heats the unburnt gases ahead of it. If the flame front moves fast enough, normal combustion occurs and the explosive mixture is ignited progressively by the flame. If it moves too slowly, ignition of the last part of the unburnt gas can occur very rapidly before the flame reaches it, producing a shock wave that travels back and forth across the combustion chamber. The result is overheating, possible damage to the plugs, an undesirable noise, and loss of power (probably due to preignition caused by overheated plugs). Knocking can be avoided by an engine design that increases turbulence in the combustion chamber and thereby increases flame speed. It also can be avoided by reducing the compression ratio, but this involves loss of efficiency. The most effective method is to use high-octane fuel (*see* octane number), which has a longer self-ignition delay than low-octane fuels. This is usually achieved by the addition of lead tetraethyl(IV) to the fuel.

Knudsen flow *See* molecular flow.

Kohlrausch's law If a salt is dissolved in water, the conductivity of the (dilute) solution is the sum of two values – one depending on the positive ions and the other on the negative ions. The law, which depends on the independent migration of ions, was deduced experimentally by the German chemist Friedrich Kohlrausch (1840–1910).

Kovar A tradename for an alloy of iron, cobalt, and nickel with an *expansivity similar to that of glass. It is therefore used in making glass-to-metal seals, especially in circumstances in which a temperature variation can be expected.

Kundt's tube An apparatus designed by August Kundt (1839–94) in 1866 to measure the speed of sound in various fluids. It consists of a closed glass tube into which a dry powder (such as lycopodium) has been sprinkled. The source of sound in the original device was a metal rod clamped at its centre with a piston at one end, which is inserted into the tube. When the rod is stroked, sound waves generated by the piston enter the tube. If the position of the piston in the tube is adjusted so that the gas column is a whole number of half wavelengths long, the dust will be disturbed by the resulting *stationary waves forming a series of striations, enabling distances between *nodes to be measured. The vibrating rod can be replaced by a small loudspeaker fed by an oscillator.

L

labelling The process of replacing a stable atom in a compound with a radioisotope of the same element to enable its path through a biological or mechanical system to be traced by the radiation it emits. In some cases a different stable isotope is used and the path is detected by means of a mass spectrometer. A compound containing either a radioactive or stable isotope is called a *labelled compound.* If a hydrogen atom in each molecule of the compound has been replaced by a tritium atom, the compound is called a *tritiated compound.* A radioactive labelled compound will behave chemically and physically in the same way as an otherwise identical stable compound, and its presence can easily be detected using a *Geiger counter. This process of *radioactive tracing* is widely used in chemistry, biology, medicine, and engineering. For example, it can be used to follow the course of the reaction of a carboxylic acid with an alcohol to give an ester, e.g.

$$CH_3COOH + C_2H_5OH \rightarrow C_2H_5COOCH_3 + H_2O$$

To determine whether the noncarbonyl oxygen in the ester comes from the acid or the alcohol, the reaction is performed with the labelled compound $CH_3CO^{18}OH$, in which the oxygen in the hydroxyl group of the acid has been 'labelled' by using the ^{18}O isotope. It is then found that the water product is $H_2{}^{18}O$; i.e. the oxygen in the ester comes from the alcohol, not the acid.

laevorotatory Designating a chemical compound that rotates the plane of plane-polarized light to the left (anticlockwise for someone facing the oncoming radiation). *See* optical activity.

lag 1. The time delay between a specified maximum value of one wave or vibration and the maximum value of another wave or vibration. A *lagging current* is an alternating current that reaches its maximum at a later instant in the cycle than the e.m.f. producing it. **2.** This time delay expressed as an angle.

lambda point Symbol λ. The temperature of 2.186 K below which helium becomes a superfluid. *See* superfluidity.

lambert A former unit of *luminance equal to the luminance of a uniformly diffusing surface that emits or reflects one lumen per square centimetre. It is approximately equal to 3.18×10^3 Cd m^{-2}. It is named after Johann H. Lambert (1728–77).

Lambert's laws (1) The *illuminance of a surface illuminated by light falling on it perpendicularly from a point source is inversely proportional to the square of the distance between the surface and the source. (2) If the rays make an angle θ with the normal to the surface, the illuminance is proportional to $\cos\theta$. (3) (Also called *Bouquer's law*) The *luminous intensity (I) of light (or other electromagnetic radiation) decreases exponentially with the distance d that it enters an absorbing medium, i.e.

$$I = I_0 \exp(-\alpha d)$$

where I_0 is the intensity of the radiation that enters the medium and α is its *linear absorption coefficient.* These laws were first stated (for light) by Johann H. Lambert (1728–77).

Lamb shift A small energy difference between two levels ($^2S_{1/2}$ and $^2P_{1/2}$) in the *hydrogen spectrum. The shift results from the quantum interaction between the atomic electron and the electromagnetic radiation. It was first explained by Willis Eugene Lamb (1913–).

laminar flow *Streamline flow of a fluid in which the fluid moves in layers without fluctuations or turbulence so that successive particles passing the same point have the same velocity. It occurs at low *Reynolds numbers, i.e. low velocities, high viscosities, or low densities. As a consequence of high pressure the flow of lubricating oil in bearings is normally laminar and it also occurs around some modern aircraft and missiles.

laminated core A core for a transformer or other electrical machine in which the ferromagnetic alloy is made into thin sheets (laminations), which are oxidized or varnished to provide a relatively high resistance between them. This has the effect of reducing *eddy currents, which occur when alternating currents are used.

Laplace equation The partial differential equation:

$$\partial^2u/\partial x^2 + \partial^2u/\partial y^2 + \partial^2u/\partial z^2 = 0$$

It may also be written in the form $\nabla^2u = 0$, where ∇^2 is called the *Laplace operator*. It was formulated by the French mathematician P. S. Laplace (1749–1827).

Larmor precession A precession of the motion of charged particles in a magnetic field. It was first deduced in 1897 by Sir Joseph Larmor (1857–1942). Applied to the orbital motion of an electron around the nucleus of an atom in a magnetic field of flux density B, the frequency of precession is given by $eB/4\pi mv\mu$, where e and m are the electronic charge and mass respectively, μ is the permeability, and v is the velocity of the electron. This is known as the *Larmor frequency*.

laser (*l*ight *a*mplification by *s*timulated *e*mission of *r*adiation) A light amplifier (also called an *optical maser*) usually used to produce monochromatic coherent radiation in the infrared, visible, and ultraviolet regions of the *electromagnetic spectrum.

Nonlaser light sources emit radiation in all directions as a result of the spontaneous emission of photons by thermally excited solids (filament lamps) or electronically excited atoms, ions, or molecules (fluorescent lamps, etc.). The emission accompanies the spontaneous return of the excited species to the *ground state and occurs randomly, i.e. the radiation is not coherent. In a laser, the atoms, ions, or molecules are first 'pumped' to an excited state and then stimulated to emit photons by collision of a photon of the same energy. This is called *stimulated emission*. In order to use it, it is first necessary to create a condition in the amplifying medium, called *population inversion*, in which the majority of the relevant entities are excited. Random emission from one entity can then trigger coherent emission from the others that it passes. In this way amplification is achieved.

The laser amplifier is converted to an oscillator by enclosing the amplifying medium within a resonator. Radiation then introduced along the axis of the resonator is reflected back and forth along its path by a mirror at one end and by a partially transmitting mirror at the other end. Between the mirrors the waves are amplified by stimulated emission. The radiation emerges through the semitransparent mirror at one end as a powerful coherent monochromatic parallel beam of light. The emitted beam is uniquely parallel because waves that do not bounce back and forth between the mirrors quickly escape through the sides of the oscillating medium without amplification.

Some lasers are solid, others are liquid or gas devices. Population inversion can be achieved by *optical pumping* with flashlights or with other lasers. It can also be achieved by such methods as chemical reactions, discharges in gases, and recombination emission in semiconducting materials (*see* recombination process).

Lasers have found many uses since their invention in 1960, including laser welding, surgery, *holography, printing, optical communications, and the reading of digital information.

latent heat Symbol L. The quantity of heat absorbed or released when a substance changes its physical phase at constant temperature (e.g. from solid to liquid at the melting point or from liquid to gas at the boiling point). For example, the latent heat of vaporization is the energy a substance absorbs from its surroundings in order to overcome the attractive forces between its molecules as it changes from a liquid to a gas and in order to do work against the external atmosphere as it expands. In thermodynamic terms the latent heat is the *enthalpy of evaporation (ΔH), i.e. $L = \Delta H = \Delta U + p\Delta V$, where ΔU is the change in the internal energy, p is the pressure, and ΔV is the change in volume.

The *specific latent heat* (symbol l) is the heat absorbed or released per unit mass of a substance in the course of its isothermal change of phase. The *molar latent heat* is the heat absorbed or released per unit amount of substance during an isothermal change of state.

lateral inversion (perversion) The type of reversal that occurs with an image formed by a plane mirror. A person with a mole on his left cheek sees an image in a plane mirror of a person with a mole on his right cheek. Since, however, that is (correctly) to the observer's left, the real reversal is of front and back; the image is 'turned through' itself to face the object – hence the alternative name.

lateral velocity The component of a celestial body's velocity that is at 90° to its *line-of-sight velocity.

latitude and longitude 1. (in geography) Imaginary lines on the earth's surface, enabling any point to be defined in terms of two angles subtended at its centre. *Parallels of latitude* are circles drawn round the earth parallel to the equator; their diameters diminish as they approach the poles. These parallels are specified by the angle subtended at the centre of the earth by the arc formed between a point on the parallel and the equator. All points on the equator therefore have a latitude of 0°, while the north pole has a latitude of 90°N and the south pole of 90°S. Parallels of latitude 1° apart are separated on the earth's surface by about 100 km.

Meridians of longitude are half *great circles passing through both poles; they cross parallels of latitude at right angles. In 1884 the meridian through Greenwich, near London, was selected as the prime meridian and designated as 0°. Other meridians are defined by the angle between the plane of the meridian and the plane of the prime meridian specifying whether it is E or W of the prime meridian. At the equator meridians 1° apart are separated by about 112 km. **2.** (in astronomy) The *celestial latitude* of a star, or other celestial body, is its angular distance north (taken as positive) or south (taken as negative) of the ecliptic measured along the great circle through the body and the poles of the ecliptic. The *celestial longitude* is the angular distance from the vernal equinox measured eastwards along the ecliptic to the intersection of the body's circle of longitude; it is measured in the same direction as the sun's apparent annual motion.

lattice The regular arrangement of atoms, ions, or molecules in a crystalline solid. *See* crystal lattice.

lattice energy A measure of the stability of a *crystal lattice, given by the energy that would be released per mole if atoms, ions, or molecules of the crystal were brought together from infinite distances

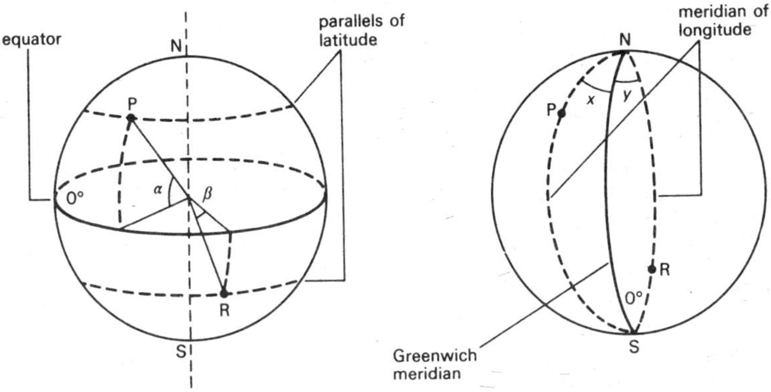

The latitude of P is given by the angle α. In this case it would be α° N. The latitude of R is β° S.

The longitude of P is given by the angle x. In this case it would be x° W. R has a longitude of y° E.

Latitude and longitude

apart to form the lattice. *See* Born–Haber cycle.

lattice vibrations The periodic vibrations of the atoms, ions, or molecules in a *crystal lattice about their mean positions. On heating, the amplitude of the vibrations increases until they are so energetic that the lattice breaks down. The temperature at which this happens is the melting point of the solid and the substance becomes a liquid. On cooling, the amplitude of the vibrations diminishes. At *absolute zero a residual vibration persists, associated with the *zero-point energy of the substance. The increase in the electrical resistance of a conductor is due to increased scattering of the free conduction electrons by the vibrating lattice particles.

latus rectum *See* ellipse; hyperbola; parabola.

launch vehicle A rocket used to launch a satellite, spaceprobe, space station, etc. Multistage rockets are usually used, the empty tanks and engine of the first two stages being jettisoned before the desired orbit is reached. The *launch window* is the time interval during which the vehicle must be launched to achieve the orbit.

lawrencium Symbol Lr. A radioactive metallic transuranic element belonging to the actinoids; a.n. 103; mass number of only known isotope 257 (half-life 8 seconds). The element was identified by A. Ghiorso and associates in 1961. The alternative name *unniltrium* has been proposed.

Lawson criterion A condition for the release of energy from a *thermonuclear reactor first laid down by J. D. Lawson. It is usually stated as the minimum value for the product of the density (n_G) of the fusion-fuel particles and the *containment time (τ) for energy breakeven, i.e. it is a measure of the density of the reacting particles required and the time for which they need to react in order to produce more energy than was used in raising the temperature of the reacting particles to the *ignition temperature. For a 50:50 mixture of deuterium and tritium at the ignition temperature, the value of $n_G\tau$ is between 10^{14} and 10^{15} cm^{-3} s.

L–D process *See* basic-oxygen process.

lead–acid accumulator An accumulator in which the electrodes are made of lead and the electrolyte consists of dilute sulphuric acid. The electrodes are usually cast from a lead alloy containing 7–12% of antimony (to give increased hardness and corrosion resistance) and a small amount of tin (for better casting properties). The electrodes are coated with a paste of lead(II) oxide (PbO) and finely divided lead; after insertion into the electrolyte a 'forming' current is passed through the cell to convert the PbO on the negative plate into a sponge of finely divided lead. On the positive plate the PbO is converted to lead(IV) oxide (PbO_2). The equation for the overall reaction during discharge is:

$$PbO_2 + 2H_2SO_4 + Pb \rightarrow 2PbSO_4 + 2H_2O$$

The reaction is reversed during charging. Each cell gives an e.m.f. of about 2 volts and in motor vehicles a 12-volt battery of six cells is usually used. The lead–acid battery produces 80–120 kJ per kilogram. *Compare* nickel–iron accumulator.

lead equivalent A measure of the absorbing power of a radiation screen, expressed as the thickness of a lead screen in millimetres that would afford the same protection as the material being considered.

Le Chatelier's principle If a system is in equilibrium, any change imposed on the system tends to shift the equilibrium to nullify the effect of the applied change. The principle, which is a consequence of the law of conservation of energy, was first stated in 1888 by Henri Le Chatelier (1850–1936). It is applied to chemical equilibria. For example, in the gas reaction

$$2SO_2 + O_2 \rightleftharpoons 2SO_3$$

an increase in pressure on the reaction mixture displaces the equilibrium to the right, since this reduces the total number of molecules present and thus decreases the pressure. The standard enthalpy change for the forward reaction is negative (i.e. the reaction is exothermic). Thus, an increase in temperature displaces the equilibrium to the left since this tends to reduce the temperature. The equilibrium constant thus falls with increasing temperature.

Leclanché cell A primary *voltaic cell consisting of a carbon rod (the anode) and a zinc rod (the cathode) dipping into an electrolyte of a 10–20% solution of ammonium chloride. *Polarization is prevented by using a mixture of manganese dioxide mixed with crushed carbon, held in contact with the anode by means of a porous bag or pot; this reacts with the hydrogen produced. This wet form of the cell, devised in 1867 by Georges Leclanché (1839–82), has an e.m.f. of about 1.5 volts. The *dry cell based on it is widely used in torches, radios, and calculators.

LED *See* light-emitting diode.

LEED Low-energy electron diffraction. *See* electron diffraction.

lens A curved, ground, and polished piece of glass, moulded plastic, or other transparent material used for the refraction of light. A *converging lens* is one that brings the rays of a parallel beam of light to a real *principal focus. They include biconvex, planoconvex, and converging meniscus lenses. *Diverging lenses* cause the rays of a parallel beam to diverge as if from a virtual principal focus; these include biconcave, planoconcave, and diverging meniscus lenses.

The *centre of curvature* of a lens face is the centre of the sphere of which the surface of the lens is a part. The *optical axis* is the line joining the two centres of curvature of a lens or, in the case of a lens with one plane surface, the line through one centre of curvature that is normal to the plane surface. The *optical centre* of a lens is the point within a lens on the optical axis through which any rays entering the lens pass without deviation. The distance between the op-

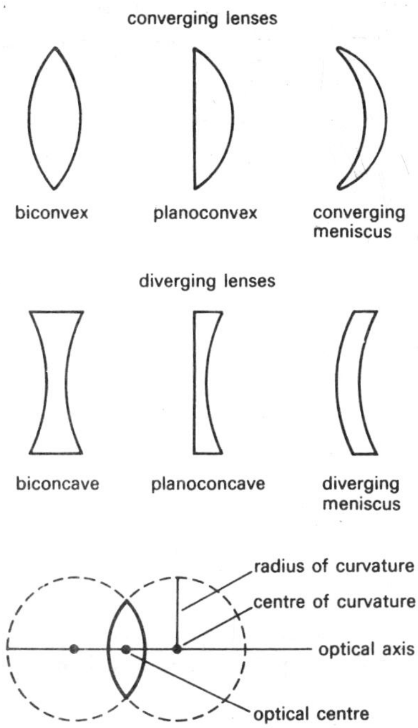

Lenses

tical centre and the principal focus of a lens is called the *focal length* (f). The distance (v) between the lens and the image it forms is related to the distance (u) between the lens and the object by:

$$1/v + 1/u = 1/f,$$

provided that the *real-is-positive convention is used. This takes distances to real objects, images, and foci as positive; those to virtual objects, images, and foci as negative. The equation does not always apply if the alternative New Cartesian convention (*see* sign convention) is used.

Lenz's law An induced electric current always flows in such a direction that it opposes the change producing it. This law, first stated by Heinrich Lenz (1804–65) in 1835, is essentially a form of the law of conservation of energy.

lepton Any of a class of *elementary particles that consists of the *electron, muon, tau particle, and three types of *neutrino (one associated with each of the other types of lepton). For each lepton there is an equivalent antiparticle. The antileptons have a charge opposite that of the leptons; the antineutrinos, like the neutrinos, have no charge. The electron, muon, and tau particle all have a charge of −1. These three particles differ from each other only in mass: the muon is 200 times more massive than the electron and the tau particle is 3500 times more massive than the electron. Leptons interact by the electromagnetic interaction and the weak interaction (*see* fundamental interactions).

Leslie's cube A metal box in the shape of a cube in which each of the four vertical sides have different surface finishes. When hot water is placed in the cube, the emissivity of the finishes can be compared. The device was first used by Sir John Leslie (1766–1832).

level An instrument used in *surveying to determine heights. It usually consists of a telescope and attached spirit level mounted on a tripod. The level is set up between a point of known height and a point for which the height is required. Before use it is adjusted until the line of sight is exactly horizontal. Sightings are then made onto a graduated levelling staff at the two points. The difference in elevation between the two points can then be calculated from the readings taken at these points.

lever A simple machine consisting of a rigid bar pivoted about a fulcrum. The mechanical advantage or *force ratio of a lever (the ratio of load to effort) is equal to the ratio of the perpendicular distance of the line of action of the effort from the fulcrum to the perpendicular distance of the line of action of the load from the fulcrum. In a first-order lever the fulcrum comes be-

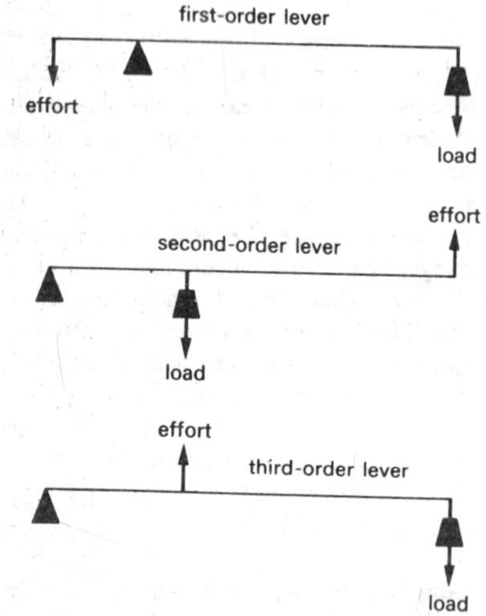

Levers

tween load and effort. In a second-order lever the load comes between the fulcrum and the effort. In a third-order lever the effort comes between the fulcrum and the load.

Leyden jar An early form of *capacitor consisting of a glass jar with a layer of metal foil on the outside and a similar layer on the inside. Contact to the inner foil is by means of a loose chain hanging inside the jar. It was invented in the Dutch town of Leyden in about 1745.

LF *See* low frequency.

***l*-form** *See* optical activity.

libration The phenomenon that enables 59% of the moon's surface to be observed from earth over a 30-year period, in spite of its *synchronous rotation. *Physical libration* arises from slight variations in the rotation of the moon on its axis, caused by minor distortions in its physical shape. *Geometric librations* are apparent oscillations arising from the fact that the moon is observed from slightly different directions at different times. The geometric *libration in longitude* results from the nonuniform orbital motion of the moon. The geometric *libration in latitude* arises because the moon's axis of rotation is not perpendicular to its orbital plane; it enables more of the lunar polar regions to be observed.

light The form of *electromagnetic radiation to which the human eye is sensitive and on which our visual awareness of the universe and its contents relies (*see* colour).

The finite velocity of light was suspected by many early experimenters in optics, but it was not established until 1676 when Ole Rømer (1644–1710) measured it. Sir Isaac Newton (1642–1727) discovered the optical *spectrum and used existing knowledge to establish a primarily *corpuscular theory* of light, in which it was regarded as a stream of particles that set up disturbances in the 'aether' of space. His successors adopted the corpuscles but ignored the wavelike disturbances until Thomas Young (1773–1829) rediscovered the *interference of light in 1801 and showed that a *wave theory* was essential to interpret this type of phenomenon. This view was accepted for most of the 19th century and it enabled James Clerk Maxwell (1831–79) to show that light forms part of the *electromagnetic spectrum. He believed that waves of electromagnetic radiation required a special medium to travel through, and revived the name 'luminiferous ether' for such a medium. The *Michelson–Morley experiment in 1887 showed that, if the medium existed, it could not be detected; it is now generally accepted that the ether is an unnecessary hypothesis. In 1905 Albert Einstein (1879–1955) showed that the *photoelectric effect could only be explained on the assumption that light consists of a stream of discrete *photons of electromagnetic energy. This renewed conflict between the corpuscular and wave theories has gradually been resolved by the evolution of the *quantum theory and *wave mechanics. While it is not easy to construct a model that has both wave and particle characteristics, it is accepted, according to Bohr's theory of *complementarity, that in some experiments light will appear wavelike, while in others it will appear to be corpuscular. During the course of the evolution of wave mechanics it has also become evident that electrons and other elementary particles have dual wave and particle properties.

light bulb *See* electric lighting.

light-emitting diode (LED) A rectifying *semiconductor device that converts electrical energy into light or infrared radiation in the range 550 nm (green light) to 1300 nm (infrared radiation). The most commonly used LED emits red light and consists of gallium arsenide–phosphide on a gallium arsenide substrate, light being emitted at a *p–n* junction, when electrons and holes recombine (*see* recombination process). LEDs are extensively used for displaying letters and numbers in digital instruments in which a self-luminous display is required.

lightning A high-energy luminous electrical discharge that passes between a charged cloud and a point on the surface of the earth, between two charged clouds, or between oppositely charged layers of the same cloud. In general, the upper parts of clouds are positively charged and the lower parts are negatively charged; the reasons for this separation of charge are complex.

Lightning usually occurs in the form of a downward step leader followed by an intensely luminous return stroke, which can produce instantaneous temperatures as high as 30 000°C. In the typical step leader a surge of electrons descends in approximately 50-metre steps with about 50-microsecond pauses between steps. When this leader reaches the earth a surge of charge returns up the preionized path taken by the leader. Cloud-to-cloud strokes also involve a leader and return stroke. The average current in a lightning stroke is about 10 000 amperes, but maximum currents in the return stroke can reach 20 000 A.

light year A unit of distance used in astronomy; the distance travelled by light in a vacuum during one year. It is equal to 9.4650×10^{15} metres or 5.8785×10^{12} miles.

limit The value that a function approaches as the independent variable approaches a specified value.

limiting friction The friction force that just balances a moving force applied to a solid body resting on a solid surface when the

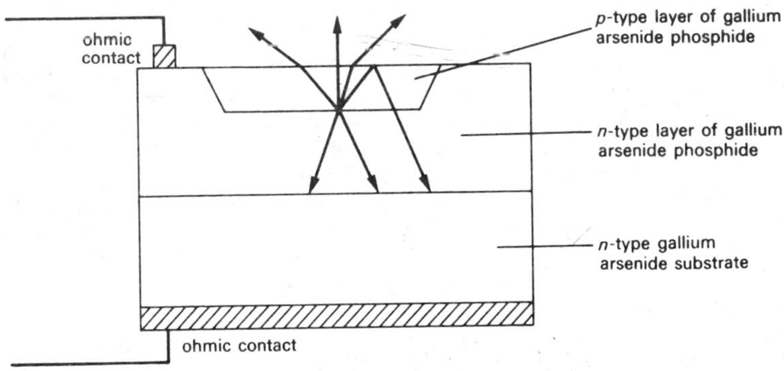

Light-emitting diode

body fails to move. If the moving force exceeds the limiting friction, the body will begin to move.

linac *See* linear accelerator.

Linde process A process for the *liquefaction of gases by the Joule–Thomson effect. In this process, devised by Carl von Linde (1842–1934) for liquefying air, the air is freed of carbon dioxide and water and compressed to 150 atmospheres. The compressed gas is passed through a copper coil to an expansion nozzle within a Dewar flask. The emerging air is cooled by the Joule–Thomson effect as it expands and then passes back within a second copper coil that surrounds the first coil. Thus the expanded gas cools the incoming gas in a process that is said to be *regenerative*. Eventually the air is reduced to its *critical temperature and, at the pressure of 150 atmospheres (well above its critical pressure), liquefies. The process is also used for other gases, especially hydrogen and helium. Hydrogen has first to be cooled below its inversion temperature (*see* Joule–Thomson effect) using liquid air; helium has first to be cooled below its inversion temperature using liquid hydrogen.

linear absorption coefficient *See* Lambert's laws.

linear accelerator (linac) A type of particle *accelerator in which charged particles are accelerated in a straight line, either by a steady electric field or by means of radiofrequency electric fields. In the latter variety the passage of the particles is synchronized with the phase of the accelerating field by means of either travelling-wave acceleration or standing-wave acceleration. In the first of these the particles are carried forward by the electric component of the field, the phase velocity of which is equal to the particle velocity. In the standing-wave type, forward and backward waves are superimposed so that the particles are alternately repelled by cylindrical electrodes through which they have passed and attracted by the cylindrical electrodes they are approaching. At room temperatures megawatt radio-frequency powers are required but in recent superconducting linacs only a few watts are needed.

linear energy transfer (LET) The energy transferred per unit path length by a moving high-energy charged particle (such as an electron or a proton) to the atoms and molecules along its path. It is of particular importance when the particles pass through living tissue as the LET modifies the effect of a specific dose of radiation. LET is proportional to the square of the charge on the particle and increases as the velocity of the particle decreases.

linear equation An equation between two variables that gives a straight line when plotted on a graph. It has the general form $y = mx + c$, where m is the gradient of the line and c is the intercept of the line on the y-axis (in Cartesian coordinates).

linear expansivity *See* expansivity.

linear momentum *See* momentum.

linear motor A form of induction motor in which the stator and armature are linear and parallel, rather than cylindrical and coaxial. In some experimental trains the magnetic force between the primary winding in the vehicle and the secondary winding on the ground support the vehicle on an air cushion thus eliminating track friction. However, because of the high cost of the installation and the low efficiency the device has not yet found commercial application.

line defect *See* defect.

line-of-sight velocity (radial velocity) The component of a celestial body's velocity along the line of sight of the observer. It is usually given in relation to the sun to avoid complications arising from the earth's orbital motion. Line-of-sight velocity is normally calculated from the *Doppler effect on the body's spectrum, a *redshift indicating a receding body (taken as a positive velocity) and a blueshift indicating an approaching body (taken as negative).

line printer An output device used with computers. It prints a line of characters at a time, typically operating at 200 to 3000 lines per minute.

lines of force Imaginary lines in a *field of force that enable the direction and strength of the field to be visualized. They are used primarily in electric and magnetic fields; in electric fields they are sometimes called *tubes of force*, to express their characteristic of being perpendicular to a conducting surface. The tangent to a line of force at any point gives the direction of the field at that point and the number of lines per unit area perpendicular to the force represents the *intensity of the field.

line spectrum *See* spectrum.

Linnz–Donnewitz process *See* basic-oxygen process.

liquation The separation of mixtures of solids by heating to a temperature at which lower-melting components liquefy.

liquefaction of gases The conversion of a gaseous substance into a liquid. This is usually achieved by one of four methods or by a combination of two of them:

(1) by vapour compression, provided that the substance is below its *critical temperature;

(2) by refrigeration at constant pressure, typically by cooling it with a colder fluid in a countercurrent heat exchanger;

(3) by making it perform work adiabatically against the atmosphere in a reversible cycle;

(4) by the *Joule–Thomson effect (*see also* Linde process).

Large quantities of liquefied gases are now used commercially, especially *liquefied petroleum gas and liquefied natural gas.

liquefied petroleum gas (LPG) Various petroleum gases, principally propane and butane, stored as a liquid under pressure. It is used as an engine fuel and has the advantage of causing very little cylinder-head deposits.

Liquefied natural gas (LNG) is a similar product and consists mainly of methane. However, it cannot be liquefied simply by pressure as it has a low critical temperature of 190 K and must therefore be cooled to below this temperature before it will liquefy. Once liquefied it has to be stored in well-insulated containers. It provides a convenient form in which to ship natural gas in bulk from oil wells to users. It is also used as an engine fuel.

liquid A phase of matter between that of a crystalline solid and a *gas. In a liquid, the large-scale three-dimensional atomic (or ionic or molecular) regularity of the solid is absent but, on the other hand, so is the total disorganization of the gas. Although liquids have been studied for many years there is still no comprehensive theory of the liquid state. It is clear, however, from diffraction studies that there is a short-range structural regularity extending over several molecular diameters. These bundles of ordered atoms, molecules, or ions move about in relation to each other, enabling liquids to have almost fixed volumes, which adopt the shape of their containers.

liquid crystal A substance that flows like a liquid but has some order in its arrangement of molecules. *Nematic crystals* have long molecules all aligned in the same direction, but otherwise randomly arranged.

Cholesteric and *smectic* liquid crystals also have aligned molecules, which are arranged in distinct layers. In cholesteric crystals, the axes of the molecules are parallel to the plane of the layers; in smectic crystals they are perpendicular.

liquid-crystal display A digital display unit used in watches, calculators, etc. It provides a source of clearly displayed digits for a very low power consumption. In the display unit a thin film of *liquid crystal is sandwiched between two transparent electrodes (glass with a thin metal or oxide coating). In the commonly used field-effect display, twisted nematic crystals are used. The nematic liquid crystal cell is placed between two crossed polarizers. Polarized light entering the cell follows the twist of the nematic liquid crystal, is rotated through 90°, and can therefore pass through the second polarizer. When an electric field is applied the molecular alignment in the liquid crystal is altered, the polarization of the entering light is unchanged, and no light is therefore transmitted. In these circumstances, a mirror placed behind the second polarizer will cause the display to appear black. One of the electrodes, shaped in the form of a digit, will then provide a black digit when the voltage is applied.

liquid-drop model A model of the atomic nucleus in which the nucleons are regarded as being analogous to the molecules in a liquid, the interactions between which maintain the droplet shape by surface tension. The model has been useful in the theory of nuclear fission.

Lissajous figures A curve in one plane traced by a point moving under the influence of two independent harmonic motions. In the common case the harmonic motions are simple, perpendicular to each other, and have a simple frequency ratio. They can be displayed by applying sinusoidal alternating potentials to the X- and Y-inputs of a *cathode-ray oscilloscope.

litre Symbol l. A unit of volume in the metric system regarded as a special name for the cubic decimetre. It was formerly defined as the volume of 1 kilogram of pure water at 4°C at standard pressure, which is equivalent to 1.000 028 dm^3.

Lloyd's mirror An optical arrangement for producing interference fringes. A slit is illuminated by monochromatic light and placed close to a plane mirror. Interference occurs between direct light from the slit and light reflected from the mirror. It was first used by Humphrey Lloyd (1800–81) in 1834.

loaded concrete Concrete containing elements (such as iron or lead) with a high mass number; it is used in making the radiation shield around nuclear reactors.

local group The cluster of *galaxies of which our own Galaxy is a member. It consists of some 25 galaxies, the most massive of which are the Galaxy and the Andromeda galaxy.

local oscillator An *oscillator in a *heterodyne or *superheterodyne radio receiver. It supplies the radio-frequency signal that beats with the incoming signal to produce the intermediate frequency.

locus A set of points whose location is specified by an equation. For example, if a point moves so that the sum of its distances from two fixed points is constant, the locus of the point is an *ellipse.

logarithm The power to which a number, called the *base*, has to be raised to give another number. Any number y can be written in the form $y = x^n$. n is then the logarithm to the base x of y, i.e. $n = \log_x y$. If the base is 10, the logarithms are called *common logarithms*. *Natural* (or *Napierian*) *logarithms* are to the base e = 2.718 28..., written $\log_e y$ or $\ln y$. Logarithms were formerly used to facilitate calculations, before the advent of electronic calculators.

A logarithm contains two parts, an integer and a decimal. The integer is called the *characteristic*, and the decimal is called the *mantissa*. For example, the logarithm to the base 10 of 210 is 2.3222, where 2 is the characteristic and 0.3222 is the mantissa.

logarithmic scale 1. A scale of measurement in which an increase or decrease of one unit represents a tenfold increase or decrease in the quantity measured. Decibels and pH measurements are common examples of logarithmic scales of measurement. **2.** A scale on the axis of a graph in which an increase of one unit represents a tenfold increase in the variable quantity. If a curve $y = x^n$ is plotted on graph paper with logarithmic scales on both axes, the result is a straight line of slope n, i.e. $\log y = n \log x$, which enables n to be determined.

logarithmic series The expansion of a logarithmic function, such as $\log_e(1 + x)$, i.e. $x \quad x^2/2 + x^3/3 - \ldots + (-1)^n x^n/n$, or $\log_e(1 - x)$, i.e. $-x - x^2/2 - x^3/3 \ldots - x^n/n$.

logic circuits The basic switching circuits or *gates used in digital computers and other digital electronic devices. The output signal, using a *binary notation, is controlled by the logic circuit in accordance with the input system. The three basic logic circuits are the *and*, *or*, and *not* circuits. The *and* circuit gives a binary 1 output if a binary 1 is present on each input circuit; otherwise the output is a binary 0. The *or* circuit gives a binary 1 output if a binary 1 is present on at least one input circuit; otherwise the output is binary 0. The *not* circuit inverts the input signal, giving a binary 1 output for a binary 0 input or a 0 output for a 1 input.

Often these basic logic circuits are used in combination, e.g. a *nand* circuit consists of *not* + *and* circuits. In terms of electronic equipment, logic circuits are now almost exclusively embodied into *integrated circuits. They formerly made use of discrete transistors, although they were originally based on thermionic valves.

longitude *See* latitude and longitude.

longitudinal wave *See* wave.

long-sightedness *See* hypermetropia.

Lorentz–Fitzgerald contraction (Fitzgerald contraction) The contraction of a moving body in the direction of its motion. It was proposed independently by H. A. Lorentz (1853–1928) and G. F. Fitzgerald (1851–1901) in 1892 to account for the null result of the *Michelson–Morley experiment. The contraction was given a theoretical background in Einstein's special theory of *relativity. In this theory, an object of length l_0 at rest in one *frame of reference will appear, to an observer in another frame moving with relative velocity v with respect to the first, to have a length

$$l_0/\sqrt{(1 - v^2/c^2)},$$

where c is the speed of light. The original hypothesis regarded this contraction as a real one accompanying the absolute motion of the body. The contraction is in any case negligible unless v is of the same order as c.

Lorentz transformations A set of equations for transforming the position and motion parameters from a frame of reference with origin at O and coordinates (x,y,z) to a frame moving relative to it with origin at O′ and coordinates (x',y',z'). They replace the *Galilean transformations used in *Newtonian mechanics and are used in relative mechanics. They are:

$$x' = \beta(x - vt)$$
$$y' = y$$
$$z' = z$$

$t' = \beta(t - vx/c^2)$
where v is the relative velocity of separation of O and O′, c is the speed of light, and $\beta = 1/\sqrt{(1 - v^2/c^2)}$.

Loschmidt's constant (Loschmidt number) The number of particles per unit volume of an *ideal gas at STP. It has the value 2.687 19 × 10^{25} m^{-3} and was first worked out by Joseph Loschmidt (1821–95).

loudness The physiological perception of sound intensity. As the ear responds differently to different frequencies, for a given intensity loudness is dependent on frequency. Sounds with frequencies between 1000 hertz and 5000 Hz are louder than sounds of the same intensity at higher or lower frequencies. Duration is also a factor in loudness, long bursts of sound being louder than short bursts. Loudness increases up to a duration of about 0.2 second; above this limit loudness does not increase with duration.

Relative loudness is usually measured on the assumption of proportionality to the logarithm of the intensity (for a given frequency), i.e. proportionality to the relative intensity on the *decibel scale. A subjective judgment is made of the relative intensity above threshold that a note of 1000 Hz must have to match the specimen sound; the loudness of this, in *phons, is then equal to that relative intensity in decibels.

loudspeaker A transducer for converting an electrical signal into an acoustic signal. Usually it is important to preserve as many characteristics of the electrical waveform as possible. The device must be capable of reproducing frequencies in the range 150–8000 hertz for speech and 20–20 000 Hz for music.

The most common loudspeaker consists of a moving-coil device. In this a cone-shaped diaphragm is attached to a coil of wire and made to vibrate in accordance with the electrical signal by the interaction between the current passing through the coil and a steady magnetic field from a permanent magnet surrounding it.

lowering of vapour pressure A reduction in the saturated vapour pressure of a pure liquid when a solute is introduced. If the solute is a solid of low vapour pressure, the decrease in vapour pressure of the liquid is proportional to the concentration of particles of solute; i.e. to the number of dissolved molecules or ions per unit volume. To a first approximation, it does not depend on the nature of the particles. *See* colligative property; Raoult's law.

low frequency (LF) A radio frequency in the range 30–300 kilohertz; i.e. having a wavelength in the range 1–10 kilometre.

lubrication The use of a substance to prevent contact between solid surfaces in relative motion in order to reduce friction, wear, overheating, and rusting. Liquid hydrocarbons (oils), either derived from petroleum or made synthetically, are the most widely used lubricants as they are relatively inexpensive, are good coolants, provide the appropriate range of viscosities, and are thermally stable. Additives include polymeric substances that maintain the desired viscosity as the temperature increases, antioxidants that prevent the formation of a sludge, and alkaline-earth phenates that neutralize acids and reduce wear.

At high temperatures, solid lubricants, such as graphite or molybdenum disulphide, are often used. Semifluid lubricants (greases) are used to provide a seal against moisture and dirt and to remain attached to vertical surfaces. They are made by adding gelling agents, such as metallic soaps, to liquid lubricants.

Recent technology has made increasing use of gases as lubricants, usually in air bearings. Their very low viscosities minimize energy losses at the bearings but necessitate some system for pumping the gas continuously to the bearings. The principle is that of the hovercraft.

lumen Symbol lm. The SI unit of *luminous flux equal to the flux emitted by a uniform point source of 1 candela in a solid angle of 1 steradian.

luminance (photometric brightness) Symbol L. The *luminous intensity of any surface in a given direction per unit projected area of the surface, viewed from that direction. It is given by the equation $L = \mathrm{d}I/(\mathrm{d}A\cos\theta)$, where I is the luminous intensity and θ is the angle between the line of sight and the normal to the surface area A being considered. It is measured in candela per square metre.

luminescence The emission of light by a substance for any reason other than a rise in its temperature. In general, atoms of substances emit *photons of electromagnetic energy when they return to the *ground state after having been in an excited state (*see* excitation). The causes of the excitation are various. If the exciting cause is a photon, the process is called *photoluminescence*; if it is an electron it is called *electroluminescence*. *Chemiluminescence* is luminescence resulting from a chemical reaction (such as the slow oxidation of phosphorus); *bioluminescence is the luminescence produced by a living organism (such as a firefly). If the luminescence persists significantly after the exciting cause is removed it is called *phosphorescence*; if it does not it is called *fluorescence*. This distinction is arbitrary since there must always be some delay; in some definitions a persistence of more than 10 nanoseconds (10^{-8} s) is treated as phosphorescence.

luminosity **1.** *Luminous intensity in a particular direction; the apparent brightness of an image. **2.** The brightness of a star defined as the total energy radiated in unit time. It is related to the surface area (A) and the *effective temperature* (t_e; the temperature of a black body having the same radius as the star and radiating the same amount of energy per unit area in one second) by a form of *Stefan's law, i.e.

$$L = A\sigma T_e^{\,4}$$

where σ is the Stefan constant and L is the luminosity.

luminous exitance *See* exitance.

luminous flux Symbol Φ_v. A measure of the rate of flow of light, i.e. the radiant flux in the wavelength range 380–760 nanometres, corrected for the dependence on wavelength of the sensitivity of the human eye. It is measured by reference to emission from a standard source, usually in lumens.

luminous intensity Symbol I_v. A measure of the light-emitting ability of a light source, either generally or in a particular direction. It is measured in candelas.

lunar eclipse *See* eclipse.

lunation *See* synodic month.

lux Symbol lx. The SI unit of *illuminance equal to the illumination produced by a *luminous flux of 1 lumen distributed uniformly over an area of 1 square metre.

Lyman series *See* hydrogen spectrum.

M

machine A device capable of making the performance of mechanical work easier, usually by overcoming a force of resistance (the load) at one point by the application of a more convenient force (the effort) at some other point. In physics, the six so-called *simple machines* are the lever, wedge, inclined plane, screw, pulley, and wheel and axle.

Mach number The ratio of the relative speeds of a fluid and a rigid body to the speed of sound in that fluid under the same conditions of temperature and pressure. If the Mach number exceeds 1 the fluid or body is moving at a *supersonic speed.* If the Mach number exceeds 5 it is said to be *hypersonic*. The number is named after Ernst Mach (1838–1916).

Mach's principle The *inertia of any particular piece of matter is attributable to the interaction between that piece of matter and the rest of the universe. A body in isolation would have zero inertia. This principle was stated by Ernst Mach in the 1870s and was made use of by Einstein in his general theory of *relativity.

Maclaurin's series *See* Taylor series.

Magellanic clouds Two small galaxies situated close to the Milky Way that are only visible from the southern hemisphere. They were first recorded by Ferdinand Magellan (1480–1521) in 1519.

magic numbers Numbers of neutrons or protons that occur in atomic nuclei to produce very stable structures. The magic numbers for both protons and neutrons are 2, 8, 14, 20, 50, and 82. For neutrons 126 and 184 are also magic numbers and for protons 114 is a magic number. The relationship between stability and magic numbers led to a nuclear *shell model in analogy to the electron shell model of the atom.

Magnadur A tradename for a ceramic material used to make permanent magnets. It consists of sintered iron oxide and barium oxide.

Magnalium A tradename for an aluminium-based alloy of high reflectivity for light and ultraviolet radiation that contains 1–2% of copper and between 5% and 30% of magnesium. Strong and light, these alloys also sometimes contain other elements, such as tin, lead, and nickel.

magnet A piece of magnetic material (*see* magnetism) that has been magnetized and is therefore surrounded by a *magnetic field. A magnet, often in the shape of a bar or horseshoe, that retains appreciable magnetization indefinitely (provided it is not heated, beaten, or exposed to extraneous magnetic fields) is called a *permanent magnet*. *See also* electromagnet.

magnetic bottle A nonuniform *magnetic field used to contain the *plasma in a thermonuclear experimental device. At the temperature of a thermonuclear reaction (10^8 K) any known substance would vaporize and the plasma has therefore to be contained in such a way that it does not come into contact with a material surface. The magnetic bottle provides a means of achieving this, by deflecting away from its boundaries the moving charged particles that make up the plasma.

magnetic bubble memory A form of computer memory in which a small magnetized region of a substance is used to store information. Bubble memories consist of materials, such as magnetic garnets, that are easily magnetized in one direction but hard to magnetize in the perpendicular direction. A thin film of these materials deposited on a nonmagnetic substrate constitutes a bubble-memory chip. When a magnetic field is applied to such a chip, by placing it between two permanent magnets, cylindrical domains (called magnetic bubbles) are formed. These bubbles constitute a magnetic region of one polarity surrounded by a magnetic region of the opposite polarity. Information is represented as the presence or absence of a bubble at a specified storage location and is retrieved by means of a rotating magnetic field. Typically a chip measures $15\,mm^2$, or $25\,mm^2$ enclosed in two permanent magnets and two rotating field coils; each chip can store up to one million bits.

magnetic circuit A closed path containing a *magnetic flux. The path is clearly delimited only if it consists mainly or wholly of ferromagnetic or other good magnetic materials; examples include transformer cores and iron parts in electrical machines. The design of these parts can often be assisted by analogy with electrical circuits, treating the *magnetomotive force as the analogue of e.m.f., the magnetic flux as current, and the *reluctance as resistance. There is, however, no actual flow around a magnetic circuit.

magnetic compass *See* compass.

magnetic constant *See* permeability.

magnetic declination *See* geomagnetism.

magnetic dip *See* geomagnetism.

magnetic disk A smooth aluminium disk, usually 35.6 cm in diameter, both surfaces of which are coated with magnetic iron oxide. The disks are used as a recording medium in computers, up to ten such disks being mounted in a disk pack. Data is recorded in concentric tracks on both surfaces with up to 236 tracks per centimetre. The disks rotate at 3600 revolutions per minute, information being put onto the disk and removed from it by means of a record-playback head. *See also* floppy disk.

magnetic domain *See* magnetism.

magnetic elements *See* geomagnetism.

magnetic equator *See* equator; geomagnetism.

magnetic field A *field of force that exists around a magnetic body (*see* magnetism) or a current-carrying conductor. Within a magnetic field a magnetic dipole may experience a torque and a moving charge may experience a force. The strength and direction of the field can be given in terms of the *magnetic flux density* (or *magnetic induction*), symbol B; it can also be given in terms of the *magnetic field strength* (*magnetizing force* or *magnetic intensity*), symbol H.

The magnetic flux density is a vector quantity and is the *magnetic flux per unit area of a magnetic field at right angles to the magnetic force. It can be defined in terms of the effects the field has, for example by $B = F/qv\sin\theta$, where F is the force that a moving charge q would experience if it was travelling at a velocity v in direction making an angle θ with the direction of the field. The unit is the tesla.

The magnetic field strength is also a vector quantity and is related to B by: $H = B/\mu$, where μ is the *permeability of the medium. The unit of field strength is the ampere per metre ($A\,m^{-1}$).

magnetic field strength *See* magnetic field.

magnetic flux Symbol Φ. A measure of quantity of magnetism, taking account of the strength and the extent of a *magnetic field. The flux $d\Phi$ through an element of area dA perpendicular to B is given by $d\Phi = BdA$. The unit of magnetic flux is the weber.

magnetic flux density *See* magnetic field.

magnetic force The attractive or repulsive force exerted on a *magnetic pole or a moving electric charge in a *magnetic field.

magnetic induction *See* magnetic field.

magnetic intensity *See* magnetic field.

magnetic meridian *See* meridian.

magnetic mirror A device used to contain *plasma in thermonuclear experimental devices. It consists of a region of high magnetic field strength at the end of a containment tube. Ions entering the region reverse their motion and return to the plasma from which they have emerged. *See also* magnetic bottle.

magnetic moment The ratio between the maximum torque (T_{max}) exerted on a magnet, current-carrying coil, or moving charge situated in a *magnetic field and the strength of that field. It is thus a measure of the strength of a magnet or current-carrying coil. In the Sommerfeld approach this quantity (also called *electromagnetic moment* or *magnetic area moment*) is T_{max}/B. In the Kennelly approach the quantity (also called *magnetic dipole moment*) is T_{max}/H.

In the case of a magnet placed in a magnetic field of field strength H, the maximum torque T_{max} occurs when the axis of the magnet is perpendicular to the field. In the case of a coil of N turns and area A carrying a current I, the magnetic moment can be shown to be $m = T/B = NIA$ or $m = T/H = \mu NIA$. Magnetic moments are measured in A m^2.

An orbital electron has an orbital magnetic moment IA, where I is the equivalent current as the electron moves round its orbit. It is given by $I = q\omega/2\pi$, where q is the electronic charge and ω is its angular velocity. The orbital magnetic moment is therefore $IA = q\omega A/2\pi$, where A is the orbital area. If the electron is spinning there is also a spin magnetic moment (*see* spin); atomic nuclei also have magnetic moments (*see* nuclear moment).

magnetic monopole A hypothetical magnetic entity consisting of an isolated elementary north or south pole. It has been postulated as a source of a *magnetic field by analogy with the way in which an electrically charged particle produces an electric field. Numerous ingenious experiments have been designed to detect the monopole but so far none has produced an unequivocal result.

magnetic permeability *See* permeability.

magnetic poles 1. *See* geomagnetism. **2.** The regions of a *magnet from which the magnetic forces appear to originate. A magnetized bar has a pole at each end; if it is freely suspended in the earth's magnetic field (*see* geomagnetism) it will rotate so that one end points approximately towards the earth's geographical north pole. This end is called the north-seeking end or the north pole of the magnet. The other end is accordingly called the south-seeking end or south pole. In the obsolete theory associated with the *c.g.s. electromagnetic system of units, a *unit magnetic pole* was treated as one of a pair, which repelled each other with a force of 1 dyne when separated by 1 cm in space.

magnetic potential *See* magnetomotive force.

magnetic quantum number *See* atom.

magnetic susceptibility *See* susceptibility.

magnetic tape A plastic tape coated with a ferromagnetic iron oxide powder. The tape is used for recording data in tape recorders and computers. To record, the tape is passed over a recording head containing a gap in a magnetic circuit whose magnetization is modulated by the information to be recorded; the information is imprinted on the tape in the form of the direction of magnetization of the individual particles of iron oxide. The particles themselves are not rotated by the magnetizing field: it is their directions of magnetization that are orientated in accordance with the information. In audio-frequency recorders a high-frequency bias (in the range 75–100 kHz) is used to reduce distortion by facilitating the re-orientation. The playback procedure is the reverse of recording; the tape containing its orientation of tiny magnets is fed over the gap of the same (now the playback) head, in whose coil corresponding e.m.f.s are generated by induction.

magnetic variation (secular magnetic variation) *See* geomagnetism.

magnetism A group of phenomena associated with *magnetic fields. Whenever an electric current flows a magnetic field is produced; as the orbital motion and the *spin of atomic electrons are equivalent to tiny current loops, individual atoms create magnetic fields around them, when their orbital electrons have a net *magnetic moment as a result of their angular momentum. The magnetic moment of an atom is the vector sum of the magnetic moments of the orbital motions and the spins of all the electrons in the atom. The macroscopic magnetic properties of a substance arise from the magnetic moments of its component atoms and molecules. Different materials have different characteristics in an applied magnetic field; there are four main types of magnetic behaviour:

(a) In *diamagnetism* the magnetization is in the opposite direction to that of the applied field, i.e. the *susceptibility is negative. Although all substances are diamagnetic, it is a weak form of magnetism and may be masked by other, stronger, forms. It results from changes induced in the orbits of electrons in the atoms of a substance by the applied field, the direction of the change (in accordance with *Lenz's law) opposing the applied flux. There is thus a weak negative susceptibility (of the order of $-10^{-8}\ m^3\ mol^{-1}$) and a relative permeability of slightly less than one.

(b) In *paramagnetism* the atoms or molecules of the substance have net orbital or spin magnetic moments that are capable of being aligned in the direction of the applied field. They therefore have a positive (but small) susceptibility and a relative permeability slightly in excess of one. Paramagnetism occurs in all atoms and molecules with unpaired electrons; e.g. free atoms, free radicals, and compounds of transition metals containing ions with unfilled electron shells. It also occurs in metals as a result of the magnetic moments associated with the spins of the conducting electrons.

(c) In *ferromagnetic* substances, within a certain temperature range, there are net atomic magnetic moments, which line up in such a way that magnetization persists after the removal of the applied field. Below a certain temperature, called the *Curie point (or Curie temperature) an increasing magnetic field applied to a ferromagnetic substance will cause increasing magnetization to a high value, called the *saturation magnetization*. This is because a ferromagnetic substance consists of small (1–0.1 mm across) magnetized regions called *domains*. The total magnetic moment of a sample of the substance is the vector sum of the magnetic moments of the component domains. Within each domain the individual atomic magnetic moments are spontaneously aligned by *exchange forces*, related to whether or not the atomic electron spins are parallel or antiparallel. However, in an unmagnetized piece of ferromagnetic material the magnetic moments of the domains themselves are not aligned; when an external field is applied those domains that are aligned with the field increase in size at the expense of the others. In a very strong field all the domains are lined up in the direction of the field and provide the high observed magnetization. Iron, nickel, cobalt, and their alloys are ferromagnetic. Above the Curie point, ferromagnetic materials become paramagnetic.

(d) Some metals, alloys, and transition-element salts exhibit another form of magnetism called *antiferromagnetism*. This occurs below a certain temperature, called the *Néel temperature, when an ordered array of atomic magnetic moments spontaneously forms in which alternate moments have opposite directions. There is therefore no net resultant magnetic moment in the absence of an applied field. In manganese fluoride, for example, this antiparallel arrangement occurs below a Néel temperature of 72 K. Below this temperature the spontaneous ordering opposes the normal tendency of the magnetic moments to align with the applied field. Above the Néel temperature the substance is paramagnetic.

A special form of antiferromagnetism is *ferrimagnetism*, a type of magnetism exhibited by the *ferrites. In these materials the magnetic moments of adjacent ions are antiparallel and of unequal strength, or the number of magnetic moments in one direction is greater than those in the opposite direction. By suitable choice of rare-earth ions in the ferrite lattices it is possible to design ferrimagnetic substances with specific magnetizations for use in electronic components. *See also* geomagnetism.

magneto An alternating-current generator used as a high-tension source in the ignition systems of petrol engines in which there are no batteries, e.g. in some tractor, marine, and aviation engines (in road vehicles the expense of a magneto is saved by using an induction coil and storage battery). Most modern magnetos consist of a permanent-magnet rotor revolving within a primary (low-voltage) winding around which a secondary winding is placed in which to induce the high voltage needed to produce the spark across the points of the plugs. Magnetos are geared to the engine shaft, the speed depending on the number of poles of the magneto and the number of engine cylinders. A make-and-break device is incorporated in the primary winding; when the primary current stops the sudden change of magnetic flux within the secondary induces in it the required large e.m.f.

magnetobremsstrahlung *See* synchrotron radiation.

magnetocaloric effect A reversible change of temperature resulting from a change in the magnetization of a ferromagnetic or paramagnetic substance (*see* magnetism). The change in temperature ΔT, accompanying an adiabatic change of magnetic field ΔH, is given by:

$$\Delta T/\Delta H = -T/C_H(\partial M/\partial T)_H$$

where C_H is the specific heat capacity per unit volume at constant H and M is the magnetization.

magnetochemistry The branch of physical chemistry concerned with measuring and investigating the magnetic properties of compounds. It is used particularly for studying transition-metal complexes, many of which are paramagnetic because they have unpaired electrons. Measurement of the magnetic susceptibility allows the magnetic moment of the metal atom to be calculated, and this gives information about the bonding in the complex.

magnetohydrodynamics (MHD) The study of the interactions between a conducting fluid and a *magnetic field. MHD is important in the study of controlled thermonuclear reactions in which the conducting fluid is a *plasma confined by a magnetic field. Other important applications include the *magnetohydrodynamic power generator*. In the open-cycle MHD generator a fossil fuel, burnt in oxygen or preheated compressed air, is seeded with an element of low *ionization potential (such as potassium or caesium). This element is thermally ionized at the combustion temperature (usually over 2500 K) producing sufficient free electrons (e.g. $K \rightarrow K^+ + e$) to provide adequate electrical conductivity. The interaction between the moving conducting fluid and the strong applied magnetic field across it generates an e.m.f. on the Faraday principle, except that the solid conductor of the con-

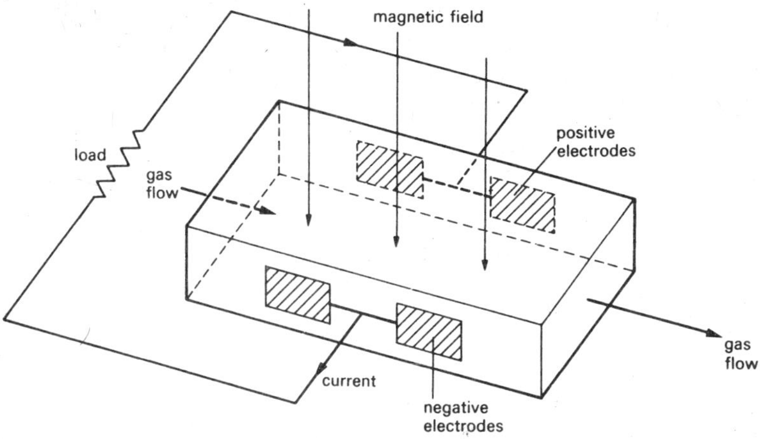

Magnetohydrodynamic generator

ventional generator is replaced by a fluid conductor. The power output per unit fluid volume (W) is given by $W = k\sigma v^2B^2$, where σ is the conductivity of the fluid, v is its velocity, B is the magnetic flux density, and K is a constant. Devices of this kind are in use in some power stations, where they are suitable for helping to meet high short-term demands and have the ability of increasing the thermal efficiency of a steam-turbine generator from about 40% to 50%. In experimental closed-cycle systems the fluid is continuously recirculated through a compressor; the fluid consists of a heated and seeded noble gas or a liquid metal.

magnetomechanical ratio *See* gyromagnetic ratio.

magnetometer An instrument for measuring the magnitude, and sometimes the direction, of a magnetic field. *Absolute magnetometers* measure the field without reference to a standard magnetic instrument. The most widely used are the *vibration magnetometer*, the *deflection galvanometer*, and the more modern *nuclear magnetometer*. The vibration instrument was devised by Gauss in 1832 and depends on the rate of oscillation of a small bar magnet suspended in a horizontal plane. The same magnet is then used as a fixed deflector to deflect a second similarly suspended magnet. The deflection galvanometer uses a Helmholtz coil system of known dimensions with a small magnet suspended at its centre. The deflected magnet comes to rest at a position controlled by the earth's magnetic field, the coil's magnetic field, and the angle through which the coil must be turned to keep the magnet and the coil in alignment. The sensitive nuclear magnetometers are based on measuring the audiofrequency voltage induced in a coil by the precessing protons in a sample of water. Various *relative magnetometers* are also in use, especially for measuring the earth's magnetic field and in calibrating other equipment.

magnetomotive force (m.m.f.) The analogue of *electromotive force in a *magnetic circuit. Mathematically, it is the circular integral of $H\cos\theta\ ds$, where $H\cos\theta$ is the component of the *magnetic field strength in the direction of a path of length ds. The m.m.f. is measured in ampere-turns. It was formerly called the *magnetic potential*.

magneton A unit for measuring *magnetic moments of nuclear, atomic, or molecular magnets. The *Bohr magneton* μ_B has the value of the classical magnetic moment of an electron, given by

$$\mu_B = eh/4\pi m_e = 9.274 \times 10^{-24}\ \text{A m}^2,$$

where e and m_e are the charge and mass of the electron and h is the Planck constant. The *nuclear magneton*, μ_N is obtained by replacing the mass of the electron by the mass of the proton and is therefore given by

$$\mu_N = \mu_B . m_e/m_p = 5.05 \times 10^{-27}\ \text{A m}^2.$$

magneto-optical effects Effects resulting from the influence of a *magnetic field upon matter that is in the process of emitting or absorbing light. Examples are the *Faraday effect and the *Zeeman effect.

magnetosphere A comet-shaped region surrounding the earth and other magnetic planets in which the charged particles of the *solar wind are controlled by the planet's magnetic field rather than the sun's magnetic field. It extends for some 60 000 km on the side facing the sun but on the opposite side it extends to a much greater extent. The boundary of the magnetosphere is known as the *magnetopause*. The magnetosphere of the earth includes the *Van Allen belts.

magnetostriction The change in length of a ferromagnetic material (*see* magnetism) when it is magnetized. It results from changes in the boundaries of the domains. A ferromagnetic rod exposed to an alternating field will vibrate along its length. This appears to be a major source of transformer hum, which can be removed by using a magnetic steel containing 6.5% silicon. Magnetostriction of a nickel transducer is used to generate and receive ultrasonic waves.

magnetron A microwave generator in which electrons, generated by a heated cathode, move under the combined force of an electric field and a magnetic field. The cathode consists of a central hollow cylinder, the outer surface of which carries the barium and strontium oxide electron emitters. The anode is also a cylinder, arranged concentrically around the cathode, and it contains a series of quarter-wavelength *resonant cavities arranged around its inner surface. The electric field is applied radially between anode and cathode, the magnetic

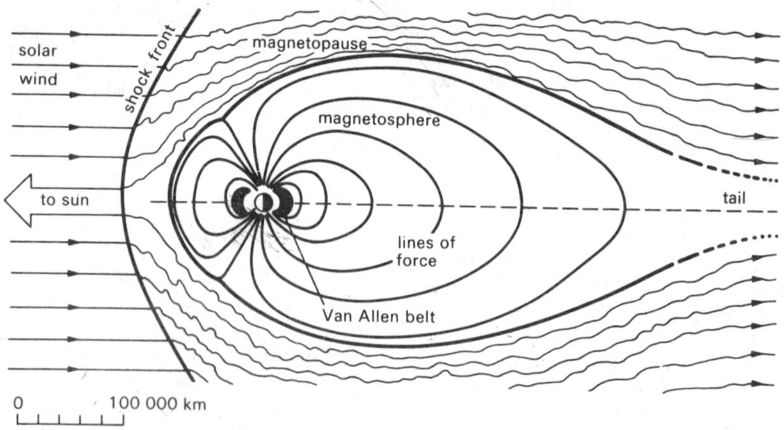

Magnetosphere and magnetopause

field is coaxial with the cathode. The whole device is maintained in a vacuum enclosure. The magnetron is extensively used as a generator for radar installations and can produce microsecond pulses of up to 10 MW.

magnification A measure of the extent to which an optical system enlarges or reduces an image. The *linear magnification*, *m*, is the ratio of the image height to the object height. If this ratio is greater than one the system is enlarging, if it is less than one, it is reducing. The *angular magnification*, *M* or γ, is the ratio of the angles formed by the final image and the object (when viewed directly, in the most favourable position available) at the eye. This is also sometimes called the *magnifying power* of an optical system.

magnifying power *See* magnification.

magnitude A measure of the relative brightness of a star or other celestial object. The *apparent magnitude* depends on the star's *luminosity, its distance, and the absorption of light between the object and the earth. In 1856 the astronomer N. R. Pogson devised a scale in which a difference of five magnitudes corresponds to a brightness ratio of 100 to 1. Two stars that differ by one magnitude therefore have a brightness ratio of $(100)^{0.2}:1 = 2.512$, known as the *Pogson ratio*. This scale is now universally adopted. Apparent magnitudes are not a measure of luminosity, which is defined in terms of the *absolute magnitude*. This is the apparent magnitude of a body if it was situated at a standard distance of 100 parsecs.

Magnox A group of magnesium alloys used to enclose uranium fuel elements in *nuclear reactors. They usually contain some aluminium as well as other elements, such as beryllium.

mainframe computer *See* computer.

main-sequence stars *See* Hertzsprung–Russell diagram.

majority carrier *See* semiconductor.

Maksutov telescope *See* telescope.

manganin A copper alloy containing 13–18% of manganese and 1–4% of nickel. It has a high electrical resistance, which is relatively insensitive to temperature changes. It is therefore suitable for use as a resistance wire.

manometer A device for measuring pressure differences, usually by the difference in height of two liquid columns. The simplest type is the U-tube manometer, which consists of a glass tube bent into the shape of a U. If a pressure to be measured is fed to one side of the U-tube and the other is open to the atmosphere, the difference in level of the liquid in the two limbs gives a measure of the unknown pressure.

mantissa *See* logarithm.

martensite A solid solution of carbon in alpha-iron (*see* iron) formed when *steel is cooled too rapidly for pearlite to form from austenite. It is responsible for the hardness of quenched steel.

mascon A gravitational anomaly on the surface of the moon resulting from a concentration of mass below the lunar surface. They occur in circular lunar maria and were caused either by the mare basalt as it flooded the basins or by uplift of high-density mantle material when the basins were formed.

maser (*m*icrowave *a*mplification by *s*timulated *e*mission of *r*adiation) A device for amplifying or generating *microwaves by means of stimulated emission (*see* laser). As oscillators, masers are used in *atomic clocks, while they are used as amplifiers in *radio astronomy, being especially suitable for amplifying feeble signals from space.

In the *ammonia gas maser* (devised in 1954) a molecular beam of ammonia passes through a small orifice into a vacuum chamber, where it is subjected to a nonuniform electric field. This field deflects ground-state ammonia molecules, shaped like a pyramid with the three hydrogen atoms forming the plane of the base and the single nitrogen atom forming the apex. The ground-state molecule has a dipole moment on account of its lack of symmetry and it is for this reason that it suffers deflection. Excited molecules, in which the nitrogen atom vibrates back and forth through the plane of the hydrogen atoms, have no resultant dipole moment and are not deflected. The beam, now consisting predominately of excited molecules, is passed to a resonant cavity fed with the microwave radiation corresponding to the energy difference between the excited and the ground states. This causes stimulated emission as the excited molecules fall to the ground state and the input microwave radiation is amplified coherently. This arrangement can also be made to oscillate and in this form is the basis of the *ammonia clock.

In the more versatile *solid-state maser* a magnetic field is applied to the electrons of paramagnetic (*see* magnetism) atoms or molecules. The energy of these electrons is quantized into two levels, depending on whether or not their spins are parallel to the magnetic field. The situation in which there are more parallel magnetic moments than antiparallel can be reversed by sudden changes in the magnetic field. This electron-spin resonance in paramagnetic materials allows amplification over broader bandwidths than gas masers.

mass A measure of a body's *inertia, i.e. its resistance to acceleration. According to Newton's laws of motion, if two unequal masses, m_1 and m_2, are allowed to collide, in the absence of any other forces both will experience the same force of collision. If the two bodies acquire accelerations a_1 and a_2 as a result of the collision, then $m_1a_1 = m_2a_2$. This equation enables two masses to be compared. If one of the masses is regarded as a standard of mass, the mass of all other masses can be measured in terms of this standard. The body used for this purpose is a 1-kg cylinder of platinum–iridium alloy, called the international standard of mass. Mass defined in this way is called the *inertial mass* of the body.

Mass can also be defined in terms of the gravitational force it produces. Thus, according to Newton's law of gravitation, $m_g = Fd^2/MG$, where M is the mass of a standard body situated a distance d from the body of mass m_g; F is the gravitational force between them and G is the *gravitational constant. The mass defined in this way is the *gravitational mass*. In the 19th century Roland Eötvös (1848–1919) showed experimentally that gravitational and inertial mass are indistinguishable, i.e. $m_i = m_g$.

Although mass is formally defined in terms of its inertia, it is usually measured by gravitation. The *weight* (W) of a body is proportional to its gravitational mass; it is the force by which a body is gravitationally attracted to the earth and is, therefore, equal to the product of the mass of the body and the *acceleration of free fall (g), i.e. $W = mg$. In the general language, weight and mass are often used synonymously; however, for scientific purposes they are different. Mass is measured in kilograms; weight, being a force, is measured in newtons. Weight, moreover, depends on where it is measured, because the value of g varies at different localities on the earth's surface. Mass, on the other hand, is constant wherever it is measured, subject to the special theory of *relativity. According to this theory, announced by Albert Einstein in 1905, the mass of a body is a measure of its total energy content. Thus, if the energy of a body increases, for example by an increase in kinetic energy or temperature, then its mass will increase. According to this law an increase in energy ΔE is accompanied by an increase in mass Δm, according to the *mass–energy equation* $\Delta m = \Delta E/c^2$, where c is the speed of light. Thus, if 1 kg of water is raised in temperature by 100 K, its internal energy will increase by 4×10^{-12} kg. This is, of course, a negligible increase and the mass–energy equation is only significant for extremely high energies. For example, the mass of an electron is increased sevenfold if it moves relative to the observer at 99% of the speed of light.

mass action The law of mass action states that the rate at which a chemical reaction takes place at a given temperature is proportional to the product of the *active masses* of the reactants. The active mass of a reactant is taken to be its molar concentration. For example, for a reaction

$$A + B \rightarrow C$$

the rate is given by

$$R = k[A][B]$$

where k is the rate constant. The principle was introduced by C. M. Guldberg and P. Waage in 1863. It is strictly correct only for ideal gases. In real cases *activities can be used.

mass decrement *See* mass defect.

mass defect 1. The difference between the rest mass of an atomic nucleus and the sum of the rest masses of its individual nucleons in the unbound state. It is thus the mass equivalent of the *binding energy on the basis of the mass–energy equation (*see* mass; relativity). **2.** (*or* **mass decrement**) The difference between the rest mass of a radioactive nucleus before decay and the total rest mass of the decay products.

mass–energy equation *See* mass; relativity.

mass number *See* nucleon number.

mass spectrum *See* spectrum.

matrix (*pl.* **matrices**) **1.** (in mathematics) A set of quantities in a rectangular array, used in certain mathematical operations. Unlike a determinant, a matrix does not have a quantitative value. The array is usually enclosed in round brackets. **2.** (in geology) The fine-grained material of rock in which the coarser-grained material is embedded.

maximum and minimum thermometer A thermometer designed to record both the maximum and minimum temperatures that have occurred over a given time period. It usually consists of a graduated capillary tube at the base of which is a bulb containing ethanol. The capillary contains a thin thread of mercury with a steel index at each end. As the temperature rises the index is pushed up the tube, where it remains in position to show the maximum temperature reached; as the temperature falls the lower index is pushed down the tube and similarly remains in position at the lowest temperature. The indexes are reset by means of a permanent magnet.

maximum permissible dose *See* dose.

maxwell A unit of magnetic flux in the *c.g.s. system, equal to the flux through 1 square centimetre perpendicular to a magnetic field of 1 gauss. 1 maxwell is equal to 10^{-8} weber. It is named after James Clerk Maxwell (1831–79).

Maxwell–Boltzmann distribution A law describing the distribution of speeds among the molecules of a gas. In a system consisting of N molecules that are independent of each other except that they exchange energy on collision, it is clearly impossible to say what velocity any particular molecule will have. However, statistical statements regarding certain functions of the molecules were worked out by James Clerk Maxwell

(1831–79) and Ludwig Boltzmann (1844–1906). One form of their law states that $n = N\exp(-E/RT)$, where n is the number of molecules with energy in excess of E, T is the thermodynamic temperature, and R is the *gas constant.

Maxwell's equations Four differential equations that describe the electric and magnetic fields at a point. From these equations James Clerk Maxwell deduced that the field vectors obey a wave equation and that light is propagated as electromagnetic waves. These equations form the basis of classical electrodynamics. They are:

$$\text{curl}\,\boldsymbol{H} = \partial\boldsymbol{D}/\partial t + \boldsymbol{j}$$
$$\text{div}\,\boldsymbol{B} = 0$$
$$\text{curl}\,\boldsymbol{E} = -\partial\boldsymbol{B}/\partial t$$
$$\text{div}\,\boldsymbol{D} = \rho$$

where $\boldsymbol{H}$ is the magnetic field strength, $\boldsymbol{D}$ is the electric displacement, $\boldsymbol{j}$ is the current density, $\boldsymbol{B}$ is the magnetic flux density, $\boldsymbol{E}$ is the electric field strength, and ρ is the volume charge density.

Mcleod gauge A vacuum pressure gauge in which a relatively large volume of a low-pressure gas is compressed to a small volume in a glass apparatus. The volume is reduced to an extent that causes the pressure to rise sufficiently to support a column of fluid high enough to read. This simple device, which relies on *Boyle's law, is suitable for measuring pressures in the range 10^3 to 10^{-3} pascal.

mean *See* average.

mean free path The average distance travelled between collisions by the molecules in a gas, the electrons in a metallic crystal, the neutrons in a moderator, etc. According to the *kinetic theory the mean free path between elastic collisions of gas molecules of diameter d (assuming they are rigid spheres) is $1/\sqrt{2}n\pi d^2$, where n is the number of molecules per unit volume in the gas. As n is proportional to the pressure of the gas, the mean free path is inversely proportional to the pressure.

mean free time The average time that elapses between the collisions of the molecules in a gas, the electrons in a crystal, the neutrons in a moderator, etc. *See* mean free path.

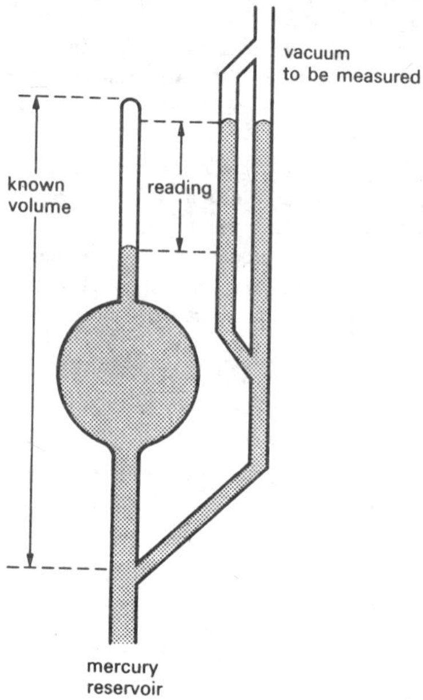

McLeod gauge

mean life *See* decay.

mean solar day *See* day.

mechanical advantage *See* force ratio.

mechanical equivalent of heat Symbol J. The ratio of a unit of mechanical energy to the equivalent unit of thermal energy, when a system of units is used in which they differ. J has the value 4.1868×10^7 ergs per calorie. The concept loses its usefulness in *SI units in which all forms of energy are expressed in joules and *J* therefore has a value of 1.

mechanics The study of the interactions between matter and the forces acting on it. *Statics is broadly concerned with the action of forces when no change of momentum is concerned, while *dynamics deals with cases in which there is a change of momentum. *Kinematics is the study of the motion of bodies without reference to the forces causing the motion. These classical sciences are concerned with macroscopic bodies in the solid state, while *fluid mechanics is the science of the interaction between forces and fluids.

median **1.** The middle number or value in a series of numbers or values. **2.** A straight line in a triangle that joins the vertex to the mid-point of the base.

medium frequency (MF) A radio frequency in the range 0.3–3 megahertz; i.e. having a wavelength in the range 100–1000 metres.

mega- Symbol M. A prefix used in the metric system to denote one million times. For example, 10^6 volts = 1 megavolt (MV).

megaton weapon A nuclear weapon with an explosive power equivalent to one million tons of TNT. *Compare* kiloton weapon.

Meissner effect The falling off of the magnetic flux within a superconducting metal when it is cooled to a temperature below the critical temperature in a magnetic field. It was discovered by Walther Meissner in 1933 when he observed that the earth's magnetic field was expelled from the interior of tin crystals below 3.72 K, indicating that as *superconductivity appeared the material became perfectly diamagnetic. *See* magnetism.

melting point (m.p.) The temperature at which a solid changes into a liquid. A pure substance under standard conditions of pressure (usually 1 atmosphere) has a single reproducible melting point. If heat is gradually and uniformly supplied to a solid the consequent rise in temperature stops at the melting point until the fusion process is complete.

mendelevium Symbol Md. A radioactive metallic transuranic element belonging to the actinoids; a.n. 101; mass number of the only known isotope 256 (half-life 1.3 hours). It was first identified by A. Ghiorso, G. T. Seaborg, and associates in 1955. The alternative name *unnilunium* has been proposed.

meniscus **1.** A concave or convex upper surface that forms on a liquid in a tube as a result of *surface tension. **2.** *See* concave.

mercury cell A primary *voltaic cell consisting of a zinc anode and a cathode of mercury(II) oxide (HgO) mixed with graphite. The electrolyte is potassium hydroxide (KOH) saturated with zinc oxide, the overall reaction being:

$$Zn + HgO \rightarrow ZnO + Hg$$

The e.m.f. is 1.35 volts and the cell will deliver about 0.3 ampere-hour per cm^3.

mercury-vapour lamp A type of discharge tube in which an arc discharge takes place in mercury vapour. The arc discharge takes place in a transparent tube of fused silica or quartz into the ends of which molybdenum and tungsten electrodes are sealed; this tube contains argon and a small amount of pure mercury. A small arc is struck between a starter electrode and one of the main electrodes causing local ionization of some argon atoms. The ionized atoms diffuse through the tube causing the main arc to strike; the heat from this arc vaporizes the mercury droplets, which become ionized current carriers. Radiation is confined to four visible wavelengths in the visible spectrum and several strong ultraviolet lines. The light is bluish but can be changed by the use of *phosphors on an outer tube. The outer tube is also usually used to filter out excessive ultraviolet radiation. The lamp is widely used for street lighting on account of its low cost and great reliability and as a source of ultraviolet radiation.

meridian 1. *See* latitude and longitude. 2. **(magnetic meridian)** An imaginary great circle on the earth's surface that passes through the north and south magnetic poles. A compass needle on the earth's surface influenced only by the earth's magnetic field (*see* geomagnetism) comes to rest along a magnetic meridian. 3. **(celestial meridian)** A great circle of the *celestial sphere that passes through the zenith and the celestial poles. It meets the horizon at the north and south points.

meso form *See* optical activity.

meson Any of a class of *elementary particles that are a subclass of the *hadrons. According to current quark theory mesons consist of quark–antiquark pairs. They exist with positive, negative, and zero charges, but when charged the charge has the same magnitude as that of the electron. They include the *kaon*, *pion*, and *psi* particles. Mesons are believed to participate in the forces that hold nucleons together in the nucleus. The muon, originally called a mu-meson, was thought to be a meson but is now recognized as a *lepton.

metal fatigue A cumulative effect causing a metal to fail after repeated applications of *stress, none of which exceeds the ultimate *tensile strength. The *fatigue strength* (or *fatigue limit*) is the stress that will cause failure after a specified number (usually 10^7) of cycles. The number of cycles required to produce failure decreases as the level of stress or strain increases. Other factors, such as corrosion, also reduce the fatigue life.

metallography The microscopic study of the structure of metals and their alloys. Both optical *microscopes and *electron microscopes are used in this work.

metallurgy The branch of engineering concerned with the production of metals from their ores, the purification of metals, the manufacture of alloys, and the use and performance of metals in engineering practice. *Process metallurgy* is concerned with the extraction and production of metals, while *physical metallurgy* concerns the mechanical behaviour of metals.

metamict state The amorphous state of a substance that has lost its crystalline structure as a result of the radioactivity of uranium or thorium. *Metamict minerals* are minerals whose structure has been disrupted by this process. The metamictization is caused by alpha-particles and the recoil nuclei from radioactive disintegration.

metastable state A condition of a system in which it has a precarious stability that can easily be disturbed. It is unlike a state of stable equilibrium in that a minor disturbance will cause a system in a metastable state to fall to a lower energy level. A book lying on a table is in a state of stable equilibrium; a thin book standing on edge is in metastable equilibrium. Supercooled water is also in a metastable state. It is liquid below 0°C; a grain of dust or ice introduced into it will cause it to freeze. An excited state of an atom or nucleus that has an appreciable lifetime is also metastable.

meteor A streak of light observable in the sky when a particle of matter enters the earth's atmosphere and becomes incandescent as a result of friction with atmospheric atoms and molecules. These particles of matter are known collectively as *meteoroids*. Meteoroids that survive their passage through the atmosphere and strike the earth's surface are known as *meteorites*.

Only some 2500 meteorites are known, excluding the *micrometeorites* (bodies less than 1 mm in diameter). Meteorites consist mainly of silicate materials (stony meteorites) or iron (iron meteorites). It is estimated that the earth collects over 10^8 kg of meteoritic material every year, mostly in the form of micrometeorites. Micrometeorites survive atmospheric friction because their small size enables them to radiate away the heat generated by friction before they vaporize.

meteorite *See* meteor.

method of mixtures A method of determining the specific heat capacities of liquids or a liquid and a solid by mixing known masses of the substances at different temperatures and measuring the final temperature of the mixture.

metre Symbol m. The SI unit of length, being the length of the path travelled by light in vacuum during a time interval of $1/(2.99\,792\,458 \times 10^8)$ second. This definition, adopted by the General Conference on Weights and Measures in October, 1983, replaced the 1967 definition based on the krypton lamp, i.e. 1 650 763.73 wavelengths in a vacuum of the radiation corresponding to the transition between the levels $2p^{10}$ and $5d^5$ of the nuclide krypton-86. This definition (in 1958) replaced the older definition of a metre based on a platinum-iridium bar of standard length. When the *metric system was introduced in 1791 in France, the metre was intended to be one ten-millionth of the earth's meridian quadrant passing through Paris. However, the original geodetic surveys proved the impractibility of such a standard and the original platinum metre bar, the *mètre des archives*, was constructed in 1793.

metre bridge *See* Wheatstone bridge.

metric system A decimal system of units originally devised by a committee of the French Academy, which included J. L. Lagrange and P. S. Laplace, in 1791. It was based on the *metre, the gram defined in terms of the mass of a cubic centimetre of water, and the second. This centimetre-gram-second system (*see* c.g.s. units) later gave way for scientific work to the metre-kilogram-second system (*see* m.k.s. units) on which *SI units are based.

metric ton (tonne) A unit of mass equal to 1000 kg or 2204.61 lb. 1 tonne = 0.9842 ton.

metrology The scientific study of measurement, especially the definition and standardization of the units of measurement used in science.

MHD *See* magnetohydrodynamics.

mho A reciprocal ohm, the former name of the unit of electrical *conductance now known as the siemens.

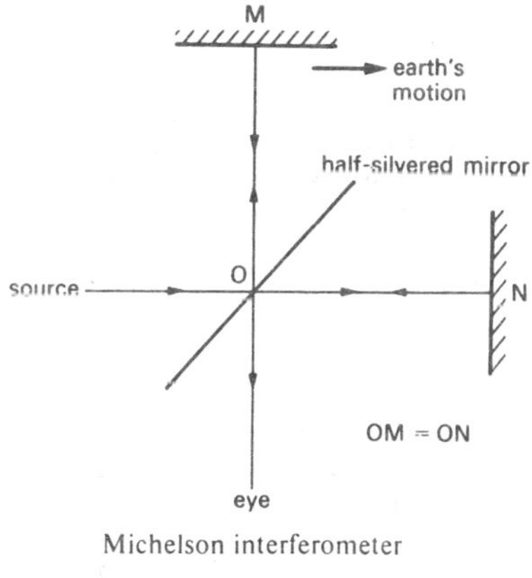

Michelson interferometer

Michelson–Morley experiment An experiment, conducted in 1887 by Albert Michelson (1852–1931) and Edward Morley (1838–1923), that attempted to measure the velocity of the earth through the *ether. Using a modified *Michelson interferometer* they expected to observe a shift in the interference fringes formed when the instrument was rotated through 90°, showing that the speed of light measured in the direction of the earth's rotation, or orbital motion, is not identical to its speed at right angles to this direction. No shift was observed. An explanation was finally provided by the *Lorentz–Fitzgerald contraction, which was an important step in the formulation of Einstein's special theory of *relativity and the abandonment of the ether concept.

micro- Symbol μ. A prefix used in the metric system to denote one millionth. For example, 10^{-6} metre = 1 micrometre (μm).

microbalance A sensitive *balance capable of weighing masses of the order 10^{-6} to 10^{-9} kg.

microelectronics The techniques of designing and making electronic circuits of very small size. As a result of these techniques a single *silicon chip measuring less than a centimetre in either direction can contain many thousands of transistors and may constitute the central processing unit of a microcomputer. In addition to an enormous drop in size, compared to an equivalent valve-operated device, these microelectronic circuits are some 100 000 times more reliable than their thermionic predecessors.

micrometeorite *See* meteor.

micrometer A gauge for measuring small diameters, thicknesses, etc., accurately. It consists of a G-shaped device in which the gap between the measuring faces is adjusted by means of an accurately calibrated screw, the end of which forms one of the measuring faces.

micron The former name for the *SI unit now called the micrometre, i.e. 10^{-6} m.

microphone A *transducer in which sound waves are converted into corresponding variations in an electrical signal for amplification, transmission to a distant point, or recording. Various types of device are used. In the *dynamic microphone* the sound waves impinge on a conductor of low mass supported in a magnetic field and cause it to oscillate at the frequency of the sound waves. These movements induce an e.m.f. in the conductor that is proportional to its velocity. The moving conductor consists of a metal ribbon, a wire, or a coil of wire. In the *moving-iron microphone*, sound waves cause a light armature to oscillate so that it varies the reluctance of a magnetic circuit. In a coil surrounding this path the varying reluctance is experienced as a variation in the magnetic flux within it, which induces a corresponding e.m.f. In the *carbon microphone*, widely used in telephones, a diaphragm constitutes a movable electrode in contact with carbon granules, which are also in contact with a fixed electrode. The movement of the diaphragm, in response to the sound waves, varies the resistance of the path through the granules to the fixed electrode. *See also* capacitor microphone; crystal microphone.

microprocessor *See* computer.

microscope A device for forming a magnified image of a small object. The *simple microscope* consists of a biconvex magnifying glass or an equivalent system of lenses, either hand-held or in a simple frame. The *compound microscope* uses two lenses or systems of lenses, the second magnifying the real image formed by the first. The lenses are usually mounted at the opposite ends of a tube that has mechanical controls to move it in relation to the object. An optical condenser and mirror, often with a separate light source, provide illumination of the object. The widely used *binocular microscope* consists of two separate instruments fastened together so that one eye looks through one while the other eye looks through the other. This gives stereoscopic

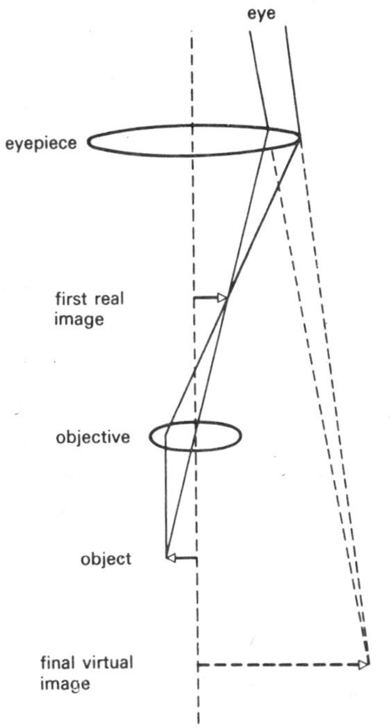

Compound microscope

vision. *See also* electron microscope; phase-contrast microscope; ultraviolet microscope.

microwave background radiation A background of radiation in the frequency range 3×10^{11} hertz to 3×10^{8} hertz discovered in space in 1965. Believed to have emanated from the primordial fireball of the big bang with which the universe is thought to have originated (*see* big-bang theory), the radiation has an energy density in intergalactic space of some 4×10^{-14} $J\,m^{-3}$.

microwave optics The study of the behaviour of microwaves by analogy with the behaviour of light waves. On the large scale microwaves are propagated in straight lines and, like light waves, they undergo reflection, refraction, diffraction, and polarization.

microwaves Electromagnetic waves with wavelengths in the range 10^{-3} to 0.03 m.

Milky Way System *See* galaxy.

milli- Symbol m. A prefix used in the metric system to denote one thousandth. For example, 0.001 volt = 1 millivolt (mV).

millibar *See* bar.

minority carrier *See* semiconductor.

minor planets *See* asteroids.

minute 1. One sixtieth of an hour. **2.** One sixtieth of a degree (angle).

mirage An optical phenomenon that occurs as a result of the bending of light rays through layers of air having very large temperature gradients. An *inferior mirage* occurs when the ground surface is strongly heated and the air near the ground is much warmer that the air above. Light rays from the sky are strongly refracted upwards near the surface giving the appearance of a pool of water. A *superior mirage* occurs if the air close to the ground surface is much colder than the air above. Light is bent downwards from the object towards the viewer so that it appears to be elevated or floating in the air.

mirror A surface that reflects most of the light falling on it. A *plane mirror* is a flat surface that produces an erect virtual *image of a real object, in which front and back are reversed. *Spherical mirrors* are formed from the surfaces of spheres and form images of real objects in much the same way as lenses. A convex mirror forms erect virtual images. They are commonly used as rear-view mirrors in road vehicles, and give a diminished wide-angle image. A concave mirror can form either inverted real images or erect virtual images. Spherical mirrors obey the *lens equation (using the

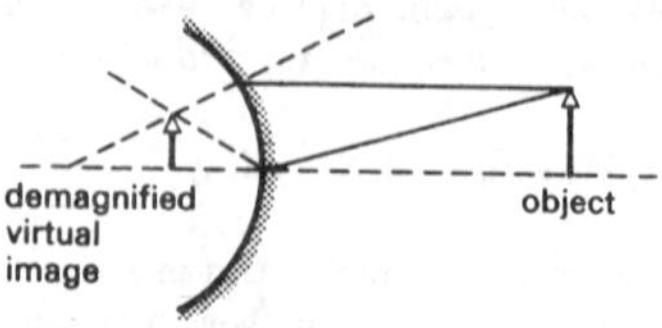

Convex mirror

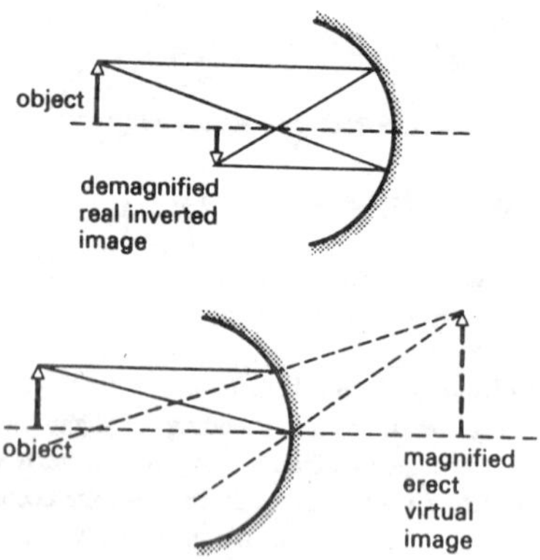

Concave mirrors

real-positive sign convention) and are subject to aberrations similar to those of lenses.

misch metal An alloy of cerium (50%), lanthanum (25%), neodymium (18%), praseodymium (5%), and other rare earths. It is used alloyed with iron (up to 30%) in lighter flints, and in small quantities to improve the malleability of iron. It is also added to copper alloys to make them harder, to aluminium alloys to make them stronger, to magnesium alloys to reduce creep, and to nickel alloys to reduce oxidation.

m.k.s. units A *metric system of units devised by A. Giorgi (and sometimes known as *Giorgi units*) in 1901. It is based on the metre, kilogram, and second and grew from the earlier *c.g.s. units. The electrical unit chosen to augment these three basic units was the ampere and the *permeability of space (magnetic constant) was taken as 10^{-7} H m^{-1}. To simplify electromagnetic calculations the magnetic constant was later changed to $4\pi \times 10^{-7}$ H m^{-1} to give the *rationalized MKSA system*. This system, with some modifications, formed the basis of *SI units, now used in all scientific work.

m.m.f. *See* magnetomotive force.

mmHg A unit of pressure equal to that exerted under standard gravity by a height of one millimetre of mercury, or 133.322 pascals.

mobility (of an ion) Symbol *u*. The velocity of an ion in an electric field divided by the field strength.

moderator A substance that slows down free neutrons in a *nuclear reactor, making them more likely to cause fissions of atoms of uranium–235 and less likely to be absorbed by atoms of uranium–238. Moderators are light elements, such as deuterium (in heavy water), graphite, and beryllium, to which neutrons can impart some of their kinetic energy on collision without being captured. Neutrons that have had their energies reduced in this way (to about 0.025 eV, equivalent to an average speed of 2200 m s^{-1}) are said to have been *thermalized* or to have become *thermal neutrons*.

modulation The process of superimposing the characteristics of a periodic signal onto a *carrier wave so that the information contained in the signal can be transmitted by the carrier wave. In *radio transmission, the simplest form of modulation is *amplitude modulation* (*AM*), in which the amplitude of the carrier is increased or diminished as the signal amplitude increases and diminishes. In *frequency modulation* (*FM*), the frequency of the carrier is increased or diminished as the signal amplitude increases and diminishes but the carrier amplitude remains constant. In *phase modulation*, the relative phase of the carrier is varied in accordance with the signal amplitude. Both frequency modu-

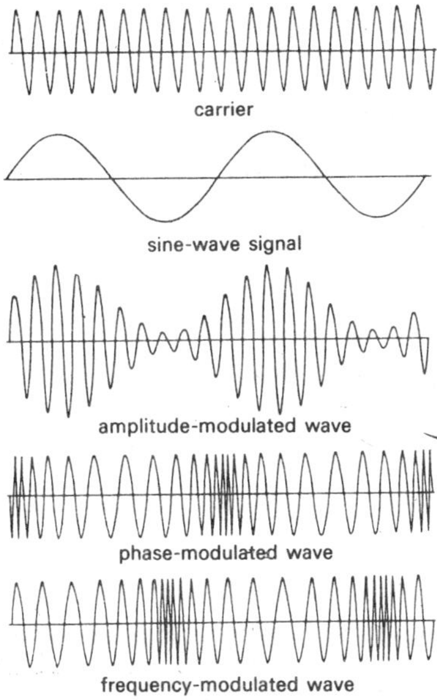

Modulation

lation and phase modulation are forms of *angle modulation*.

In *pulse modulation* the information is transmitted by controlling the amplitude, duration, position, or presence of a series of pulses. Morse code is a simple form of a pulse modulation.

modulus *See* absolute value.

modulus of elasticity *See* elastic modulus.

Moho (Mohorovičić discontinuity) A discontinuity within the *earth that marks the junction between the crust and the underlying mantle. Below the discontinuity earthquake seismic waves undergo a sudden increase in velocity, a feature that was first observed in 1909 by the Yugoslavian geophysicist Andrija Mohorovičić (1857–1936), after whom the discontinuity was named. The Moho lies at a depth of about 10–12 km below the oceans and about 33–35 km below the continents.

Mohs' scale A hardness scale in which a series of ten minerals are arranged in order, each mineral listed being scratched by and therefore softer than those below it. The minerals are: (1) talc; (2) gypsum; (3) calcite; (4) fluorite; (5) apatite; (6) orthoclase; (7) quartz; (8) topaz; (9) corundum; (10) diamond. As a rough guide a mineral with a value up to 2.5 on this scale can be scratched by a fingernail, up to 4 can be scratched by a coin, and up to 6 by a knife. The scale was devised by Friedrich Mohs (1773–1839).

molar Denoting that an extensive physical property is being expressed per *amount of substance, usually per mole. For example, the molar heat capacity of a compound is the heat capacity of that compound per unit amount of substance, i.e. it is usually expressed in J K^{-1} mol^{-1}.

molar conductivity Symbol Λ. The conductivity of that volume of an electrolyte that contains one mole of solution between electrodes placed one metre apart.

molar heat capacity *See* heat capacity.

molar latent heat *See* latent heat.

molar volume (molecular volume) The volume occupied by a substance per unit amount of substance.

mole Symbol mol. The SI unit of *amount of substance. It is equal to the amount of substance that contains as many elementary units as there are atoms in 0.012 kg of carbon-12. The elementary units may be atoms, molecules, ions, radicals, electrons, etc., and must be specified. 1 mole of a compound has a mass equal to its *relative molecular mass expressed in grams.

molecular beam A beam of atoms, ions, or molecules at low pressure, in which all the particles are travelling in the same direction and there are few collisions between them. They are formed by allowing a gas or vapour to pass through an aperture into an enclosure, which acts as a collimator by containing several additional apertures and vacuum pumps to remove any particles that do not pass through the apertures. Molecular beams are used in studies of surfaces and chemical reactions and in spectroscopy.

molecular distillation Distillation in high vacuum (about 0.1 pascal) with the condensing surface so close to the surface of the evaporating liquid that the molecules of the liquid travel to the condensing surface without collisions. This technique enables very much lower temperatures to be used than are used with distillation at atmospheric pressure and therefore heat-sensitive substances can be distilled. Oxidation of the distillate is also eliminated as there is no oxygen present.

molecular flow (Knudsen flow) The flow of a gas through a pipe in which the mean free path of gas molecules is large compared to the dimensions of the pipe. This occurs at low pressures; because most collisions are with the walls of the pipe rather than other gas molecules, the flow characteristics depend on the relative molecular mass of the gas rather than its viscosity. The effect was studied by M. H. C. Knudsen (1871–1949).

molecular formula *See* formula.

molecular orbital *See* orbital.

molecular volume *See* molar volume.

molecular weight *See* relative molecular mass.

molecule One of the fundamental units forming a chemical compound; the smallest part of a chemical compound that can take part in a chemical reaction. In most covalent compounds, molecules consist of groups of atoms held together by covalent or coordinate bonds. Covalent substances that form macromolecular crystals have no discrete molecules (in a sense, the whole crystal is a molecule). Similarly, ionic compounds do not have single molecules, being collections of oppositely charged ions.

mole fraction Symbol X. A measure of the amount of a component in a mixture. The mole fraction of component A is given by $X_A = n_A/N$, where n_A is the amount of substance of A (for a given entity) and N is the total amount of substance of the mixture (for the same entity).

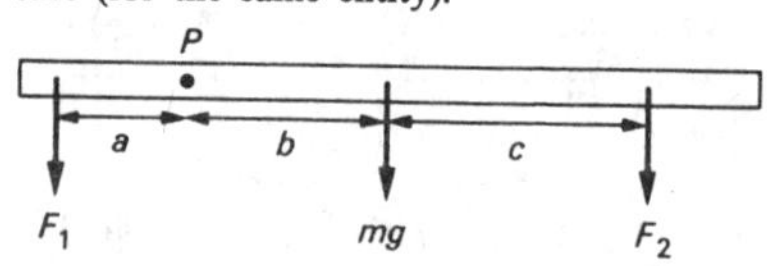

Moment of a force. For equilibrium $mgb + F_2(b+c) = F_1 a$, where mg is the weight of the beam acting through its centre of mass

moment of a force A measure of the turning effect produced by a force about an axis. The magnitude of the moment is the product of the force and the perpendicular distance from the axis to the line of action of the force. An object will be in rotational equilibrium if the algebraic sum of all the moments of the forces on it about any axis is zero.

moment of inertia Symbol I. The moment of inertia of a massive body about an axis is the sum of all the products formed by multiplying the magnitude of each element of mass (δm) by the square of its distance (r) from the line, i.e. $I_m = \Sigma r^2 \delta m$. It is the analogue in rotational dynamics of mass in linear dynamics. The basic equation is $T = I\alpha$, where T is the torque causing angular acceleration α about the specified axis.

momentum The *linear momentum* (p) of a body is the product of its mass (m) and its velocity (v), i.e. $p = mv$. *See also* angular momentum.

monochromatic radiation Electromagnetic radiation, especially visible radiation, of only one frequency or wavelength. Completely monochromatic radiation cannot be produced, but *lasers produce radiation within a very narrow frequency band. *Compare* polychromatic radiation.

monochromator A device that provides monochromatic radiation from a polychromatic source. In the case of visible radiation, for example, a prism can be used together with slits to select a small range of wavelengths.

monoclinic *See* crystal system.

moon The earth's only natural satellite, which orbits the earth at a mean distance of 384 400 km. It has a diameter of 3476 km. It has no atmosphere or surface water. Its surface temperature varies between 80 K (night minimum) and 400 K (noon at the equator). It is the only celestial body outside the earth to have been reached by man (1969).

Moseley's law The frequencies of the lines in the *X-ray spectra of the elements are related to the atomic numbers of the elements. If the square roots of the frequencies of corresponding lines of a set of elements are plotted against the atomic numbers a straight line is obtained. The law was discovered by H. G. Moseley (1887–1915).

Mössbauer effect The emission without recoil of a gamma-ray photon from a nucleus embedded in a solid. The emission of a gamma ray by a single atom in a gas causes the atom to recoil and reduces the energy of the gamma ray from its usual transition energy E_0 to $E_0 - R$, where R is the recoil energy. In 1957 R. L. Mössbauer (1929–) discovered that if the emitting nucleus is held by strong forces in the lattice of a solid, the recoil energy is shared by all the nuclei in the lattice. As there may typically be 10^{10}–10^{20} atoms in the lattice the recoil will be negligible and the gamma-ray photon has the energy E_0. This effect is used in *Mössbauer-effect spectroscopy* to elucidate problems in nuclear physics, solid-state physics, and chemistry.

motion A change in the position of a body or system with respect to time, as measured by a particular observer in a particular *frame of reference. Only relative motion can be measured; absolute motion is meaningless. *See also* equation of motion; Newton's laws of motion.

motor Any device for converting chemical energy or electrical energy into mechanical energy. *See* electric motor; internal-combustion engine; linear motor.

motor generator An electric motor mechanically coupled to an electric generator. The motor is driven by a supply of specified voltage, frequency, or number of phases and the generator provides an output in which one or more of these parameters is different to suit a particular purpose.

moving-coil instrument A measuring instrument in which current or voltage is determined by the force on a small coil pivoted between the poles of a magnet with curved poles, giving a radial magnetic field. When a current flows through the coil it turns against a return spring. If the angle through which it turns is α, the current I is given by $I = k\alpha/BAN$, where B is the magnetic flux density, A is the area of the coil and N is its number of turns; k is a constant depending on the strength of the return spring. The instrument is suitable for measuring d.c. but can be converted for a.c. by means of a rectifier network. It is usually made as a *galvanometer and converted to an ammeter or voltmeter by means of a *shunt or a *multiplier.

moving-iron instrument A measuring instrument in which current or voltage is determined by the force of attraction on a bar of soft iron pivoted within the magnetic field of a fixed coil or by the repulsion between the poles induced in two soft iron rods within the coil. As the deflection caused by the passage of a current through the coil does not depend on the direction of the current, moving-iron instruments can be used with either d.c. or a.c. without a rectifier. They are, however, less sensitive than *moving-coil instruments.

moving-iron microphone *See* microphone.

multimeter An electrical measuring instrument designed to measure potential differences or currents over a number of ranges. It also usually has an internal dry cell to enable resistances to be measured. Most multimeters are moving-coil instruments with a switch to enable series resistors or parallel resistors to be incorporated into the circuit.

multiplet 1. A spectral line formed by more than two (*see* doublet) closely spaced lines. **2.** A group of *elementary particles that are identical in all respects except that of electric charge.

multiplication factor Symbol k. The ratio of the average number of neutrons produced in a *nuclear reactor per unit time to the number of neutrons lost by absorption or leakage in the same time. If $k = 1$, the reactor is said to be *critical.* If $k > 1$ it is *supercritical* and if $k < 1$ it is *subcritical.* *See also* critical reaction.

multiplier A fixed resistance used with a voltmeter, usually a *moving-coil instrument, to vary its range. Many voltmeters are provided with a series of multipliers from which the appropriate value can be selected. If the original instrument requires i amperes for full-scale deflection and the resistance of the moving coil is r ohms, the value R of the resistance of the multiplier required to give a full-scale deflection when a voltage V is applied across the terminals is given by $R = V/i - r$.

multivibrator An electronic *oscillator consisting of two active devices, usually transistors, interconnected in an electrical network. The purpose of the device is to generate a continuous square wave with which to store information in binary form in a logic circuit. This is achieved by applying a portion of the output voltage or current of each active device to the input of the other with the appropriate magnitude and polarity, so that the devices are conducting alternately for controllable periods.

Mumetal The original trade name for a ferromagnetic alloy, containing 78% nickel, 17% iron, and 5% copper, that had a high *permeability and a low *coercive force. More modern versions also contain chromium and molybdenum. These alloys are used in some transformer cores and for shielding various devices from external magnetic fields.

Muntz metal A form of *brass containing 60% copper, 39% zinc, and small amounts of lead and iron. Stronger than alpha-brass, it is used for hot forgings, brazing rods, and large nuts and bolts. It is named after G. F. Muntz (1794–1857).

muon *See* lepton; elementary particles.

mutarotation Change of optical activity with time as a result of spontaneous chemical reaction.

mutual inductance *See* inductance.

myopia Short-sightedness. It results from the lens of the eye refracting the parallel rays of light entering it to a focus in front of the retina generally because of an abnormally long eyeball. The condition is corrected by using diverging spectacle lenses to move the image back to the retina.

N

nadir The point opposite the *zenith on the *celestial sphere.

nand circuit *See* logic circuit.

nano- Symbol n. A prefix used in the metric system to denote 10^{-9}. For example, 10^{-9} second = 1 nanosecond (ns).

Napierian logarithm *See* logarithm.

natural abundance *See* abundance.

natural frequency **1.** The frequency of the free oscillation of a system. **2.** The frequency at which resonance occurs in an electrical circuit.

natural logarithm *See* logarithm.

nautical mile A measure of distance used at sea. In the UK it is defined as 6080 feet but the international definition is 1852 metres. 1 international nautical mile is therefore equivalent to 1.15078 land (statute) miles.

near point The nearest point at which the human eye can focus an object. As the lens becomes harder with age, the extent to which accommodation can bring a near object into focus decreases. Therefore with advancing age the near point recedes – a condition known as *presbyopia.

nebula Originally a fixed, extended, and somewhat fuzzy white haze observed in the sky with a telescope. Many of these objects can now be resolved into clouds of individual stars and have been identified as *galaxies. They are still sometimes referred to as *extragalactic nebulae*. The *gaseous nebulae*, however, cannot be resolved into individual stars and consist, for the most part, of interstellar dust and gas. In some of these gaseous nebulae the gas atoms have been ionized by ultraviolet radiation from nearby stars and light is emitted as these ions interact with the free electrons in the gas. These are called *emission nebulae*. In the *dark nebulae*, there are no nearby stars and these objects are consequently dark; they can only be detected by what they obscure.

Néel temperature The temperature above which an antiferromagnetic substance becomes paramagnetic (*see* magnetism). The *susceptibility increases with temperature reaching a maximum at the Néel temperature, after which it abruptly declines. The phenomenon was discovered around 1930 by L. E. F. Néel (1904–).

negative charge *See* charge.

negative feedback *See* feedback.

nematic crystal *See* liquid crystal.

neon lamp A small lamp consisting of a pair of electrodes, treated to emit electrons freely, sealed in a glass bulb containing neon gas at low pressure. When a minimum voltage of between 60 and 90 volts is applied across the electrodes, the kinetic energy of the electrons is sufficient to ionize the neon atoms around the cathode, causing the emission of a reddish light. With d.c. the glow is restricted to the cathode; with a.c. both electrodes act alternately as cathodes and a glow appears to emanate from both electrodes. The device consumes a very low power and is widely used as an indicator light showing that a circuit is live.

neper A unit used to express a ratio of powers, currents, etc., used especially in telecommunications to denote the attenuation of an amplitude A_1 to an amplitude A_2 as N nepers, where

$$N = \ln(A_2/A_1).$$

1 neper = 8.686 decibels. The unit is named after John Napier (1550–1617), the inventor of natural logarithms.

neptunium Symbol Np. A radioactive metallic transuranic element belonging to the actinoids; a.n. 93; r.a.m. 237.0482. The most stable isotope, neptunium–237, has a half-life of 2.2×10^6 years and is produced in small quantities as a by-product by nuclear reactors. Other isotopes have mass numbers 229–236 and 238–241. The only other relatively long-lived isotope is neptunium–236 (half-life 5×10^3 years). The element was first produced by McMillan and Abelson in 1940.

neptunium series *See* radioactive series.

Nernst effect An effect in which a temperature gradient along an electric conductor or semiconductor placed in a perpendicular magnetic field, causes a potential difference to develop in the third perpendicular direction between opposite edges of the conductor. This effect, an analogue of the *Hall effect, was discovered in 1886 by Walter Nernst (1864–1941).

Nernst heat theorem A statement of the third law of *thermodynamics in a restricted form: if a chemical change takes place between pure crystalline solids at *absolute zero there is no change of entropy.

Neumann's law The magnitude of an electromagnetically induced e.m.f. (E) is given by $E = -d\Phi/dt$, where Φ is the magnetic flux. This is a quantitative statement of *Faraday's second law of electromagnetic induction and is sometimes known as the Faraday–Neumann law.

neutrino A *lepton (*see also* elementary particles) that exists in three forms, one in association with the electron, one with the muon, and one with the tau particle. Each form has its own antiparticle. The neutrino, which was postulated in 1931 to account for the 'missing' energy in *beta decay, was identified tentatively in 1953 and, more definitely, in 1956. Neutrinos have no charge, are thought to have zero rest mass, and travel at the speed of light.

neutron A neutral hadron (*see* elementary particles) that is stable in the atomic nucleus but decays into a proton, an electron, and an antineutrino with a mean life of 12 minutes outside the nucleus. Its rest mass is slightly greater than that of the proton, being $1.674\,92 \times 10^{-27}$ kg. Neutrons occur in all atomic nuclei except normal hydrogen. The neutron was first reported in 1932 by James Chadwick (1891–1974).

neutron diffraction The scattering of neutrons by atoms in solids, liquids, or gases. This process has given rise to a technique, analogous to *X-ray diffraction techniques, using a flux of thermal neutrons from a nuclear reactor to study solid-state phenomena. Thermal neutrons have an average kinetic energy of about 0.025 eV (4×10^{-21} J) giving them an equivalent wavelength of 0.1 nanometre, which is suitable for the study of interatomic interference. There are two types of interaction in the scattering of neutrons by atoms: one is the interaction between the neutrons and the atomic nucleus, the other is the interaction between the *magnetic moments of the neutrons and the spin and orbital magnetic moments of the atoms. The latter interaction has provided valuable information on antiferromagnetic and ferrimagnetic materials (*see* magnetism). Interaction with the atomic nucleus gives diffraction patterns that complement those from X-rays. X-rays, which interact with the extranuclear electrons, are not suitable for investigating light elements (e.g. hydrogen), whereas neutrons do give diffraction patterns from such atoms because they interact with nuclei.

neutron excess *See* isotopic number.

neutron number Symbol N. The number of neutrons in an atomic nucleus of a particular nuclide. It is equal to the difference between the *nucleon number and the *atomic number.

neutron star A star that has reached the end of its evolutionary life (*see* stellar evolution) and run out of nuclear fuel. It then suffers gravitational collapse and becomes so compressed that within the star itself electrons are forced into atomic nuclei, where they combine with protons to form neutrons. The current model of a neutron star envisages that it has a gaseous surface a few metres thick below which it is believed there is a solid crust some 1 km in depth. Underlying the crust is thought to be a *superfluid layer consisting almost exclusively of neutrons. At the centre of the star there is an immensely hard core. The whole star is probably no more than 10 km in diameter, but has a mass one and a half times that of the sun, i.e. 3×10^{30} kg. This gives the star an average density some 10^{15} times that of water – a density enormously greater than anything met elsewhere in the universe. Neutron stars were postulated in the 1930s and many astronomical observations confirm their existence, though the accuracy of the current model is by no means assured. Neutron stars are believed to be *pulsars if they rotate sufficiently rapidly and have sufficiently strong magnetic fields.

neutron temperature A concept used to express the energies of neutrons that are in thermal equilibrium with their surroundings, assuming that they behave like a monatomic gas. The neutron temperature T, on the Kelvin scale, is given by $T = 2E/3k$, where E is the neutron energy and k is the *Boltzmann constant.

newton Symbol N. The *SI unit of force, being the force required to give a mass of one kilogram an acceleration of 1 m s^{-2}. It is named after Sir Isaac Newton (1642–1727).

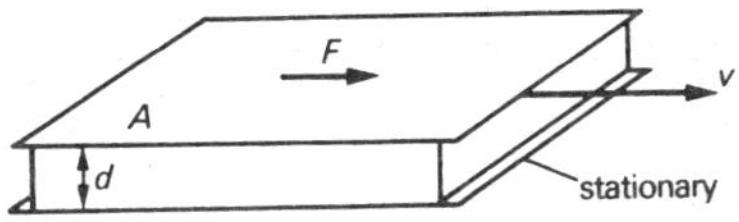

Newtonian fluid

Newtonian fluid A fluid in which the velocity gradient is directly proportional to the shear stress. If two flat plates of area A are separated by a layer of fluid of thickness d and move relative to each other at a velocity v, then the rate of shear is v/d and the shear stress is F/A, where F is the force applied to each. For a Newtonian fluid $F/A = \mu v/d$, where μ is the constant of proportionality and is called the Newtonian *viscosity. Many liquids and gases are Newtonian over a wide range of temperatures and pressures. However, some are not; these are called *non-Newtonian fluids*. In such fluids there is a departure from the simple Newtonian relationships. For example, in some liquids the viscosity increases as the velocity gradient increases, i.e. the faster the liquid moves the more viscous it becomes. Such liquids are said to be *dilatant* and the phenomenon they exhibit is called *dilatancy*. It occurs in some pastes and suspensions. More common, however, is the opposite effect in which the viscosity depends not only on the velocity gradient but also on the time for which it has been applied. These liquids are said to exhibit *thixotropy*. The faster a *thixotropic liquid* moves the less viscous it becomes. This property is used in nondrip paints (which are more viscous on the brush than on the wall) and in lubricating oils (which become thinner when the parts they are lubricating start to move). Another example is the non-Newtonian flow of macromolecules in solution or in polymer melts. In this case the shearing force F is not parallel to the shear planes and the linear relationship does not apply. In general, the many types of non-Newtonian fluid are somewhat complicated and no theory has been developed to accommodate them fully.

Newtonian mechanics The system of *mechanics that relies on *Newton's laws of motion. Newtonian mechanics is applicable to bodies moving at speeds relative to the observer that are small compared to the speed of light. Bodies moving at speeds comparable to the speed of light require an approach based on *relativistic mechanics, in which the mass of a body changes with its speed.

Newtonian telescope *See* telescope.

Newton's formula For a lens, the distances p and q between two conjugate points is given by $pq = f^2$, where f is the focal length of the lens.

Newton's law of cooling The rate at which a body loses heat is proportional to the difference in temperature between the body and the surroundings. It is an empirical law that is only true for substantial temperature differences if the heat loss is by forced convection or conduction.

Newton's law of gravitation There is a force of attraction between any two massive particles in the universe. For any two point masses m_1 and m_2, separated by a distance d, the force of attraction F is given by $F = m_1 m_2 G/d^2$, where G is the *gravitational constant. Real bodies having spherical symmetry act as point masses positioned at their centres of mass.

Newton's laws of motion The three laws of motion on which *Newtonian mechanics is based. (1) A body continues in a state of rest or uniform motion in a straight line unless it is acted upon by external forces. (2) The rate of change of momentum of a moving body is proportional to and in the same direction as the force acting on it, i.e. $F = \mathrm{d}(mv)/\mathrm{d}t$, where F is the applied force, v is the velocity of the body, and m its mass. If the mass remains constant, $F = m\mathrm{d}v/\mathrm{d}t$ *or* $F = ma$, where a is the acceleration. (3) If one body exerts a force on another, there is an equal and opposite force, called a *reaction*, exerted on the first body by the second.

Newton's rings 1. (in optics) *Interference fringes formed by placing a slightly convex lens on a flat glass plate. If monochromatic light is reflected by the two close surfaces into the observer's eye at a suitable angle, the point of contact of the lens is seen as a dark spot surrounded by a series of bright and dark rings. The radius of the nth dark ring is in an ideal case given by $r_n = \sqrt{nR\lambda}$, where λ is the wavelength of the monochromatic light and R is the radius of curvature of the lens. The phenomenon is used in the quality testing of lens surfaces. **2.** (in photography) The irregular patterns produced by thin film interference between a projected transparency and its cover glass.

Nichrome Trade name for a group of nickel–chromium alloys used for wire in heating elements as they possess good resistance to oxidation and have a high resistivity. Typical is Nichrome V containing 80% nickel and 19.5% chromium, the balance consisting of manganese, silicon, and carbon.

nickel–iron accumulator (Edison cell; NIFE cell) A *secondary cell devised by Thomas Edison (1847–1931) having a positive plate of nickel oxide and a negative plate of iron both immersed in an electrolyte of potassium hydroxide. The reaction on discharge is

$$2NiOOH.H_2O + Fe \rightarrow 2Ni(OH)_2 + Fe(OH)_2,$$

the reverse occurring during charging. Each cell gives an e.m.f. of about 1.2 volts and produces about 100 kJ per kilogram during each discharge. *Compare* lead–acid accumulator.

nickel silver *See* German silver.

Nicol prism A device for producing *plane-polarized light. It consists of two pieces of calcite cut with a 68° angle and stuck together with Canada balsam. The extraordinary ray (*see* double refraction) passes through the prism while the ordinary ray suffers total internal reflection at the interface between the two crystals, as the refractive index of the calcite is 1.66 for the ordinary ray and that of the Canada balsam is 1.53. Modifications of the prism using different shapes and cements are used for special purposes. It was devised in 1828 by William Nicol (1768–1851).

NIFE cell *See* nickel–iron accumulator.

nit A unit of *luminance equal to one *candela per square metre.

NMR *See* nuclear magnetic resonance.

nobelium Symbol No. A radioactive metallic transuranic element belonging to the actinoids; a.n. 102; mass number of most stable element 254 (half-life 55 seconds). Seven isotopes are known. The element was first identified with certainty by A. Ghiorso and G. T. Seaborg in 1966. The alternative name *unnilbium* has been proposed.

nodal points Two points on the axis of a system of lenses; if the incident ray passes through one, the emergent ray will pass through the other.

node 1. (in physics) A point of minimum disturbance in a *stationary-wave system. **2.** (in astronomy) Either of two points at which the orbit of a celestial body intersects a reference plane, usually the plane of the *ecliptic or the celestial equator (*see* celestial sphere).

noise 1. Any undesired sound. It is measured on a *decibel scale ranging from the threshold of hearing (0 dB) to the threshold of pain (130 dB). Between these limits a whisper registers about 20 dB, heavy urban traffic about 90 dB, and a heavy hammer on steel plate about 110 dB. A high noise level (industrial or from overamplified music, for example) can cause permanent hearing impairment. **2.** Any unwanted disturbance within a useful frequency band in a communication channel.

nomogram A graph consisting of three lines, each with its own scale, each line representing the values of a variable over a specified range. A ruler laid between two points on two of the lines enables the value of the third variable to be read off the third line.

non-Euclidean geometry A type of geometry that does not comply with the basic postulates of *Euclidean geometry, particularly a form of geometry that does not accept Euclid's postulate that only one straight line can be drawn through a point in space parallel to a given straight line. Several types of non-Euclidean geometry exist.

non-Newtonian fluid *See* Newtonian fluid.

nonpolar compound A compound that has covalent molecules with no permanent dipole moment. Examples of nonpolar compounds are methane and benzene.

nonrelativistic quantum theory *See* quantum theory.

normal 1. (in mathematics) A line drawn at right angles to a surface. **2.** (in chemistry) Having a concentration of one gram equivalent per dm^3.

normalizing The process of heating steel to above an appropriate critical temperature followed by cooling in still air. The process promotes the formation of a uniform internal structure and the elimination of internal stress.

not circuit *See* logic circuit.

note 1. A musical sound of specified pitch. **2.** A representation of such a sound in a musical score. Such a representation has a specified duration as well as a specified pitch.

nova A star that, over a period of only a few days, becomes 10^3–10^4 times brighter than it was. Some 10–15 such events occur in the Milky Way each year. Novae are believed to be close *binaries, one component of which is usually a *white dwarf and the other a *red giant. The outer envelope of the red giant engulfs the white dwarf and the two stars spiral together causing the subsequent nova explosion. *See also* supernova.

N.T.P. *See* s.t.p.

***n*-type conductivity** *See* semiconductor; transistor.

nuclear battery A single cell, or battery of cells, in which the energy of particles emitted from the atomic nucleus is converted internally into electrical energy. In the high-voltage type, a beta-emitter, such as strontium–90, krypton–85, or tritium, is sealed into a shielded glass vessel, the electrons being collected on an electrode that is separated from the emitter by a vacuum or by a solid dielectric. A typical cell delivers some 160 picoamperes at a voltage proportional to the load resistance. It can be used to maintain the voltage of a charged capacitor. Of greater use, especially in space technology, are the various types of low-voltage nuclear batteries. Typical is the gas-ionization device in which a beta-emitter ionizes a gas in an electric field. Each beta-particle produces about 200 ions, thus multiplying the current. The electric field is obtained by the contact potential difference between two electrodes, such as lead dioxide and magnesium. Such a cell, containing argon and tritium, gives about 1.6 nanoamperes at 1.5 volts. Other types use light from a phosphor receiving the beta-particles to operate photocells or heat from the nuclear reaction to operate a thermopile.

nuclear energy Energy obtained as a result of *nuclear fission or *nuclear fusion. The nuclear fission of one uranium atom yields about 3.2×10^{-11} joule, whereas the combustion of one carbon atom yields about 6.4×10^{-19} joule. Mass for mass, uranium yields about 2 500 000 times more energy by fission than carbon does by combustion. The nuclear fusion of deuterium to form helium releases about 400 times as much energy as the fission of uranium (on a mass basis).

nuclear fission A nuclear reaction in which a heavy nucleus (such as uranium) splits into two parts (*fission products*), which subsequently emit either two or three neutrons, releasing a quantity of energy equivalent to the difference between the rest mass of the neutrons and the fission products and that of the original nucleus. Fission may occur spontaneously or as a result of irradiation by neutrons. For example, the fission of a uranium–235 nucleus by a *slow neutron may proceed thus:

$$^{235}U + n \rightarrow {}^{148}La + {}^{85}Br + 3n$$

The energy released is approximately 3×10^{-11} J per ^{235}U nucleus. For 1 kg of ^{235}U this is equivalent to 20 000 megawatt-hours – the amount of energy produced by the combustion of 3×10^6 tonnes of coal. Nuclear fission is the process used in *nuclear reactors and atom bombs (*see* nuclear weapons).

nuclear force A strong attractive force between *nucleons in the atomic nucleus that holds the nucleus together. At close range (up to about 2×10^{-15} metre) these forces are some 100 times stronger than electromagnetic forces. *See* fundamental interactions.

nuclear fuel A substance that will sustain a fission chain reaction so that it can be used as a source of *nuclear energy. The *fissile isotopes uranium–235, uranium–233, and plutonium–239 are the only ones known. The first occurs in nature as 1 part in 140 of natural uranium, the other two have to be made artificially. ^{233}U is produced when thorium–232 captures a neutron and ^{239}Pu is produced by neutron capture in ^{238}U. ^{232}Th and ^{238}U are called *fertile isotopes.

nuclear fusion A type of *nuclear reaction in which atomic nuclei of low atomic number fuse to form a heavier nucleus with the release of large amounts of energy. In *nuclear fission reactions a neutron is used to break up a large nucleus, but in nuclear fusion the two reacting nuclei themselves have to be brought into collision. As both nuclei are positively charged there is a strong repulsive force between them, which can only be overcome if the reacting nuclei have very high kinetic energies. These high kinetic energies imply temperatures of the order of 10^8 K. As the kinetic energy required increases with the nuclear charge (i.e. atomic number), reactions involving low atomic-number nuclei are the easiest to produce. At these elevated temperatures, however, fusion reactions are self-sustaining; the reactants at these temperatures are in the form of a *plasma (i.e. nuclei and free electrons) with the nuclei possessing sufficient energy to overcome electrostatic repulsion forces. The fusion bomb (*see* nuclear weapons) and the stars generate energy in this way. It is hoped that the method will be harnessed in the *thermonuclear reactor as a source of energy for man's use.

Typical fusion reactions with the energy release in joules are:

$$D + D = T + p + 6.4 \times 10^{-19}\ J$$

$$T + D = {}^{4}He + n + 28.2 \times 10^{-13}\ J$$

$$^{6}Li + D = 2\,{}^{4}He + 35.8 \times 10^{-13}\ J$$

By comparison the formation of a water molecule from hydrogen and oxygen is accompanied by the release of 1.5×10^{-19} J. *See also* plasma physics.

nuclear isomerism A condition in which atomic nuclei with the same number of neutrons and protons have different lifetimes. This occurs when nuclei exist in different unstable quantum states, from which they decay to lower excited states or to the ground state, with the emission of gamma-ray photons. If the lifetime of a particular excited state is unusually long it is said to be isomeric, although there is no fixed limit separating isomeric decays from normal decays.

nuclear magnetic resonance (NMR) The absorption of electromagnetic radiation at a suitable precise frequency by a nucleus with a nonzero magnetic moment in an external magnetic field. The phenomenon occurs if the nucleus has nonzero *spin, in which case it behaves as a small magnet. In an external magnetic field, the nucleus's magnetic moment vector precesses about the field direction but only certain orientations are allowed by quantum rules. Thus, for hydrogen (spin of ½) there are two possible states in the presence of a field, each with a slightly different energy. Nuclear magnetic resonance is the absorption of radiation at a photon energy equal to the difference between these levels, causing a transition from a lower to a higher energy state. For practical purposes, the difference in energy levels is small and the radiation is in the radiofrequency region of the electromagnetic spectrum. It depends on the field strength.

NMR can be used for the accurate determination of nuclear moments. It can also be used in a sensitive form of magnetometer to measure magnetic fields. In medicine, NMR *tomography is being developed, in which images of tissue are produced by magnetic-resonance techniques.

The main application of NMR is as a technique for chemical analysis and structure determination, known as *NMR spectroscopy*. It depends on the fact that the electrons in a molecule shield the nucleus to some extent from the field, causing different atoms to absorb at slightly different frequencies (or at slightly different fields for a fixed frequency). Such effects are known as *chemical shifts*. In an NMR spectrometer, the sample is subjected to a strong field, which can be varied in a controlled way over a small region. It is irradiated with radiation at a fixed frequency, and a detector monitors the field at the sample. As the field changes, absorption corresponding to transitions occurs at certain values, and this causes oscillations in the field, which induce a signal in the detector. The most common nucleus studied is ^{1}H. For instance, an NMR spectrum of ethanol (CH_3CH_2OH) has three peaks in the ratio 3:2:1, corresponding to the three different hydrogen-atom environments. The peaks also have a fine structure caused by interaction between spins in the molecule. Other nuclei can also be used for NMR spectroscopy (e.g. ^{13}C, ^{14}N, ^{19}F) although these generally have lower magnetic moment and natural abundance than hydrogen. *See also* electron spin resonance.

nuclear moment A property of atomic nuclei in which lack of spherical symmetry of the nuclear charge gives rise to electric moments and the intrinsic spin and rotational motion of the component nucleons give rise to magnetic moments.

nuclear physics The physics of atomic nuclei and their interactions, with particular reference to the generation of *nuclear energy.

nuclear power Electric power or motive power generated by a *nuclear reactor.

nuclear reaction Any reaction in which there is a change to an atomic nucleus. This may be a natural spontaneous disintegration or an artificial bombardment of a nucleus with an energetic particle, as in a *nuclear reactor. Nuclear reactions are commonly represented by enclosing within a bracket the symbols for the incoming and outgoing particles; the initial nuclide is shown before the bracket and the final nuclide after it. For example, the reaction:

$$^{12}C + {}^{2}H \rightarrow {}^{13}N + {}^{1}n$$

is shown as $^{12}C(d,n)^{13}N$, where d is the symbol for deuterium.

nuclear reactor A device in which a *nuclear fission *chain reaction is sustained and controlled in order to produce *nuclear energy, radioisotopes, or new nuclides. The fuels available for use in a fission reactor are uranium–235, uranium–233, and plutonium–239; only the first occurs in nature (as 1 part in 140 of natural uranium), the others have to be produced artificially (*see* nuclear fuel). When a uranium–235 nucleus is made to undergo fission by the impact of a neutron it breaks into two roughly equal fragments, which release either two or three very high-energy neutrons. These *fast neutrons need to be slowed down to increase the probability that they will cause further fissions of ^{235}U nuclei and thus sustain the chain reaction. This slowing down process occurs naturally to a certain extent when the neutrons collide with other nuclei; unfortunately, however, the predominant uranium isotope, ^{238}U, absorbs fast neutrons to such an extent that in natural uranium the fission reaction is not self-sustaining. In order to create a controlled self-sustaining chain reaction it is necessary either to slow down the neutrons (using a *moderator in a *thermal reactor*) to greatly reduce the number absorbed by ^{238}U, or to reduce the predominance of ^{238}U in natural uranium by enriching it with more ^{235}U than it normally contains. In a *fast reactor* the fuel used is enriched uranium and no moderator is employed.

In thermal reactors, neutrons are slowed down by collisions with light moderator atoms (such as graphite, deuterium, or beryllium); they are then in thermal equilibrium with the surrounding material and are known as *thermal neutrons*. In a *heterogeneous thermal reactor* the fuel and moderator are in separate solid and liquid phases (e.g. solid uranium fuel and a heavy water moderator). In the *homogeneous thermal reactor* the fuel and moderator are mixed together, for example in a solution, molten dispersion, slurry, or suspension.

In the reactor *core* the *fuel elements* encase the fuel; in a heterogeneous reactor the fuel elements may fit into a lattice that also contains the moderator. The progress of the reaction is controlled by *control rods, which when lowered into the core absorb neutrons and so slow down or stop the chain reaction. The heat produced by the nuclear reaction in the core is used to generate electricity by the same means as in a conventional power station, i.e. by raising steam to drive a steam turbine that turns a generator. The heat is transferred to the steam-raising boiler or heat-exchanger by the *coolant. Water is frequently used as the coolant; in the case of the *boiling-water reactor* (BWR) and the *pressurized-water reactor* (PWR) water is both coolant and moderator. In the BWR the primary coolant drives the turbine; in the PWR the primary coolant raises steam in a secondary circuit for driving the turbine. In the *gas-cooled reactor* the coolant is a gas, usually carbon dioxide with an outlet temperature of about 350°C, or 600°C in the case of the *advanced gas-cooled reactor* (AGR).

In fast reactors, in which there is no moderator, the temperature is higher and a liquid-metal coolant is used, usually liquid sodium. Some fast reactors are used as converters or breeders. A *converter reactor* is one that converts *fertile material (such as ^{238}U) into *fissile material (such as ^{239}Pu). A *breeder reactor* produces the same fissile material as it uses. For example, a *fast breeder reactor* using uranium enriched with ^{239}Pu as the fuel can produce more ^{239}Pu than it uses by converting ^{238}U to ^{239}Pu. *See also* thermonuclear reactor.

nuclear waste *See* radioactive waste.

nuclear weapons Weapons in which an explosion is caused by *nuclear fission, *nuclear fusion, or a combination of both. In the fission bomb (*atomic bomb* or *A-bomb*) two subcritical masses (*see* critical mass) of a *fissile material (uranium-235 or plutonium-239) are brought together by a chemical explosion to produce one supercritical mass. The resulting nuclear explosion is typically in the *kiloton range with temperatures of the order 10^8 K being reached. The fusion bomb (*thermonuclear weapon, hydrogen bomb*, or *H-bomb*) relies on a nuclear-fusion reaction, which becomes self-sustaining at a critical temperature of about 35 $\times 10^6$ K. This is achieved by means of an inner fission bomb, which is surrounded by a hydrogenous material, such as heavy hydrogen (deuterium) or lithium deuteride. The *megaton explosion produced in such a thermonuclear reaction has not yet been used in war, the largest explosion so far produced being the 58 megaton weapon exploded in the Arctic by the USSR in 1961.

nucleon A *proton or a *neutron.

nucleonics The technological aspects of *nuclear physics, including the design of nuclear reactors, devices to produce and detect radiation, and nuclear transport systems. It is also concerned with the technology of *radioactive waste disposal and with radioisotopes.

nucleon number (mass number) Symbol A. The number of *nucleons in an atomic nucleus of a particular nuclide.

nucleus (of atom) The central core of an atom that contains most of its mass. It is positively charged and consists of one or more nucleons (protons or neutrons). The positive charge of the nucleus is determined by the number of protons it contains (*see* atomic number) and in the neutral atom this is balanced by an equal number of electrons, which move around the nucleus. The simplest nucleus is the hydrogen nucleus, consisting of one proton only. All other nuclei also contain one or more neutrons. The neutrons contribute to the atomic mass (*see* nucleon number) but not to the nuclear charge. The most massive nucleus that occurs in nature is uranium-238, containing 92 protons and 146 neutrons. The symbol used for this *nuclide is $^{238}_{92}U$, the upper figure being the nucleon number and the lower figure the atomic number. In all nuclei the nucleon number (A) is equal to the sum of the atomic number (Z) and the neutron number (N), i.e. $A = Z + N$.

nuclide 1. An atomic nucleus as characterized by its *atomic number and its *neutron number. An *isotope refers to a series of different atoms that have the same atomic number but different neutron numbers (e.g. uranium-238 and uranium-235 are isotopes of uranium), whereas a nuclide refers only to a particular nuclear species (e.g. the nuclides uranium-235 and plutonium-239 are fissile). **2.** The atom to which a specified nucleus belongs.

null method A method of making a measurement in which the quantity to be measured is balanced by another similar reading by adjusting the instrument to read zero (*see* Wheatstone bridge).

nutation An irregular periodic oscillation of the earth's poles. It causes an irregularity of the precessional circle traced by the celestial poles and results from the varying distances and relative directions of the sun and the moon.

O

objective The *lens or system of lenses nearest to the object being examined through an optical instrument.

occlusion 1. The trapping of small pockets of liquid in a crystal during crystallization. **2.** The absorption of a gas by a solid such that atoms or molecules of the gas occupy interstitial positions in the solid lattice. Palladium, for example, can occlude hydrogen.

occultation The disappearance of a star, planet, or other celestial body behind the moon or another planet, while it is being observed. A solar *eclipse is a form of occultation.

octane number A number that provides a measure of the ability of a fuel to resist *knocking when it is burnt in a spark-ignition engine. It is the percentage by volume of iso-octane (C_8H_{18}; 2,2,4-trimethylpentane) in a blend with normal heptane (C_7H_{16}) that matches the knocking behaviour of the fuel being tested in a single cylinder four-stroke engine of standard design. *Compare* cetane number.

octave The interval between two musical notes that have fundamental frequencies in the ratio 2:1; the word describes the interval in terms of the eight notes of the diatonic scale.

octet A stable group of eight electrons in the outer shell of an atom (as in an atom of an inert gas).

ocular *See* eyepiece.

odd–even nucleus An atomic nucleus containing an odd number of protons and an even number of neutrons.

odd–odd nucleus An atomic nucleus containing an odd number of protons and an odd number of neutrons.

oersted Symbol Oe. The unit of magnetic field strength in the *c.g.s. electromagnetic system. A field has a strength of one oersted if it exerts a force of one dyne on a unit magnetic pole placed in it. It is equivalent to $10^3/4\pi$ A m^{-1}. The unit was named after Hans Christian Oersted (1777–1851).

ohm Symbol Ω. The derived *SI unit of electrical resistance, being the resistance between two points on a conductor when a constant potential difference of one volt, applied between these points, produces a current of one ampere in the conductor. The former *international ohm* (sometimes called the 'mercury ohm') was defined in terms of the resistance of a column of mercury. The unit is named after Georg Ohm (1787–1854).

ohmmeter Any direct-reading instrument for measuring the value of a resistance in ohms. The instrument commonly used is a *multimeter capable of measuring also both currents and voltages. To measure resistance a dry cell and resistor are switched in series with the moving coil *galvanometer and the unknown resistance is connected across the instrument's terminals. The value of the resistance is then read off an ohms scale. Such instruments are increasingly being replaced by electronic digital multimeters.

Ohm's law The ratio of the potential difference between the ends of a conductor to the current flowing through it is constant. This constant is the *resistance of the conductor, i.e. $V = IR$, where V is the potential difference in volts, I is the current in amperes, and R is the resistance in ohms. The law was discovered in 1827 by Georg Ohm. Most materials do not obey this simple linear law; those that do are said to be *ohmic* but remain so only if physical conditions, such as temperature, remain constant. Metals are the most accurately ohmic conductors.

oil-immersion lens *See* immersion objective.

Olbers' paradox If the universe is infinite, uniform, and unchanging the sky at night would be bright, as in whatever direction one looked one would eventually see a star. The number of stars would increase in proportion to the square of the distance from the earth; the intensity of light reaching the earth from a given star is inversely proportional to the square of the distance. Consequently, the whole sky should be about as bright as the sun. The paradox, that this is not the case, was stated by Heinrich Olbers (1758–1840) in 1826. (It had been discussed earlier, in 1744, by J. P. L. Chesaux.) The paradox is resolved by the fact that, according to the *big-bang theory, the universe is not infinite, not uniform, and not unchanging. For instance, light from the most distant galaxies displays an extreme *redshift and ceases to be visible.

opacity The extent to which a medium is opaque to electromagnetic radiation, especially to light. It is the reciprocal of the *transmittance. A medium that is opaque to X-rays and gamma rays is said to be *radio-opaque*.

open cluster A diffuse group of between 20 and a few hundred stars that move through the Milky Way together. About 1000 such clusters are known; some of them closely resemble *globular clusters although they are less tightly bound and disperse more easily (as a result of rotation of parts of the Galaxy and the effects of interstellar dust clouds).

open-hearth process A traditional method for manufacturing steel by heating together scrap, pig iron, hot metal, etc., in a refractory-lined shallow open furnace heated by burning producer gas in air.

opera glasses *See* binoculars.

operator A mathematical symbol indicating that a specified operation should be carried out. For example, the operator $\sqrt{}$ in $\sqrt{x}$ indicates that the square root of x should be taken; the operator d/dx in dy/dx indicates that y should be differentiated with respect to x, etc.

opposition The moment at which a planet having its orbit outside that of the earth is in a line with the earth and the sun. When a planet is in opposition it can be observed during the night and is near to its closest point to the earth; it is therefore a favourable opportunity for observation.

optical activity The ability of certain substances to rotate the plane of plane-polarized light as it passes through a crystal, liquid, or solution. It occurs when the molecules of the substance are asymmetric, so that they can exist in two different structural forms each being a mirror image of the other. The two forms are *optical isomers* or *enantiomers*. The existence of such forms is also known as *enantiomorphism* (the mirror images being *enantiomorphs*). One form will rotate the light in one direction and the other will rotate it by an equal amount in the other. The two possible forms are described as *dextrorotatory or *laevorotatory according to the direction of rotation, and prefixes are used to designate the isomer, as in *d*-tartaric and *l*-tartaric acids. An equimolar mixture of the two forms is not optically active. It is called a *racemic mixture* (or *racemate*) and designated by *dl*-. In addition, certain molecules can have a *meso form* in which one part of the molecule is a mirror image of the other. Such molecules are not optically active.

Molecules that show optical activity have no plane of symmetry. The commonest case of this is in organic compounds in which a carbon atom is linked to four different groups. An atom of this type is said to be a *chiral centre*. Asymmetric molecules showing optical activity can also occur in inorganic compounds. For example, an octahedral complex in which the central ion coordinates to eight different ligands would be optically active. Many naturally occurring

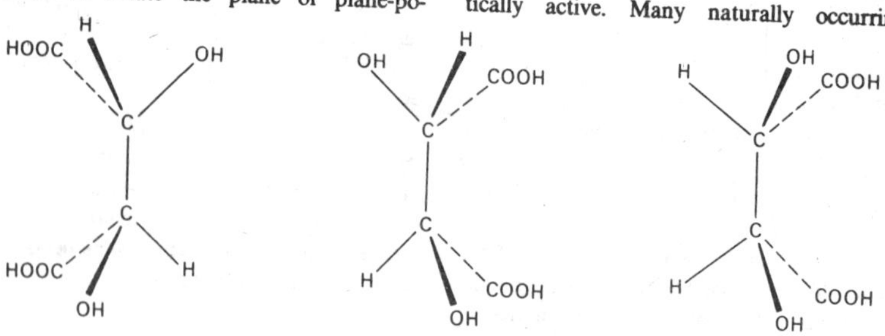

D-form L-form *meso*-form

Isomers of tartaric acid

compounds show optical isomerism and usually only one isomer occurs naturally. For instance, glucose is found in the dextrorotatory form. The other isomer, *l*-glucose, can be synthesized in the laboratory, but cannot be synthesized by living organisms. *See also* absolute configuration.

optical axis (principal axis; optic axis) The line passing through the *optical centre and the centre of a curvature of a *lens or spherical *mirror.

optical centre The point at the geometrical centre of a *lens through which a ray of light entering the lens passes without deviation.

optical fibre A *waveguide through which light can be transmitted with very little leakage through the sidewalls. In the *step-index fibre* a pure glass core, with a diameter between 6 and 250 micrometres, is surrounded by a coaxial glass or plastic cladding of lower refractive index. The cladding is usually between 10 and 150 micrometres thick. The interface between core and cladding acts as a cylindrical mirror at which *total internal reflection of the transmitted light takes place. This structure enables a beam of light to travel through many kilometres of fibre. In the *graded-index fibre*, each layer of glass, from the fibre axis to its outer wall, has a slightly lower refractive index than the layer inside it. This arrangement also prevents light from escaping through the fibre walls by a combination of refraction and total internal reflection, and can be made to give the same transit time for rays at different angles.

Fibre-optic systems use optical fibres to transmit information, in the form of coded pulses or fragmented images (using bundles of fibres), from a source to a receiver. Over moderate distances they are used in telecommunications, for which purpose they are becoming competitive with electric cables. They are also used in medical instruments (*fibrescopes*) to examine internal body cavities, such as the stomach and bladder.

optical flat A flat glass disc having very accurately polished surfaces so that the deviation from perfect flatness does not exceed (usually) 50 nanometres. It is used to test the flatness of such plane surfaces as gauge anvils by means of the *interference patterns formed when parallel beams of light pass through the flat and are reflected by the surface being inspected.

Surfaces are said to be *optically flat* if the deviation from perfect flatness is smaller than the wavelength of light.

optical glass Glass used in the manufacture of lenses, prisms, and other optical parts. It must be homogeneous and free from bubbles and strain. Optical *crown glass* may contain potassium or barium in place of the sodium of ordinary crown glass and has a refractive index in the range 1.51 to 1.54. *Flint glass* contains lead oxide and has a refractive index between 1.58 and 1.72. Higher refractive indexes are obtained by adding lanthanoid oxides to glasses; these are now known as lanthanum crowns and flints.

optical isomers *See* optical activity.

optical maser *See* laser.

optical microscope *See* microscope.

optical pumping *See* laser.

optical pyrometer *See* pyrometer.

optical rotary dispersion (ORD) The effect in which the amount of rotation of plane-polarized light by an optically active compound depends on the wavelength. A graph of rotation against wavelength has a characteristic shape showing peaks or troughs.

optical rotation Rotation of plane-polarized light. *See* optical activity.

optical telescope *See* telescope.

optical temperature *See* radiation temperature.

optic axis 1. The direction in a doubly refracting crystal in which light is transmitted without double refraction. **2.** *See* optical axis.

optics The study of *light and the phenomena associated with its generation, transmission, and detection. In a broader sense, optics includes all the phenomena associated with infrared and ultraviolet radiation. *Geometrical optics* assumes that light travels in straight lines and is concerned with the laws controlling the reflection and refraction of rays of light. *Physical optics* deals with phenomena that depend on the wave nature of light, e.g. diffraction, interference, and polarization.

orbit 1. (in astronomy) The path through space of one celestial body about another. For one small body moving in the gravitational field of another the orbit is a *conic section. Most such orbits are elliptical and most planetary orbits in the solar system are nearly circular. The shape and size of an elliptical orbit is specified by its *eccentricity, *e*, and the length of its semimajor axis, *a*. **2.** (in physics) The path of an electron as it travels round the nucleus of an atom. *See* orbital.

orbital A region in which an electron may be found in an atom or molecule. In the original *Bohr theory of the atom the electrons were assumed to move around the nucleus in circular orbits, but further advances in quantum mechanics led to the view that it is not possible to give a definite path for an electron. According to *wave mechanics, the electron has a certain probability of being in a given element of space. Thus for a hydrogen atom the electron can be anywhere from close to the nucleus to out in space but the maximum probability occurs in a spherical shell around the nucleus with a radius equal to the Bohr radius of the atom. The probabilities of finding an electron in different regions can be obtained by solving the Schrödinger wave equation to give the wave function ψ, and the probability is then proportional to $|\psi|^2$. Thus the idea of electrons in fixed orbits has been replaced by that of a probability distribution around the nucleus – an *atomic orbital*. Alternatively, the orbital can be thought of as an electric charge distribution (averaged over time). In representing orbitals it is convenient to take a surface enclosing the space in which the electron is likely to be found with a high probability.

The possible atomic orbitals correspond to subshells of the atom. Thus there is one *s*-orbital for each shell (orbital quantum number $l = 0$). This is spherical. There are three *p*-orbitals (corresponding to the three values of l) and five *d*-orbitals. The shapes of orbitals depend on the value of l. For instance, *p*-orbitals each have two lobes; most *d*-orbitals have four lobes.

In molecules, the valence electrons move under the influence of two nuclei (in a bond involving two atoms) and there are corresponding *molecular orbitals* for electrons.

orbital quantum number *See* atom.

orbital velocity (orbital speed) The speed of a satellite, spacecraft, or other body travelling in an *orbit around the earth or around some other celestial body. If the orbit is elliptical, the orbital speed, *v*, is given by:

$$v = \sqrt{[gR^2(2/r - 1/a)]},$$

where *g* is the accceleration of free fall, *R* is the radius of the orbited body, *a* is the semimajor axis of the orbit, and *r* is the distance between the orbiting body and the

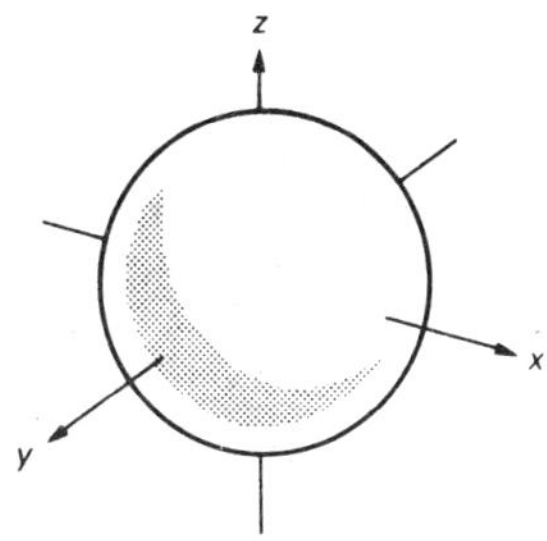

symmetrical
s–orbital

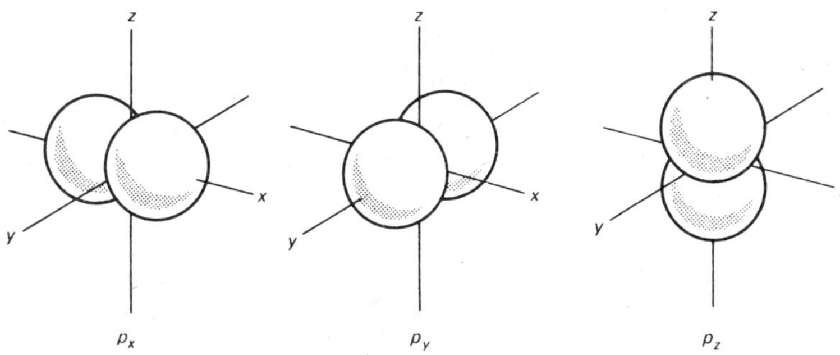

three equivalent *p*-orbitals, each having 2 lobes

Atomic orbitals

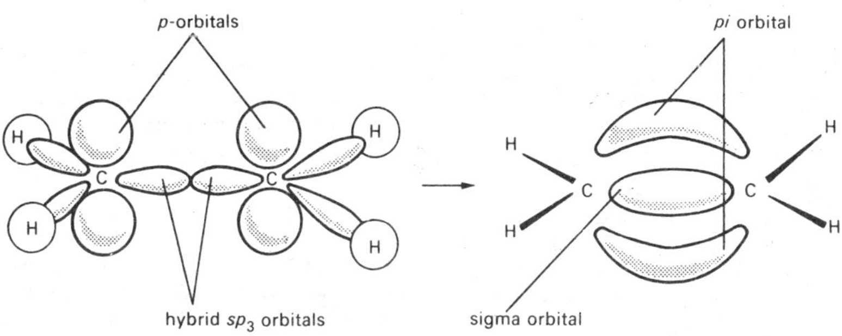

Molecular orbitals: formation of the double bond in ethene

Orbitals

centre of mass of the system. If the orbit is circular, $r = a$ and $v = \sqrt{(gR^2/r)}$.

or circuit *See* logic circuits.

order The number of times a variable is differentiated. dy/dx represents a first-order derivative, d^2y/dx^2 a second-order derivative, etc. In a *differential equation the order of the highest derivative is the order of the equation. $d^2y/dx^2 + 2dy/dx = 0$ is a second-order equation of the first degree. *See also* degree.

order of magnitude A value expressed to the nearest power of ten.

ordinary ray *See* double refraction.

ordinate *See* Cartesian coordinates.

origin *See* Cartesian coordinates.

oscillator An electronic device that produces an alternating output of known frequency. If the output voltage or current has the form of a sine wave with respect to time, the device is called a *sinusoidal* (or *harmonic*) *oscillator*. If the output voltage changes abruptly from one level to another (as in a *square wave or *sawtooth wave) it is called a *relaxation oscillator*. A harmonic oscillator consists of a frequency-determining circuit or device, such as a *resonant circuit, maintained in oscillation by a source of power that by positive feedback also makes up for the resistive losses. In some relaxation oscillators the circuit is arranged so that in each cycle energy is stored in a reactive element (a capacitor or inductor) and subsequently discharged over a different time interval. *See also* multivibrator.

oscilloscope *See* cathode-ray oscilloscope.

osmiridium A hard white naturally occurring alloy consisting principally of osmium (17–48%) and iridium (49%). It also contains small quantities of platinum, rhodium, and ruthenium. It is used for making small items subject to wear, e.g. electrical contacts or the tips of pen nibs.

osmometer *See* osmosis.

osmosis The passage of a solvent through a *semipermeable membrane* separating two solutions of different concentrations. A semipermeable membrane is one through which the molecules of a solvent can pass but the molecules of most solutes cannot. There is a thermodynamic tendency for solutions separated by such a membrane to become equal in concentration, the water (or other solvent) flowing from the weaker to the stronger solution. Osmosis will stop when the two solutions reach equal concentration, and can also be stopped by applying a pressure to the liquid on the stronger-solution side of the membrane. The pressure required to stop the flow from a pure solvent into a solution is a characteristic of the solution, and is called the *osmotic pressure* (symbol Π). Osmotic pressure depends only on the concentration of particles in the solution, not on their nature (i.e. it is a *colligative property). For a solution of n moles in volume V at thermodynamic temperature T, the osmotic pressure is given by $\Pi V = RT$, where R is the gas constant. Osmotic-pressure measurements are used in finding the relative molecular masses of compounds, particularly macromolecules. A device used to measure osmotic pressure is called an *osmometer*.

The distribution of water in living organisms is dependent to a large extent on osmosis, water entering the cells through their membranes. A cell membrane is not truly semipermeable as it allows the passage of certain solute molecules; it is described as *differentially permeable*. Animals have evolved various means to counteract the effects of osmosis (*see* osmoregulation); in plant cells, excessive osmosis is prevented by the pressure exerted by the cell wall, which opposes the osmotic pressure.

osmotic pressure *See* osmosis.

Ostwald's dilution law An expression for the degree of dissociation of a weak electrolyte. For example, if a weak acid dissociates in water

$$HA \rightleftharpoons H^+ + A^-$$

the dissociation constant K_a is given by

$$K_a = \alpha^2 n/(1 - \alpha)V$$

where α is the degree of dissociation, n the initial amount of substance (before dissociation), and V the volume. If α is small compared with 1, then $\alpha^2 = KV/n$; i.e. the degree of dissociation is proportional to the square root of the dilution. The law was first put forward by W. Ostwald (1853–1932) to account for electrical conductivities of electrolyte solutions.

Otto engine *See* internal-combustion engine.

ounce 1. One sixteenth of a pound (avoirdupois), equal to 0.028 349 kg. **2.** Eight drachms (Troy), equal to 0.031 103 kg. **3. (fluid ounce)** Eight fluid drachms, equal to 0.028 413 dm^3.

overdamped *See* damping.

overpotential A potential that must be applied in an electrolytic cell in addition to the theoretical potential required to liberate a given substance at an electrode. The value depends on the electrode material and on the current density. It occurs because of the significant activation energy for electron transfer at the electrodes, and is particularly important for the liberation of such gases as hydrogen and oxygen. For example, in the electrolysis of a solution of zinc ions, hydrogen ($E^{\ominus} = 0.00$ V) would be expected to be liberated at the cathode in preference to zinc ($E^{\ominus} = -0.76$ V). In fact, the high overpotential of hydrogen on zinc (about 1 V under suitable conditions) means that zinc can be deposited instead.

overtones *See* harmonic.

P

packing density 1. The number of devices (such as *logic circuits) or integrated circuits per unit area of a *silicon chip. **2.** The quantity of information stored in a specified space of a storage system associated with a computer, e.g. *bits per inch of magnetic tape.

packing fraction The algebraic difference between the relative atomic mass of an isotope and its mass number, divided by the mass number.

pair production The creation of an electron and a positron from a photon in a strong electric field, such as that surrounding an atomic nucleus. The electron and the positron each have a mass of about 9×10^{-31} kg, which is equivalent on the basis of the mass–energy equation ($E = mc^2$) to a total of 16×10^{-14} J. The frequency, ν, associated with a photon of this energy (according to $E = h\nu$) is 2.5×10^{20} Hz. As a photon of this energy is in the gamma-ray range, it is only a gamma-ray photon that can cause pair production. Any energy in excess of this minimum becomes kinetic energy of the products.

palaeomagnetism The study of magnetism in rocks, which provides information on variations in the direction and intensity of the earth's magnetic field with time. During the formation of an igneous or sedimentary rock containing magnetic minerals the polarity of the earth's magnetic field at that time becomes 'frozen' into the rock. Studies of this fossil magnetism in samples of rocks have enabled the former positions of magnetic poles at various geological times to be located. It has also revealed that periodic reversals in the geomagnetic field have taken place (i.e. the N-pole becomes the S-pole and vice versa). This information has been important in plate tectonics in establishing the movements of lithospheric plates over the earth's surface. The magnetic reversals provided crucial evidence for the sea-floor spreading hypothesis proposed in the early 1960s.

para- Prefix denoting the form of diatomic molecules in which the nuclei have opposite spins, e.g. parahydrogen. *Compare* ortho-.

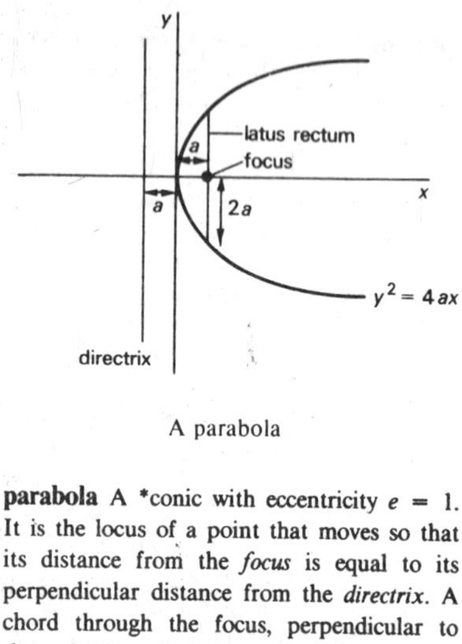

A parabola

parabola A *conic with eccentricity $e = 1$. It is the locus of a point that moves so that its distance from the *focus* is equal to its perpendicular distance from the *directrix*. A chord through the focus, perpendicular to the axis, is called the *latus rectum*. For a parabola with its vertex at the origin, lying symmetrically about the x-axis, the equation is $y^2 = 4ax$, where a is the distance from the vertex to the focus. The directrix is the line $x = -a$, and the latus rectum is $4a$.

parabolic reflector (paraboloidal reflector) A reflector having a section that is a parabola. A concave parabolic reflector will reflect a parallel beam of radiation through its focus and, conversely, will produce a parallel beam if the source of the radiation is placed at its focus. Parabolic mirrors are used in reflecting optical *telescopes to collect the light and in some light sources that require a parallel beam of light. In radio telescopes a dish aerial may also consist of a parabolic reflector.

paraboloid A solid formed by rotating a parabola about its axis of symmetry.

parallax 1. An apparent displacement of a distant object (with respect to a more distant background) when viewed from two different positions. If such an object is viewed from two points at either end of a base line, the angle between the lines joining the object to the ends of the base line is the *angle of parallax*. If the base line is the distance between the two eyes of an observer the angle is called the *binocular parallax*. **2.** The angular displacement in the apparent position of a celestial body when observed from two different points. *Diurnal parallax* results from the earth's daily rotation, the celestial body being viewed from the surface of the earth rather than from its centre. *Annual parallax* is caused by the earth's motion round the sun, the celestial body being viewed from the earth rather than from the centre of the sun. *Secular parallax* is caused by the motion of the solar system relative to the fixed stars.

parallel circuits A circuit in which the circuit elements are connected so that the current divides between them. For resistors in parallel, the total resistance, R, is given by $1/R = 1/r_1 + 1/r_2 + 1/r_3 \ldots$, where r_1, r_2, and r_3 are the resistances of the individual elements. For capacitors in parallel, the total capacitance, C, is given by $C = c_1 + c_2 + c_3 \ldots$.

parallelepiped (parallelopiped) A solid with six faces, all of which are parallelograms.

parallelogram of forces *See* parallelogram of vectors.

parallelogram of vectors A method of determining the *resultant of two *vector quantities. The two vector quantities are represented by two adjacent sides of a parallelogram and the resultant is then the diagonal through their point of intersection. The magnitude and direction of the resultant is found by scale drawing or by trigonometry. The method is used for such vectors as forces (*parallelogram of forces*) and velocities (*parallelogram of velocities*).

parallelogram of velocities *See* parallelogram of vectors.

paramagnetism *See* magnetism.

parametric equation An equation of a curve expressed in the form of the parameters that locate points on the curve. The parametric equations of a straight line are $x = a + bt$, $y = c + dt$. For a circle, they are $x = a\cos\theta$, $y = a\sin\theta$.

parasitic capture The absorption of a neutron by a nuclide that does not result in either fission or a useful artificial element.

paraxial ray A ray of light that falls on a reflecting or refracting surface close to and almost parallel to the axis. It is for such rays that simple lens theory can be developed, by means of making small angle approximations.

parent *See* daughter.

parity Symbol P. The property of a *wave function that determines the behaviour of the wave function when all its spatial coordinates are reversed in direction, i.e. when x,y,z are replaced by $-x,-y,-z$. If a wave function ψ satisfies the equation $\psi(x,y,z) = \psi(-x,-y,-z)$ it is said to have even parity, if it satisfies $\psi(x,y,z) = -\psi(-x,-y,-z)$ it has odd parity. In general, $\psi(x,y,z) = P\psi(-x,-y,-z)$, where P is a quantum number called parity that can have the value $+1$ or -1. The principle of *conservation of parity* (or *space-reflection symmetry*) would hold if all physical laws could be stated in a coordinate system independent of left- or right-handedness. If parity was conserved there would therefore be no fundamental way of distinguishing between left and right. In electromagnetic and strong interactions, parity is, in fact, conserved. In 1956, however, it was shown that parity is not conserved in weak interactions. In the beta decay of cobalt-60, for example, the electrons from the decay are emitted preferentially in a direction opposite to that of the cobalt spin. This experiment provides a fundamental distinction between left and right.

parsec A unit of length used to express astronomical distance. The distance at which

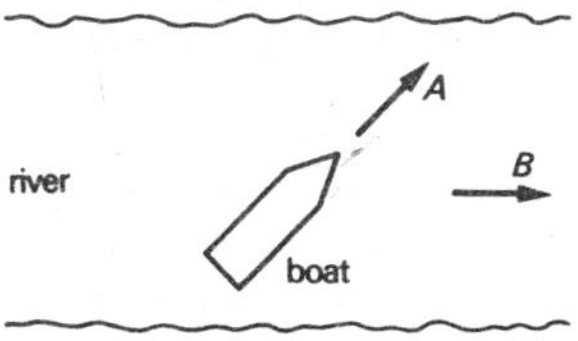

A is the velocity of the boat in the direction indicated. *B* is the velocity of the river

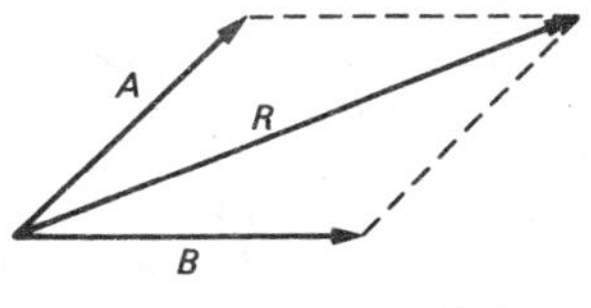

parallelogram of velocities. *R* is the resultant velocity of the boat in the direction indicated

Parallelogram of vectors

the mean radius of the earth's orbit subtends an angle of one second of arc. One parsec is equal to 3.0857×10^{16} metres or 3.2616 light years.

partial A simple component of a complex tone. When a musical instrument produces a note, say, middle C, it will produce a complex tone in which the fundamental frequency is mixed with a number of partials. Some of these partials, for example, if the note is produced by bowing a taut string, will be *harmonics, i.e. integral multiples of the fundamental. If the string is struck, however, some of the partials can be inexact multiples of the fundamental. Partials are not therefore identical with harmonics.

partial differential The infinitesimal change in a function consisting of two or more variables when one of the variable changes and the others remain constant. If $z = f(x,y)$, $\partial z/\partial x$ is the partial differential of z with respect to x, while y remains unchanged. A *partial differential equation*, such as *Laplace's equation, is an equation containing partial derivatives of a function.

partial eclipse *See* eclipse.

partial pressure *See* Dalton's law.

particle 1. (in physics) One of the fundamental components of matter. *See* elementary particles. **2.** (in mechanics) A hypothetical body that has mass but no physical extension. As it is regarded as having no volume, a particle is incapable of rotation and therefore can only have translational motion. Thus a real body may often, for translational purposes, be regarded as a particle located at the body's centre of mass and having a mass equal to that of the whole body.

particle physics The study of *elementary particles.

partition If a substance is in contact with two different phases then, in general, it will have a different affinity for each phase. Part of the substance will be absorbed or dissolved by one and part by the other, the relative amounts depending on the relative affinities. The substance is said to be *partitioned* between the two phases. For example, if two immiscible liquids are taken and a third compound is shaken up with them, then an equilibrium is reached in which the concentration in one solvent differs from that in the other. The ratio of the concentrations is the *partition coefficient* of the system. The *partition law* states that this ratio is a constant for given liquids.

partition coefficient *See* partition.

pascal The *SI unit of pressure equal to one newton per square metre.

Pascal's law In a confined fluid, externally applied pressure is transmitted uniformly in all directions. In a static fluid, force is transmitted at the speed of sound throughout the fluid and acts at right angles to any surface in or bounding the fluid. This principle is made use of in the hydraulic jack, the pneumatic tyre, and similar devices. The law was discovered in 1647 by Blaise Pascal (1623–62).

Paschen series *See* hydrogen spectrum.

passive device 1. An electronic component, such as a capacitor or resistor, that is incapable of *amplification. **2.** An artificial *satellite that reflects an incoming signal without amplification. **3.** A solar-power device that makes use of an existing structure to collect and utilize solar energy without the use of pumps, fans, etc. **4.** A radar device that provides information for navigation, guidance, surveillance, etc., by receiving the microwave radiation emitted by a warm body or reflected by a body from some other source. Such a passive device emits no microwave energy itself and therefore does not disclose its position. *Compare* active device.

Pauli exclusion principle The principle that no two particles in a system, such as electrons in an atom, can have an identical set of quantum numbers. This principle, which was first formulated by Wolfgang Pauli (1900–58) in 1925, is now known to apply to all *fermions but not to *bosons.

p.d. (potential difference) *See* electric potential.

pearlite *See* steel.

Peltier effect The change in temperature produced at a junction between two dissimilar metals or semiconductors when an electric current passes through the junction. The direction of the current determines whether the temperature rises or falls. The first metals to be investigated were bismuth and copper; if the current flows from bismuth to copper the temperature rises. If the current is reversed the temperature falls. The effect was discovered in 1834 by J. C. A. Peltier (1785–1845) and has been used recently for small-scale refrigeration. *Compare* Seebeck effect.

pendulum Any rigid body that swings about a fixed point. The *ideal simple pendulum* consists of a bob of small mass oscillating back and forth through a small angle at the end of a string or wire of negligible mass. Such a device has a period $2\pi\sqrt{(l/g)}$, where l is the length of the string or wire and g is the *acceleration of free fall. This type of pendulum moves with *simple harmonic motion.

The *compound pendulum* consists of a rigid body swinging about a point within it. The period of such a pendulum is given by $T = 2\pi\sqrt{[(h^2 + k^2)/hg]}$, where k is the radius of gyration about an axis through the centre of mass and h is the distance from the pivot to the centre of mass. *See also* Kater's pendulum.

pentode A *thermionic valve with a *suppressor grid* between the anode and the screen grid of a tetrode. Its purpose is to suppress the loss of electrons from the anode as a result of secondary emission. The suppressor grid is maintained at a negative potential relative to the anode and to the screen grid.

penumbra *See* shadow.

perfect gas *See* ideal gas; gas.

perfect pitch *See* absolute pitch.

perfect solution *See* Raoult's law.

pericynthion The point in the orbit around the moon of a satellite launched from the earth that is nearest to the moon. For a satellite launched from the moon the equivalent point is the *perilune*. *Compare* apocynthion.

perigee *See* apogee.

perihelion The point in the orbit of a planet, comet, or artificial satellite in solar orbit at which it is nearest to the sun. The earth is at perihelion on about 3 January. *Compare* aphelion.

period *See* periodic motion.

periodic law The principle that the physical and chemical properties of elements are a periodic function of their proton number. The concept was first proposed in 1869 by the Russian chemist Dimitri Mendeleev (1834–1907), using relative atomic mass rather than proton number, as a culmination of efforts to rationalize chemical properties by J. W. Döbereiner (1817), J. A. R. Newlands (1863), and Lothar Meyer (1864). One of the major successes of the periodic law was its ability to predict chemical and physical properties of undiscovered elements and unknown compounds that were later confirmed experimentally.

periodic motion Any motion of a system that is continuously and identically repeated. The time T that it takes to complete one cycle of an oscillation or wave motion is called the *period*, which is the reciprocal of the *frequency. *See* pendulum; simple harmonic motion.

peripheral device Any device, such as an input or output device, connected to the central processing unit of a *computer. Backing store is also usually regarded as a peripheral.

periscope An optical device that enables an observer to see over or around opaque objects. The simplest type consists of a long tube with mirrors at each end set at 45° to the direction to be viewed. A better type uses internally reflecting prisms instead of plane mirrors. Periscopes are used in tanks (to enable the observer to see over obstacles without being shot at) and in submarines (when the vessel is submerged). Such periscopes are usually quite complicated instruments and include telescopes.

Permalloys A group of alloys of high magnetic permeability consisting of iron and nickel (usually 40–80%) often with small amounts of other elements (e.g. 3–5% molybdenum, copper, chromium, or tungsten). They are used in thin foils in electronic transformers, for magnetic shielding, and in computer memories.

permanent gas A gas, such as oxygen or nitrogen, that was formerly thought to be impossible to liquefy. A permanent gas is now regarded as one that cannot be liquefied by pressure alone at normal temperatures (i.e. a gas that has a critical temperature below room temperature).

permanent magnet *See* magnet.

permeability (magnetic permeability) Symbol μ. The ratio of the magnetic flux density, B, in a substance to the external field strength, H; i.e. $\mu = B/H$. The permeability of free space, μ_0, is also called the *magnetic constant* and has the value $4\pi \times 10^{-7}\ \mathrm{H\,m^{-1}}$. The relative permeability of a substance, μ_r, is given by μ/μ_0 and is therefore dimensionless. *See* magnetism.

permittivity Symbol ε. The ratio of the *electric displacement in a medium to the intensity of the electric field producing it. It is important for electrical insulators used as *dielectrics.

If two charges Q_1 and Q_2 are separated by a distance r in a vacuum, the force F between the charges is given by:

$$F = Q_1 Q_2 / r^2 4\pi\varepsilon_0$$

In this statement of *Coulomb's law using SI units, ε_0 is called the absolute permittivity of free space, which is now known as the *electric constant*. It has the value $8.854 \times 10^{-12}\ \mathrm{F\,m^{-1}}$.

If the medium between the charges is anything other than a vacuum the equation becomes:

$$F = Q_1 Q_2 / r^2 4\pi\varepsilon$$

and the force between the charges is reduced. ε is the *absolute permittivity* of the new medium. The *relative permittivity* (ε_r) of a medium, formerly called the *dielectric constant*, is given by $\varepsilon_r = \varepsilon/\varepsilon_0$.

permutations and combinations A combination is any subset of a particular set of objects, regardless of the order of selection. If the set consists of n objects, r objects can be selected giving $n!/r!(n - r)!$ different combinations. This can be written ${}_nC_r$.

A permutation is an ordered subset (i.e. attention is paid to the order of selection or arrangement) of a particular set of objects. If the set consists of n objects, r objects can be selected in this way to give $n!/(n - r)!$ permutations. This is written ${}_nP_r$.

perpetual motion 1. Perpetual motion of the first kind. Motion in which a mechanism, once started, would continue indefinitely to perform useful work without being supplied with energy from an outside source. Such a device would contravene the first law of *thermodynamics and is therefore not feasible. Many historical attempts, exercising great ingenuity, were constructed before the concept of energy and its conservation were understood. Some attempts have been made, since the first law of thermodynamics became generally accepted, by inventors seeking to establish loopholes in the laws of nature. **2.** Perpetual motion of the second kind. Motion in which a mechanism extracts heat from a source and converts all of it into some other form of energy. An example of such a mechanism would be a ship that utilized the internal energy of the oceans for propulsion. Such a device does not contravene the first law of thermodynamics but it does contravene the second law. In the case of the ship, the sea would have to be at a higher temperature than the ship to establish a useful flow of heat. This could not occur without an external energy source. **3.** Perpetual motion of the third kind. A form of motion that continues indefinitely but without doing any useful work. An example is the random molecular motion in a substance. This type postulates the complete elimination of friction. A mechanism consisting of frictionless bearings maintained in a vacuum could turn indefinitely, once started, without contravening the first or second laws of thermodynamics, provided it did no external work. Experience indicates that on the macroscopic scale such a condition cannot be achieved. On the microscopic scale, however, a superconducting ring of wire will apparently sustain a perpetual current flow without the application of an external force. This could be considered a form of perpetual motion of the third kind, if the energy required to cool the wire to superconducting temperatures is ignored.

perturbation A departure by a celestial body from the trajectory or orbit it would follow if it moved only under the influence of a single central force. According to *Kepler's law, for example, a single planet orbiting the sun would move in an elliptical orbit. In fact, planets are perturbed from elliptical orbits by the gravitational forces exerted on them by other planets. Similarly, the moon's orbit round the earth is perturbed by the gravitational effect of the sun and the trajectories of comets are perturbed when they pass close to planets.

perversion *See* lateral inversion.

peta- Symbol P. A prefix used in the metric system to denote one thousand million million times. For example, 10^{15} metres = 1 petametre (Pm).

pewter An alloy of lead and tin. It usually contains 63% tin; pewter tankards and food containers should have less than 35% of lead so that the lead remains in solid solution with the tin in the presence of weak acids in the food and drinks. Copper is sometimes added to increase ductility and antimony is added if a hard alloy is required.

pH *See* pH scale.

phase 1. A homogeneous part of a heterogeneous system that is separated from other parts by a distinguishable boundary. A mixture of ice and water is a two-phase system. A solution of salt in water is a single-phase system. **2.** A description of the stage that a periodic motion has reached, usually by comparison with another such motion of the same frequency. Two varying quantities are said to be *in phase* if their maximum and minimum values occur at the same instants; otherwise, there is said to be a *phase difference*. *See also* phase angle. **3.** One of the circuits in an electrical system or device in which there are two or more alternating currents that are not in phase with each other. In a three-phase system the displacement between the currents is one third of a period. **4.** *See* phases of the moon.

phase angle The difference in *phase between two sinusoidally varying quantities. The displacement x_1 of one quantity at time t is given by $x_1 = a\sin\omega t$, where ω is the angular frequency and a is the amplitude. The displacement x_2 of a similar wave that reaches the end of its period T, a fraction β of the period before the first is said to lead the first quantity by a time βT. The value of x_2 is then given by $x_2 = a\sin(\omega t + \phi)$. ϕ is called the phase angle and it is equal to $2\pi\beta$.

phase-contrast microscope A type of *microscope that is widely used for examining such specimens as biological cells and tissues. It makes visible the changes in phase that occur when nonuniformly transparent specimens are illuminated. In passing through an object the light is slowed down and becomes out of phase with the original light. With transparent specimens having some structure *diffraction occurs, causing a larger phase change in light outside the central maximum of the pattern. The phase-contrast microscope provides a means of combining this light with that of the central maximum by means of an annular diaphram and a *phase-contrast plate*, which produces a matching phase change in the light of the central maximum only. This gives greater contrast to the final image, due to constructive interference between the two sets of light waves. This is *bright contrast*; in *dark contrast* a different phase-contrast plate is used to make the same structure appear dark, by destructive interference of the same waves.

phase diagram A graph showing the relationship between solid, liquid, and gaseous *phases over a range of conditions (e.g. temperature and pressure).

phase modulation *See* modulation.

phase rule For any system at equilibrium, the relationship $P + F = C + 2$ holds, where P is the number of distinct phases, C the number of components, and F the number of degrees of freedom of the system.

phases of the moon The shapes of the illuminated surface of the moon as seen from the earth. The shape changes as a result of the relative positions of the earth, sun, and moon.

New moon occurs when the nearside is totally unilluminated by the sun. As the moon moves eastwards in its orbit the sunrise *terminator crosses the nearside from east to west producing a *crescent moon*. The moon is half illuminated at *first quarter*. When it is more than half-phase but less than full phase it is said to be a *gibbous moon*. When the moon is at *opposition the nearside is fully illuminated producing a *full moon*. The sunset terminator then follows to produce a waning gibbous moon, *last quarter*, a waning crescent moon, and eventually the next new moon.

phase space *See* statistical mechanics.

phase speed (phase velocity) Symbol V_p. The speed of propagation of a pure sine wave. $V_p = \lambda f$, where λ is the wavelength and f is the frequency. The value of the phase speed depends on the nature of the medium through which it is travelling and may also depend on the mode of propagation. For electromagnetic waves travelling through space the phase speed c is given by $c^2 = 1/\varepsilon_0\mu_0$, where ε_0 and μ_0 are the *electric constant and the *magnetic constant respectively.

phasor A rotating *vector that represents a sinusoidally varying quantity. Its length represents the amplitude of the quantity and it is imagined to rotate with angular velocity equal to the angular frequency of the quantity, so that the instantaneous value of the quantity is represented by its projection upon a fixed axis. The concept is convenient for representing the *phase angle between two quantities; it is shown on a diagram as the angle between their phasors.

phlogiston theory A former theory of combustion in which all flammable objects were supposed to contain a substance called *phlogiston*, which was released when the object burned. The existence of this hypothetical substance was proposed in 1669 by Johann Becher, who called it 'combustible earth' (*terra pinguis*: literally 'fat earth'). For example, according to Becher, the conversion of wood to ashes by burning was explained on the assumption that the original wood consisted of ash and *terra pinguis*, which was released on burning. In the early 18th century Georg Stahl renamed the substance *phlogiston* (from the Greek for 'burned') and extended the theory to include the calcination (and corrosion) of metals. Thus, metals were thought to be composed of *calx* (a powdery residue) and phlogiston; when a metal was heated, phlogiston was set free and the calx remained. The process could be reversed by heating the metal over charcoal (a substance believed to be rich in phlogiston, because combustion almost totally consumed it). The calx would absorb the phlogiston released by the burning charcoal and become metallic again.

The theory was finally demolished by Antoine Lavoisier, who showed by careful experiments with reactions in closed containers that there was no *absolute* gain in mass – the gain in mass of the substance was matched by a corresponding loss in mass of the air used in combustion. After experiments with Priestley's dephlogisticated air, Lavoisier realized that this gas, which he named oxygen, was taken up to form a calx (now called an oxide). The role of oxygen in the new theory was almost exactly the opposite of phlogiston's role in the old. In combustion and corrosion phlogiston was released; in the modern theory, oxygen is taken up to form an oxide.

phon A unit of loudness of sound that measures the intensity of a sound relative to a reference tone of defined intensity and frequency. The reference tone usually used has a frequency of 1 kilohertz and a root-mean-square sound pressure of 2×10^{-5} pascal. The observer listens with both ears to the reference tone and the sound to be measured alternately. The reference tone is then increased until the observer judges it to be of equal intensity to the sound to be measured. If the intensity of the reference tone has been increased by n *decibels to achieve this, the sound being measured is said to have an intensity of n phons. The decibel and phon scales are not identical as the phon scale is subjective and relies on the sensitivity of the ear to detect changes of intensity with frequency.

phonon A quantum of sound-wave energy having a magnitude hf, where h is the *Planck constant and f is the frequency of the sound wave. Analogous to the light quantum, or *photon, it is useful in the treatment of the scattering of electrons and other particles by thermal waves (short sound waves) in matter.

phosphor bronze An alloy of copper containing 4% to 10% of tin and 0.05% to 1% of phosphorus as a deoxidizing agent. It is used particularly for marine purposes and where it is exposed to heavy wear, as in gear wheels. *See also* bronze.

phosphorescence *See* luminescence.

phot A unit of illuminance equal to 10^4 lux or one lumen per square centimetre.

photocathode A *cathode that emits electrons when light falls upon it, as a result of the *photoelectric effect. *See* photoelectric cell.

photocell *See* photoelectric cell.

photochemical reaction A chemical reaction caused by light or ultraviolet radiation. The incident photons are absorbed by reactant molecules to give excited molecules or free radicals, which undergo further reaction.

photochemistry The branch of chemistry concerned with *photochemical reactions.

photochromism A change of colour occurring in certain substances when exposed to light. Photochromic materials are used in sunglasses that darken in bright sunlight.

photoconductive effect *See* photoelectric effect.

photodiode A *semiconductor diode used to detect the presence of light or to measure its intensity. It usually consists of a *p–n* junction device in a container that focuses any light in the environment close to the junction. The device is usually biased in reverse so that in the dark the current is small; when it is illuminated the current is proportional to the amount of light falling on it. *See* photoelectric effect.

photodisintegration The decay of a nuclide as a result of the absorption of a gamma-ray photon.

photoelasticity An effect in which certain materials exhibit double refraction when subjected to stress. It is used in a technique for detecting strains in transparent materials (e.g. Perspex, celluloid, and glass). When polarized white light is passed through a stressed sample, the birefringence causes coloured patterns to be seen on the viewing screen of a suitable *polariscope. If monochromatic polarized light is used, a complex pattern of light and dark fringes is produced.

photoelectric cell (photocell) Any of several devices that produce an electric signal in response to exposure to electromagnetic radiation. The original photocells utilized photoemission from a photosensitive cathode (*photocathode*). The electrons emitted are attracted to an anode. A positive potential on the anode enables a current to flow through an external circuit, the current being proportional to the intensity of the illumination on the cathode. The electrodes are enclosed in an evacuated glass tube (*see also* photomultiplier).

More modern light-sensitive devices utilize the photoconductive effect and the photovoltaic effect (*see* photoelectric effect; photodiode; phototransistor; solar cell).

photoelectric effect The liberation of electrons (*see* photoelectron) from a metal surface exposed to electromagnetic radiation. The number of electrons emitted depends on the intensity of the radiation. The kinetic energy of the electrons emitted depends on the frequency of the radiation. The effect is a quantum process in which the radiation is regarded as a stream of *photons, each having an energy hf, where h is the Planck constant and f is the frequency of the radiation. A photon can only eject an electron if the photon energy exceeds the *work function, ϕ, of the solid, i.e. if $hf_0 = \phi$ an electron will be ejected; f_0 is the minimum frequency (or *threshold frequency*) at which ejection will occur. For many metals the photoelectric effect occurs at ultraviolet frequencies or above, but for some materials (having low work functions) it occurs with light. The maximum kinetic energy, E_m, of the photoelectron is given by *Einstein's equation: $E_m = hf - \phi$ (*see also* photoionization).

Apart from the liberation of electrons from metals other phenomena are also referred to as photoelectric effects. These are the *photoconductive effect* and the *photovoltaic effect*. In the photoconductive effect, an increase in the electrical conductivity of a semiconductor is caused by radiation as a result of the excitation of additional free charge carriers by the incident photons. *Photoconductive cells*, using such photosensitive materials as cadmium sulphide, are widely used as radiation detectors and light switches (e.g. to switch on street lighting).

In the photovoltaic effect, an e.m.f. is produced between two layers of different materials as a result of irradiation. The effect is made use of in *photovoltaic cells*, most of which consist of $p-n$ *semiconductor junctions (*see also* photodiode; phototransistor). When photons are absorbed near a $p-n$ junction new free charge carriers are produced (as in photoconductivity); however, in the photovoltaic effect the electric field in the junction region causes the new charge carriers to move, creating a flow of current in an external circuit without the need for a battery. *See also* photoelectric cell.

photoelectron An electron emitted from a substance by irradiation as a result of the *photoelectric effect or *photoionization.

photoelectron spectroscopy A technique for determining the *ionization potentials of molecules. The sample is a gas or vapour irradiated with a narrow beam of ultraviolet radiation (usually from a helium source at 58.4 nm, 21.21 eV photon energy). The photoelectrons produced in accordance with *Einstein's equation are passed through a slit into a vacuum region, where they are deflected by magnetic or electrostatic fields to give an energy spectrum. The photoelectron spectrum obtained has peaks corresponding to the ionization potentials of the molecule (and hence the orbital energies). The technique also gives information on the vibrational energy levels of the ions formed. *ESCA* (electron spectroscopy for chemical analysis) is a similar analytical technique in which a beam of X-rays is used. In this case, the electrons ejected are from the inner shells of the atoms. Peaks in the electron spectrum for a particular element show characteristic chemical shifts, which depend on the presence of other atoms in the molecule.

photoemission The process in which electrons are emitted by a substance as a result of irradiation. *See* photoelectric effect; photoionization.

photofission A *nuclear fission that is caused by a gamma-ray photon.

photographic density A measure of the opacity of a photographic emulsion (negative or transparency). *See* densitometer.

photography The process of forming a permanent record of an image on specially treated film or paper. In normal black-and-white photography a camera is used to expose a film or plate to a focused image of the scene for a specified time. The film or plate is coated with an emulsion containing silver salts and the exposure to light causes the silver salts to break down into silver atoms; where the light is bright dark areas of silver are formed on the film after development (by a mild reducing agent) and fixing. The negative so formed is printed, either by a contact process or by projection. In either case light passing through the negative film falls on a sheet of paper also coated with emulsion. Where the negative is dark, less light passes through and the resulting positive is light in this area, corresponding with a light area in the original scene. As photographic emulsions are sensitive to ultraviolet and X-rays, they are widely used in studies involving these forms of electromagnetic radiation. *See also* colour photography.

photoionization The *ionization of an atom or molecule as a result of irradiation by electromagnetic radiation. For a photoionization to occur the incident photon of the radiation must have an energy in excess of the *ionization potential of the species being irradiated. The ejected photoelectron will have an energy, E, given by $E = hf - I$, where h is the Planck constant, f is the frequency of the incident radiation, and I is the ionization potential of the irradiated species.

photolithography A technique used in the manufacture of semiconductor components, integrated circuits, etc. It depends on the principle of masking selected areas of a surface and exposing the unmasked areas to such processes as the introduction of impurities, deposition of thin films, removal of material by etching, etc. The technique has been developed for use on tiny structures (typically measured in micrometres), which can only be examined by means of an electron microscope.

photoluminescence *See* luminescence.

photolysis Chemical reaction produced by exposure to light or ultraviolet radiation. Photolytic reactions often involve free radicals, the first step being homolytic fission of a chemical bond.

photometer An instrument used to measure *luminous intensity, illumination, and other photometric quantities. The older types rely on visual techniques to compare a source of light with a standard source. More modern photometers use *photoelectric cells, of the photoconductive, photoemissive, or photovoltaic typess. The photovoltaic types do not require an external power source and are therefore very convenient to use but are relatively insensitive. The photoemissive type usually incorporates a *photomultiplier, especially for use in astronomy and with other weak sources. Photoconductive units require only low-voltage supplies, which makes them convenient for commercial illumination meters and photographers' exposure meters.

photometric brightness *See* luminance.

photometry The study of visual radiation, especially the calculations and measurements of *luminous intensity, *luminous flux, etc. In some cases photometric calculations and measurements extend into the near infrared and the near ultraviolet.

In photometry, two types of measurement are used: those that measure *luminous* quantities rely on the use of the human eye (for example, to compare the illuminance of two surfaces); those called *radiant* quantities rely on the use of photoelectric devices to measure electromagnetic energy. *See also* photometer.

photomicrography The use of photography to obtain a permanent record (a *photomicrograph*) of the image of an object as viewed through a microscope.

photomultiplier A sensitive type of *photoelectric cell in which electrons emitted from a photocathode are accelerated to a second electrode where several electrons are liberated by each original photoelectron, as a result of *secondary emission. The whole process is repeated as many times as necessary to produce a useful electric current by secondary emission from the last electrode. A photomultiplier is thus a photocathode with the output amplified by an electron multiplier. The initial photocurrent can be amplified by a factor of 10^8. Photomultipliers are thus useful when it is necessary to detect low intensities of light, as in stellar photometry, star and planet tracking in guidance systems, and more mundanely in process control.

photon A particle with zero *rest mass consisting of a *quantum of electromagnetic radiation. The photon may also be regarded as a unit of energy equal to hf, where h is the *Planck constant and f is the frequency of the radiation in hertz. Photons travel at the speed of light. They are required to explain the photoelectric effect and other phenomena that require *light to have particle character.

photoneutron A neutron emitted by an atomic nucleus undergoing a *photonuclear reaction.

photonuclear reaction A *nuclear reaction that is initiated by a (gamma-ray) photon.

photopic vision The type of vision that occurs when the cones in the eye are the principal receptors, i.e. when the level of illumination is high. Colours can be identified with photopic vision. *Compare* scotopic vision.

photoreceptor A sensory cell or group of cells that reacts to the presence of light. It usually contains a pigment that undergoes a chemical change when light is absorbed, thus stimulating a nerve. *See* eye.

photosensitive substance **1.** Any substance that when exposed to electromagnetic radiation produces a photoconductive, photoelectric, or photovoltaic effect. **2.** Any substance, such as the emulsion of a photographic film, in which electromagnetic radiation produces a chemical change.

photosphere The visible surface of the *sun or other star and the source of its absorption spectrum. It is a gaseous layer several hundreds of kilometres thick with an average temperature of 5780 K. Where the photosphere merges with the *chromosphere the temperature is 4000 K.

phototransistor A junction *transistor that is photosensitive. When radiation falls on the emitter-base junction, new free charge carriers are created in the base region and the collector current is increased. Phototransistors are similar to *photodiodes except that the primary photoelectric current is amplified internally and it is therefore more sensitive to light than the photodiode. Some types can be used as switching or bistable devices, a small intensity of radiation switching them from a low to high current state.

photovoltaic effect *See* photoelectric effect.

pH scale A logarithmic scale for expressing the acidity or alkalinity of a solution. To a first approximation, the pH of a solution can be defined as $-\log_{10}c$, where c is the concentration of hydrogen ions in moles per cubic decimetre. A neutral solution at 25°C has a hydrogen-ion concentration of 10^{-7} mol dm^{-3}, so the pH is 7. A pH below 7 indicates an acid solution; one above 7 indicates an alkaline solution. More accurately, the pH depends not on the concentration of hydrogen ions but on their *activity, which cannot be measured experimentally. For practical purposes, the pH scale is defined by using a hydrogen electrode in the solution of interest as one half of a cell, with a reference electrode (e.g. a calomel electrode) as the other half cell. The pH is then given by

$$(E - E_R)F/2.303RT,$$

where E is the e.m.f. of the cell and E_R the standard electrode potential of the reference electrode. In practice, a glass electrode is more convenient than a hydrogen electrode.

pH stands for 'potential of hydrogen'. The scale was introduced by S. P. Sørensen in 1909.

physical chemistry The branch of chemistry concerned with the effect of chemical structure on physical properties. It includes chemical thermodynamics and electrochemistry.

physics The study of the laws that determine the structure of the universe with reference to the matter and energy of which it consists. It is concerned not with chemical changes that occur but with the forces that exist between objects and the interrelationship between matter and energy. Traditionally, the study was divided into separate fields: heat, light, sound, electricity and magnetism, and mechanics. Since the turn of the century, however, quantum mechanics and relativistic physics have become increasingly important; the growth of modern physics has been accompanied by the studies of atomic physics, nuclear physics, and particle physics. The physics of astronomical bodies and their interactions is known as *astrophysics*, the physics of the earth is known as *geophysics*, and the study of the physical aspects of biology is called *biophysics*.

physisorption *See* adsorption.

pi Symbol π. The ratio of the circumference of any circle to its diameter. It is a *transcendental number with the value 3.141 592 . . . .

pico- Symbol p. A prefix used in the metric system to denote 10^{-12}. For example, 10^{-12} farad = 1 picofarad (pF).

pie chart A diagram in which percentages are shown as sectors of a circle. If x percent of the electorate vote for party X, y percent for party Y, and z percent for party Z, a pie chart would show three sectors having central angles $3.6x°$, $3.6y°$, and $3.6z°$.

pi electron An electron in a pi orbital. *See* orbital.

piezoelectric effect The generation of a potential difference across the opposite faces of certain nonconducting crystals (*piezoelectric crystals*) as a result of the application of mechanical stress between these faces. The *electric polarization produced is proportional to the stress and the direction of the polarization reverses if the stress changes from compression to tension. The *reverse piezoelectric effect* is the opposite phenomenon: if the opposite faces of a piezoelectric crystal are subjected to a potential difference, the crystal changes its shape. Rochelle salt and quartz are the most frequently used piezoelectric materials. While Rochelle salt produces the greater polarization for a given stress, quartz is more widely used as its crystals have greater strength and are stable at temperatures in excess of 100°C.

If a quartz plate is subjected to an alternating electric field, the reverse piezoelectric effect causes it to expand and contract at the field frequency. If this field frequency is made to coincide with the natural elastic frequency of the crystal, the plate resonates; the direct piezoelectric effect then augments the applied electric field. This is the basis of the *crystal oscillator and the *quartz clock. *See also* crystal microphone; crystal pickup.

pig iron The impure form of iron produced by a blast furnace, which is cast into pigs (blocks) for converting at a later date into cast iron, steel, etc. The composition depends on the ores used, the smelting procedure, and the use to which the pigs will later be put.

pi-meson *See* pion.

pinch effect A magnetic attraction between parallel conductors carrying currents flowing in the same direction. The force was noticed in early induction furnaces. Since the late 1940s it has been widely studied as a means of confining the hot plasma in a *thermonuclear reactor. In an experimental toroidal thermonuclear reactor a large electric current is induced in the plasma by electromagnetic induction; this current both heats the plasma and draws it away from the walls of the tube as a result of the pinch effect.

pion (pi-meson) An *elementary particle classified as a *meson. It exists in three forms: neutral, positively charged, and negatively charged. The charged pions decay into muons and electrons; the neutral pion converts directly into gamma-ray photons.

Pirani gauge An instrument used to measure low pressures ($1-10^{-4}$ torr; 100–0.01 Pa). It consists of an electrically heated filament, which is exposed to the gas whose pressure is to be measured. The extent to which heat is conducted away from the filament depends on the gas pressure, which thus controls its equilibrium temperature. Since the resistance of the filament is dependent on its temperature, the pressure is related to the resistance of the filament. The filament is arranged to be part of a *Wheatstone bridge circuit and the pressure is read from a microammeter calibrated in pressure units. As the effect depends on the thermal conductivity of the gas, the calibration has to be made each time the pressure of a different gas is measured.

pitch 1. (in physics) The property of a sound that characterizes its highness or lowness to an observer. It is related to, but not identical with, frequency. Below about 1000 Hz the pitch is slightly higher than the frequency and above 1000 the position is reversed. The loudness of a sound also affects the pitch. Up to 1000 Hz an increase in loudness causes a decrease in pitch. From about 1000 to 3000 Hz the pitch is independent of loudness, while above 3000 Hz an increase in loudness seems to cause a raising of pitch. Pitch is usually measured in mels; a note of 1000 Hz frequency with a loudness of 40 decibels above the absolute threshold of hearing has a pitch of 1000 mels. **2.** (in mechanics) *See* screw.

Pitot tube A device for measuring the speed of a fluid. It consists of two tubes, one with an opening facing the moving fluid and the other with an opening at 90° to the direction of the flow. The two tubes are connected to the opposite sides of a manometer so that the difference between the dynamic pressure in the first tube and the static pressure in the second tube can be measured. The speed v of the flow of an incompressible fluid is then given by: $v^2 = 2(P_2 - P_1)/\rho$, where P_2 is the dynamic pressure, P_1 is the static pressure, and ρ is the density of the fluid. The device has a wide variety of applications. It was devised by Henri Pitot (1695–1771).

Planck constant Symbol h. The fundamental constant equal to the ratio of the energy of a quantum of energy to its frequency. It has the value $6.626\,196 \times 10^{-34}$ J s. It is named after Max Planck (1858–1947).

Planck's radiation law A law giving the distribution of energy radiated by a *black body. It introduced into physics the novel concept of energy as a quantity that is radiated by a body in small discrete packets rather than as a continuous emission. These small packets became known as quanta and the law formed the basis of *quantum theory. The *Planck formula* giving the energy M radiated in unit time by unit area of a surface at thermodynamic temperature T for wavelength λ is given by:

$$M = C_1\lambda^{-5}/(\exp C_2\lambda T - 1),$$

where $C_1 = 2\pi hc^2$ and $C_2 = hc/k$. In the constants C_1 and C_2, h is the *Planck constant, c is the speed of light, and k is the Boltzmann constant.

plane A flat surface defined by the condition that any two points in the plane are joined by a straight line that lies entirely in the surface.

plane-polarized light *See* polarization of light.

planet A body that revolves around a central astronomical body, especially one of the nine bodies (Mercury; Venus; earth; Mars;

Jupiter; Saturn; Uranus; Neptune; Pluto) that revolve in elliptical orbits around the sun. *See* solar system.

planimeter An instrument used to measure the area of a closed curve. The outline of the curve is followed by a pointer on the instrument and the area is given on a graduated disc.

plano-concave lens *See* concave; lens.

plano-convex lens *See* convex; lens.

plasma A highly ionized gas in which the number of free electrons is approximately equal to the number of positive ions. Sometimes described as the fourth state of matter, plasmas occur in interstellar space, in the atmospheres of stars (including the sun), in discharge tubes, and in experimental thermonuclear reactors.

Because the particles in a plasma are charged, its behaviour differs in some respects from that of a gas. Plasmas can be created in the laboratory by heating a low-pressure gas until the mean kinetic energy of the gas particles is comparable to the *ionization potential of the gas atoms or molecules. At very high temperatures, from about 50 000 K upwards, collisions between gas particles cause cascading ionization of the gas. However, in some cases, such as a fluorescent lamp, the temperature remains quite low as the plasma particles are continually colliding with the walls of the container causing cooling and recombination. In such cases ionization is only partial and requires a large energy input. In *thermonuclear reactors an enormous plasma temperature is maintained by confining the plasma away from the container walls using electromagnetic fields (*see* pinch effect). The study of plasmas is known as *plasma physics*.

plasticity The property of solids that causes them to change permanently in size or shape as a result of the application of a stress in excess of a certain value, called the *yield point*.

pleochroic Denoting a crystal that appears to be of different colours, depending on the direction from which it is viewed. It is caused by polarization of light as it passes through an anisotropic medium.

plutonium Symbol Pu. A dense silvery radioactive metallic transuranic element belonging to the actinoids; a.n. 94; mass number of most stable isotope 244 (half-life 7.6×10^7 years); r.d. 19.84; m.p. 641°C; b.p. 3232°C. Thirteen isotopes are known, by far the most important being plutonium–239 (half-life 2.44×10^4 years), which undergoes *nuclear fission with slow neutrons and is therefore a vital power source for *nuclear weapons and some *nuclear reactors. About 20 tonnes of plutonium are produced annually by the world's nuclear reactors, a detailed inventory of every gram of which is kept in order to prevent its military misuse. The element was first produced by Seaborg, McMillan, Kennedy, and Wahl in 1940.

point-contact transistor *See* transistor.

point defect *See* defect.

point discharge *See* corona.

poise A *c.g.s. unit of viscosity equal to the tangential force in dynes per square centimetre required to maintain a difference in velocity of one centimetre per second between two parallel planes of a fluid separated by one centimetre. 1 poise is equal to 10^{-1} N s m^{-2}.

Poiseuille's equation An equation relating the volume flow rate, V, of a fluid through a cylindrical tube to the pressure difference, p, between the ends of the tube: $V = \pi pr^4/8l\eta$, where r is the radius and l the length of the tube; η is the viscosity of the fluid. It applies if the Reynolds number is less than 2000 and was first stated by Jean Louis Poiseuille (1799–1869).

poison A substance that absorbs neutrons in a nuclear reactor and therefore slows down the reaction. It may be added intentionally for this purpose or may be formed as a fission product and need to be periodically removed.

Poisson's ratio The ratio of the lateral strain to the longitudinal strain in a stretched rod. If the original diameter of the rod is d and the contraction of the diameter under stress is Δd, the lateral strain $\Delta d/d = s_d$; if the original length is l and the extension under stress Δl, the longitudinal strain is $\Delta l/l = s_l$. Poisson's ratio is then s_d/s_l. For steels the value is between 0.28 and 0.30 and for aluminium alloys it is about 0.33. It was first introduced by Simeon Poisson (1781–1840).

polar coordinates A system used in analytical geometry to locate a point P, with reference to two or three axes. The distance of P from the origin is r, and the angle between the x-axis and the *radius vector OP* is θ, thus in two-dimensional polar coordinates the coordinates of P are (r,θ). If the *Cartesian coordinates of P are (x,y) then, $x = r\cos\theta$ and $y = r\sin\theta$.

In three dimensions the point P may be regarded as lying on the surface of a cylinder, giving *cylindrical polar coordinates*, or on the surface of a sphere, giving *spherical polar coordinates*. In the former the coordinates of P would be (r,θ,z); in the latter they would be (r,θ,ϕ) (see illustration).

polariscope (polarimeter) A device used to study optically active substances (*see* optical activity). The simplest type of instrument consists of a light source, collimator, polarizer, and analyser. The specimen is placed between polarizer and analyser, so that any rotation of the plane of polarization of the light can be assessed by turning the analyser. *See also* photoelasticity.

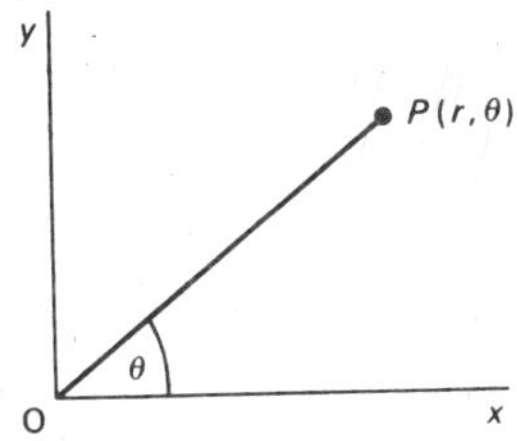

two-dimensional coordinates

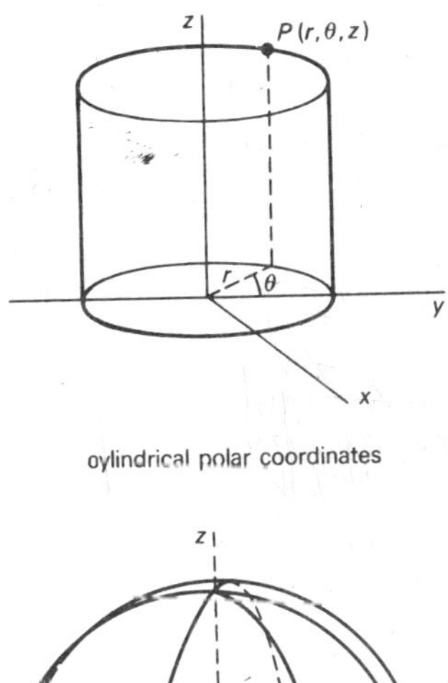

cylindrical polar coordinates

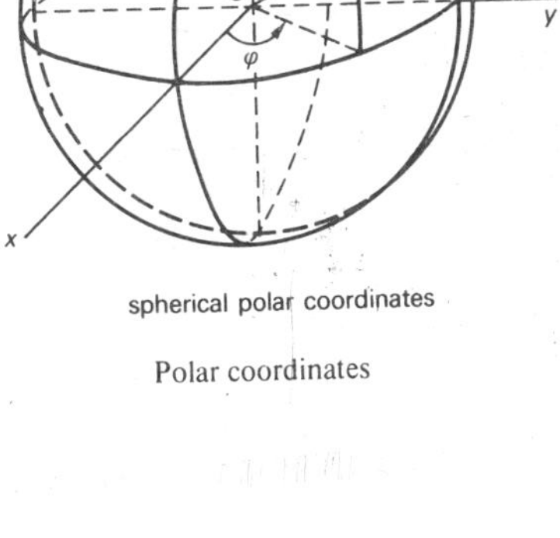

spherical polar coordinates

Polar coordinates

polarization **1.** The process of confining the vibrations of the vector constituting a transverse wave to one direction. In unpolarized radiation the vector oscillates in all directions perpendicular to the direction of propagation. *See* polarization of light. **2.** The formation of products of the chemical reaction in a *voltaic cell in the vicinity of the electrodes resulting in increased resistance to current flow and, frequently, to a reduction in the e.m.f. of the cell. *See also* depolarization. **3.** The partial separation of electric charges in an insulator subjected to an electric field.

polarization of light The process of confining the vibrations of the electric vector of light waves to one direction. In unpolarized light the electric field vibrates in all directions perpendicular to the direction of propagation. After reflection or transmission through certain substances (*see* Polaroid) the electric field is confined to one direction and the radiation is said to be *plane-polarized light*. The plane of plane-polarized light can be rotated when it passes through certain substances (*see* optical activity).

In *circularly polarized light*, the tip of the electric vector describes a circular helix about the direction of propagation with a frequency equal to the frequency of the light. The magnitude of the vector remains constant. In *elliptically polarized light*, the vector also rotates about the direction of propagation but the amplitude changes; a projection of the vector on a plane at right angles to the direction of propagation describes an ellipse. Circularly and elliptically polarized light are produced using a *retardation plate.

polarizing angle *See* Brewster's law.

polar molecule A molecule that has a dipole moment; i.e. one in which there is some separation of charge in the chemical bonds, so that one part of the molecule has a positive charge and the other a negative charge.

Polaroid A doubly refracting material that plane-polarizes unpolarized light passed through it. It consists of a plastic sheet strained in a manner that makes it birefringent by aligning its molecules. Sunglasses incorporating a Polaroid material absorb light that is vibrating horizontally – produced by reflection from horizontal surfaces – and thus reduce glare.

pole **1.** *See* magnetic poles; magnetic monopole. **2.** The *optical centre of a curved mirror.

polonium Symbol Po. A rare radioactive metallic element; a.n. 84; r.a.m. 210; r.d. 9.32; m.p. 254°C; b.p. 962°C. The element occurs in uranium ores to an extent of about 100 micrograms per 1000 kilograms. It has 27 isotopes, more than any other element. The longest-lived isotope is polonium–209 (half-life 103 years). Polonium has attracted attention as a possible heat source for spacecraft as the energy released as it decays is 1.4×10^5 J kg^{-1} s^{-1}. It was discovered by Marie Curie in 1898 in a sample of pitchblende.

polychromatic radiation Electromagnetic radiation that consists of a mixture of different wavelengths. This need not refer only to visible radiation. *Compare* monochromatic radiation.

polygon A plane figure with a number of sides. In a *regular polygon* all the sides and internal angles are equal. For a regular polygon with n sides, the interior angle is $(180 - 360/n)$ degrees and the sum of the interior angles is $(180n - 360)$ degrees.

polygon of forces A polygon in which the sides represent, in magnitude and direction, all forces acting on a rigid body. If such a polygon can be constructed, taking the forces in order, the body will be in equilibrium.

polyhedron A solid bounded by polygonal faces. In a *regular polyhedron* all the faces are congruent regular polygons. The cube is one of five possible regular polyhedrons. The others are the *tetrahedron* (four triangular faces), the *octahedron* (eight triangular faces), the *dodecahedron* (twelve pentagonal faces), and the *icosahedron* (twenty triangular faces).

polynomial A mathematical expression containing three or more terms. It has the general form $a_0x^n + a_1x^{n-1} + a_2x^{n-2} + \ldots a_n$, where a_0, a_1, etc., are constants and n is the highest power of the variable, called the *degree* of the polynomial.

population inversion *See* laser.

population type A method of classifying stars as either population I or population II bodies, devised in 1944 by Wilhelm Baade (1893–1960). Population I stars are the young metal-rich highly luminous stars found in the spiral arms of galaxies. Population II stars are older metal-deficient stars that occur in the centres of galaxies.

positive charge *See* charge.

positive feedback *See* feedback.

positron The antiparticle of the *electron. *See also* annihilation; elementary particles; pair production.

potassium–argon dating A *dating technique for certain rocks that depends on the decay of the radioisotope potassium–40 to argon–40, a process with a half-life of about 1.27×10^{10} years. It assumes that all the argon–40 formed in the potassium-bearing mineral accumulates within it and that all the argon present is formed by the decay of potassium–40. The mass of argon–40 and potassium–40 in the sample is estimated and the sample is then dated from the equation:

$$^{40}\text{Ar} = 0.1102\ ^{40}\text{K}(e^{\lambda t} - 1),$$

where λ is the decay constant and t is the time in years since the mineral cooled to about 300°C, when the ^{40}Ar became trapped in the crystal lattice. The method is effective for micas, feldspar, and some other minerals.

potential barrier A region containing a maximum of potential that prevents a particle on one side of it from passing to the other side. According to classical theory a particle must possess energy in excess of the height of the potential barrier to pass it. However, in quantum theory there is a finite probability that a particle with less energy will pass through the barrier (*see* tunnel effect). A potential barrier surrounds the atomic nucleus and is important in nuclear physics; a similar but much lower barrier exists at the interface between *semiconductors and metals and between differently doped semiconductors. These barriers are important in the design of electronic devices.

potential difference *See* electric potential.

potential divider *See* voltage divider.

potential energy *See* energy.

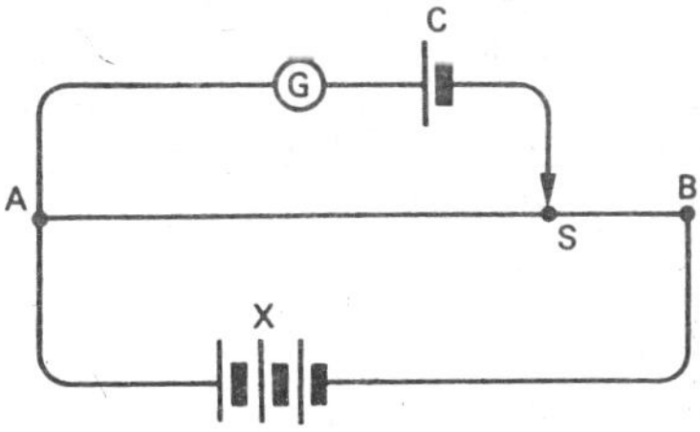

Potentiometer

potentiometer 1. *See* voltage divider. **2.** An instrument for measuring, comparing, or dividing small potential differences. A typical example of its use is the measurement of the e.m.f. (E_1) of a cell by comparing it with the e.m.f. (E_2) of a standard cell. In this case a circuit is set up as illustrated, in which AB is a wire of uniform resistance and S is a sliding contact onto this wire. An accumulator X maintains a steady current through the wave. To measure the e.m.f. of a cell C, it is connected up as shown in the diagram and the sliding contact moved until the e.m.f. of C exactly balances the p.d. from the accumulator, as indicated by a zero reading on the galvanometer G. If the length AS is then l_1, the value of E_1 is given by $E_1/E_2 = l_1/l_2$, where l_2 is the length AS when the standard cell is used as the cell C.

pound The unit of mass in the *f.p.s. system of units defined as 0.453 592 37 kilogram. Before 1963 it was defined in terms of a platinum cylinder called the Imperial Standard Pound.

poundal The unit of force in the *f.p.s. system of units equal to the force required to impart to a mass of one pound an acceleration of one foot per second per second.

powder metallurgy A process in which powdered metals or alloys are pressed into a variety of shapes at high temperatures. The process started with the pressing of powdered tungsten into incandescent lamp filaments in the first decade of this century and is now widely used for making self-lubricating bearings and cemented tungsten carbide cutting tools.

The powders are produced by atomization of molten metals, chemical decomposition of a compound of the metal, or crushing and grinding of the metal or alloy. The parts are pressed into moulds at pressures ranging from 140×10^6 Pa to 830×10^6 Pa after which they are heated in a controlled atmosphere to bond the particles together (*see* sintering).

power 1. (in physics) Symbol *P*. The rate at which work is done or energy is transferred. In SI units it is measured in watts (joules per second). *See also* horsepower. **2.** (in mathematics) The number of times a quantity is multiplied by itself. x^5 is the fifth power of *x*. A *power series* is one in which the power of the variable increases with each term, e.g. $a_0 + a_1x + a_2x^2 + a_3x^3 + \ldots + a_nx^n$.

power factor *See* electric power.

power reactor A *nuclear reactor designed to produce electrical power.

preamplifier An *amplifier in a radio, record player, etc., providing a first stage of amplification. It is usually located close to the signal source (i.e. the aerial or pick-up) and the signal is then transmitted by cable to the main amplifier. Preamplification at this early stage improves the signal-to-noise ratio of the whole system.

precessional motion A form of motion that occurs when a torque is applied to a rotating body in such a way that it tends to change the direction of its axis of rotation. It arises because the resultant of the angular velocity of rotation and the increment of angular velocity produced by the torque is an angular velocity about a new direction; this commonly changes the axis of the applied torque and leads to sustained rotation of the original axis of rotation.

A spinning top, the axis of which is not exactly vertical, has a torque acting on it as a result of gravity. Instead of falling over, the top precesses about a vertical line through the pivot. The earth also experiences a torque and undergoes a slow precession, primarily as a result of the gravitational attraction of the sun and the moon on its equatorial bulge (*see* precession of the equinoxes).

precession of the equinoxes The slow westward motion of the *equinoxes about the ecliptic as a result of the earth's *precessional motion. The equinoxes move round the ecliptic with a period of 25 800 years.

presbyopia A loss of accommodation that normally develops in human eyes over the age of 45–50 years. Vision of distant objects remains unchanged but accommodation of the eye to near objects is reduced as a result of loss of elasticity in the lens of the eye. The defect is corrected by reading glasses using weak converging lenses.

pressure The force acting on unit area of a surface or the ratio of force to area. It is measured in *pascals in SI units. *Absolute pressure* is pressure measured on a gauge that reads zero at zero pressure rather than at atmospheric pressure. *Gauge pressure* is measured on a gauge that reads zero at atmospheric pressure.

pressure gauge Any device used to measure pressure. Three basic types are in use: the liquid-column gauge (e.g. the mercury *barometer and the *manometer), the expanding-element gauge (e.g. the *Bourdon gauge and the aneroid *barometer), and the electrical transducer. In the last category the *strain gauge is an example. Capacitor pressure gauges also come into this category. In these devices, the pressure to be measured displaces one plate of a capacitor and thus alters its capacitance.

pressurized-water reactor *See* nuclear reactor.

Prévost's theory of exchanges A body emits and absorbs radiant energy at equal rates when it is in equilibrium with its surroundings. Its temperature then remains constant. If the body is not at the same temperature as its surroundings there is a net flow of energy between the surroundings and the body because of unequal emission and absorption. The theory was proposed by Pierre Prévost (1751–1839) in 1791.

primary cell A *voltaic cell in which the chemical reaction producing the e.m.f. is not satisfactorily reversible and the cell cannot therefore be recharged by the application of a charging current. *See* Daniell cell; Leclanché cell; Weston cell; mercury cell. *Compare* secondary cell.

primary colour Any one of a set of three coloured lights that can be mixed together to give the sensation of white light as well as approximating all the other colours of the spectrum. An infinite number of such sets exists, the condition being that none of the individual colours of a set should be able to be matched by mixing the other two; however, unless the colours are both intense and very different the range that they can match well will be limited. The set of primary colours most frequently used is red, green, and blue. *See also* colour.

primary winding The winding on the input side of a *transformer or *induction coil. *Compare* secondary winding.

principal axis *See* optical axis.

principal focus A point through which rays close to and parallel to the axis of a lens or spherical mirror pass, or appear to pass, after refraction or reflection. A mirror has one principal focus, a lens has a principal focus on both sides.

principal plane The plane that is perpendicular to the optical axis of a lens and that passes through the optical centre. A thick lens has two principal planes, each passing through a *principal point.

principal point Either of two points on the principal axis of a thick lens from which simply related distances can be measured, as from the optical centre of a thin lens.

principle of superposition The resultant displacement at any point in a region through which two waves of the same type pass is the algebraic sum of the displacements that the two would separately produce at that point. Both waves leave the region of superposition unaltered.

printed circuit An electronic circuit consisting of a conducting material deposited (printed) onto the surface of an insulating sheet. These devices are now common in all types of electronic equipment, facilitating batch production and eliminating the unreliability of the hand-soldered joint.

prism **1.** (in mathematics) A polyhedron with two parallel congruent polygons as bases and parallelograms for all other faces. A *triangular prism* has triangular bases. **2.** (in optics) A block of glass or other transparent material, usually having triangular bases. Prisms have several uses in optical systems: they can be used to deviate a ray, to disperse white light into the visible spectrum, or to erect an inverted image (*see* binoculars). Prisms of other materials are used for different kinds of radiation. *See also* Nicol prism; Wollaston prism.

probability The likelihood of a particular event occurring. If there are n equally likely outcomes of some experiment, and a ways in which event E could occur, then the probability of event E is a/n. For instance, if a die is thrown there are 6 possible outcomes and 3 ways in which an odd number may occur. The probability of throwing an odd number is $3/6 = 1/2$.

program *See* computer.

progressive wave *See* wave.

projectile Any body that is thrown or projected. If the projectile is discharged on the surface of the earth at an angle θ to the horizontal it will describe a parabolic flight path (if $\theta < 90°$ and the initial velocity < the *escape velocity). Neglecting air resistance, the maximum height of this flight path will be $(v^2\sin^2\theta)/2g$, where v is the velocity of discharge and g is the acceleration of free fall. The horizontal distance covered will be $(v^2\sin 2\theta)/g$ and the time of the flight will be $(2v\sin\theta)/g$.

projector An optical device for throwing a large image of a two-dimensional object onto a screen. In an *episcope*, light is reflected from the surface of an opaque two-dimensional object (such as a diagram or photographic print) and an enlarged image is thrown onto a distant screen by means of a system of mirrors and lenses. The *diascope* passes light through the two-dimensional object (such as a photographic transparency, slide, or film) and uses a converging projection lens to form an enlarged image on a distant screen. An *epidiascope* is a device that can be used as both episcope and diascope. An *overhead projector* is a form of diascope that throws its image on a wall or screen behind and above the operator. In a *motion-picture projector* (or *ciné projector*) the film, consisting of a long sequence of transparent pictures, is driven by a motor past the light source in such a way that each picture comes to rest for a brief period in front of the light source. The illusion of motion is created as each image on the screen is replaced by the next; during the picture change the light is interrupted.

prompt neutrons The neutrons emitted during a nuclear fission process within less than a microsecond of fission. *Compare* delayed neutrons.

proper motion The apparent angular motion of a star on the *celestial sphere. This is motion in a direction that is perpendicular to the line of sight.

proportional counter A type of detector for *ionizing radiation in which the size of the output pulse is proportional to the number of ions formed in the initial ionizing event. It operates in a voltage region, called the *proportional region*, intermediate between that of an *ionization chamber and a *Geiger counter, avalanche ionization being limited to the immediate vicinity of the primary ionization rather than the entire length of the central wire electrode.

proportional limit *See* elasticity.

prostar *See* stellar evolution.

protactinium Symbol Pa. A radioactive metallic element belonging to the actinoids; a.n. 91; r.a.m. 231.036; r.d. 15.37 (calculated); m.p. <1600°C (estimated). The most stable isotope, protactinium-231, has a half-life of 3.43×10^4 years; ten other radioisotopes are known. Protactinium-231 occurs in all uranium ores as it is a member of the uranium series. Protactinium has no practical applications; it was discovered by Lise Meitner and Otto Hahn in 1917.

proton An *elementary particle that is stable, bears a positive charge equal in magnitude to that of the *electron, and has a mass of $1.672\,614 \times 10^{-27}$ kg, which is 1836.12 times that of the electron. The proton is a hydrogen ion and occurs in all atomic nuclei.

proton number *See* atomic number.

psi particle (J particle) A *meson discovered in 1974, which led to the extension of the quark model and the hypothesis that a fourth quark existed with the property of charm (*see* elementary particles). The psi particle is believed to consist of a charmed quark and its antiquark.

psychrometer *See* hygrometer.

Ptolemaic astronomy The system of astronomy originally proposed by Apollonius of Perga in the third century BC and completed by Claudius Ptolemaeus of Alexandria (100–178 AD). It assumed that the earth was at the centre of the universe and that each known planet, the moon, and the sun moved round it in a circular orbit, called the deferent. In addition to this motion the orbiting bodies also described epicycles, small circles about points on the deferent. The system gave moderately good predictions, but was completely replaced by the heliocentric astronomy of Copernicus in the 16th century. The Ptolemaic system was published by Ptolemaeus in the work known by its Arabic name, the *Almagest*.

***p*-type conductivity** *See* semiconductor; transistor.

pulley A simple machine consisting of a wheel with a flat, crowned, or grooved rim to take a belt, rope, or chain with which a

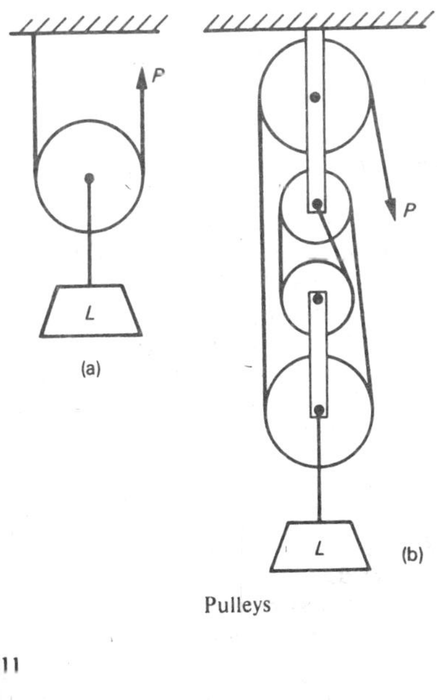

Pulleys

load can be raised.

In fig (a), assuming the system is frictionless, the force P in any part of the rope is constant, therefore $2P = L$, where L is the load. In general, $nP = L$, where n is the number of supporting ropes. In fig (b), the number of supporting ropes is 4. The mechanical advantage of a pulley system is the ratio of the load, L, to the effort applied to the free end of the rope, P, i.e. mechanical advantage = L/P = $L(L/n)^{-1} = n$. Thus in fig (b) the mechanical advantage is 4. A combination of ropes and pulleys as in fig (b) is called a *block and tackle.*

pulsar A celestial source of radiation emitted in brief (0.03 second to 4 seconds) regular pulses. First discovered in 1968, a pulsar is believed to be a rotating *neutron star. The strong magnetic field of the neutron star concentrates charged particles in two regions and the radiation is emitted in two directional beams. The pulsing effect occurs as the beams rotate. Most pulsars are radio sources (emit electromagnetic radiation of radio frequencies) but a few that emit light or X-rays have been detected. Over 300 pulsars are now known, but it is estimated that there are over one million in the Milky Way.

pulsatance *See* angular frequency.

pulse a. A brief variation in a quantity, usually for a finite time, especially in a quantity that is normally constant. **b.** A series of such variations having a regular waveform in which the variable quantity rises sharply from a base value to a maximum value and then falls back to the base value in a relatively short time.

pulse jet A type of ramjet (*see* jet propulsion) in which a louvred valve at the front of the projectile is blown open by the ram effect of the moving projectile and remains open until pressure has built up in the combustion chamber. Fuel is then admitted and the mixture exploded by spark ignition. This closes the louvred valve and produces thrust at the open rear end of the projectile. The German flying bombs of World War II were powered by pulse jets.

pulse modulation *See* modulation.

pump A device that imparts energy to a fluid in order to move it from one place or level to another or to raise its pressure (*compare* vacuum pump). *Centrifugal pumps* and *turbines have rotating impellers, which increase the velocity of the fluid, part of the energy so acquired by the fluid then being converted to pressure energy. Displacement pumps act directly on the fluid, forcing it to flow against a pressure. They include piston, plunger, gear, screw, and cam pumps. *See also* electromagnetic pump.

pyramid A solid having a polygonal base with n sides, each side forming the base of a triangle. The n triangles so formed have a common vertex. The *axis* of the pyramid is a line joining the vertex to the centre of symmetry of the base. If the axis is perpendicular to the base the solid is a *right pyramid.* A *square pyramid* has a square base and a *triangular pyramid* has a triangular base (*see* tetrahedron). The volume of a pyramid is one third of the base area multiplied by the height.

pyroelectricity The property of certain crystals, such as tourmaline, of acquiring opposite electrical charges on opposite faces when heated. In tourmaline a rise in temperature of 1 K at room temperature produces a polarization of some 10^{-5} C m^{-2}.

pyrometric cones *See* Seger cones.

pyrometry The measurement of high temperatures (beyond the range of thermometers) using a *pyrometer.* Modern *narrow-band* or *spectral* pyrometers use infrared-sensitive *photoelectric cells behind filters that exclude visible light. In the *optical pyrometer* (or disappearing filament pyrometer) the image of the incandescent source is focused in the plane of a tungsten filament that is heated electrically. A variable resistor is used to adjust the current through the filament until it blends into the image of the source, when viewed through a red filter and an eyepiece. The temperature is then read from a calibrated ammeter or a calibrated dial on the variable resistor. In the *total-radiation pyrometer* radiation emitted by the source is focused by a concave mirror onto a blackened foil to which a thermopile is attached. From the e.m.f. produced by the thermopile the temperature of the source can be calculated.

pyrophoric Igniting spontaneously in air. *Pyrophoric alloys* are alloys that give sparks when struck. *See* misch metal.

Q

QSG *See* quasars.

QSO *See* quasars.

QSS *See* quasars.

quadratic equation An equation of the second degree having the form $ax^2 + bx + c = 0$. Its roots are:

$$[-b \pm \sqrt{(b^2 - 4ac)}]/2a.$$

quadrature The position of the moon or an outer planet when the line joining it to the earth makes a right angle with a line joining the earth to the sun.

quality of sound (timbre) The quality a musical note has as a result of the presence of *harmonics. A pure note consists only of the *fundamental; however, a note from a musical instrument will have several harmonics present, depending on the type of instrument and the way in which it is played. For example, a plucked string (as in a guitar) produces a series of harmonics of diminishing intensity, whereas a struck string (as in a piano) produces a series of harmonics of more nearly equal intensity.

quantum The minimum amount by which certain properties, such as energy or angular momentum, of a system can change. Such properties do not, therefore, vary continuously, but in integral multiples of the relevant quantum. This concept forms the basis of the *quantum theory. In waves and fields the quantum can be regarded as an excitation, giving a particle-like interpretation to the wave or field. Thus, the quantum of the electromagnetic field is the *photon and the *graviton is the quantum of the gravitational field. *See* quantum mechanics.

quantum chromodynamics *See* elementary particles.

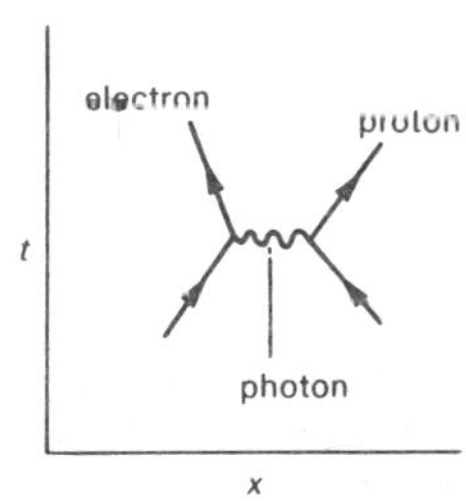

An electron-proton collision

quantum electrodynamics The study of the properties of electromagnetic radiation and the way in which it interacts with charged matter in terms of *quantum mechanics. The collision of a moving electron with a proton, in this theory, can be visualized by a space–time diagram (*Feynman diagram*) in which photons are exchanged (see illustration).

quantum jump A change in a system (e.g. an atom or molecule) from one quantum state to another.

quantum mechanics A system of mechanics that was developed from *quantum theory and is used to explain the properties of atoms and molecules. Using the energy *quantum as a starting point it incorporates Heisenberg's *uncertainty principle and the *de Broglie wavelength to establish the wave–particle duality (*see* complementarity) on which *Schrödinger's equation is based. This form of quantum mechanics is called *wave mechanics*. An alternative but equivalent formalism, *matrix mechanics*, is based on mathematical operators.

quantum number *See* atom; spin.

quantum state The state of a quantized system as described by its quantum numbers. For instance, the state of a hydrogen *atom is described by the four quantum numbers n, l, m, m_s. In the ground state they have values 1, 0, 1, and ½ respectively.

quantum statistics A statistical description of a system of particles that obeys the rules of *quantum mechanics rather than classical mechanics. In quantum statistics, energy states are considered to be quantized; if the particles are treated as indistinguishable, *Bose–Einstein statistics* apply and the particles are called *bosons*. All known bosons have an angular momentum $n\hbar$, where n is zero or an integer and $\hbar$ is the Planck constant. For identical bosons the *wave function is always symmetric. For the particular case in which not more than one particle may appear in any of the cells into which the particles are distributed, *Fermi–Dirac statistics* apply and the particles are called *fermions*. All known fermions have a total angular momentum $(n + ½)\hbar$ and any wave function that involves identical fermions is always antisymmetric.

quantum theory The theory devised by Max Planck (1858–1947) in 1900 to account for the emission of the black-body radiation from hot bodies. According to this theory energy is emitted in quanta (*see* quantum), each of which has an energy equal to $h\nu$, where h is the *Planck constant and ν is the frequency of the radiation. This theory led to the modern theory of the interaction between matter and radiation known as *quantum mechanics, which generalizes and replaces classical mechanics and Maxwell's electromagnetic theory. In *nonrelativistic quantum theory* particles are assumed to be neither created nor destroyed, to move slowly relative to the speed of light, and to have a mass that does not change with velocity. These assumptions apply to atomic and molecular phenomena and to some aspects of nuclear physics. *Relativistic quantum theory* applies to particles that have zero *rest mass or travel at or near the speed of light.

quark *See* elementary particles.

quarter-wave plate *See* retardation plate.

quartz The most abundant and common mineral, consisting of crystalline silica (silicon dioxide, SiO_2), crystallizing in the trigonal system. It has a hardness of 7 on the Mohs' scale. The mineral has the property of being piezoelectric and hence is used to make oscillators for clocks (*see* quartz clock), radios, and radar instruments. It is also used in optical instruments and in glass, glaze, and abrasives.

quartz clock A clock based on a *piezoelectric crystal of quartz. Each quartz crystal has a natural frequency of vibration, which depends on its size and shape. If such a crystal is introduced into an oscillating electronic circuit that resonates at a frequency very close to that of the natural frequency of the crystal, the whole circuit (including the crystal) will oscillate at the crystal's natural frequency and the frequency will remain constant over considerable periods (a good crystal will maintain oscillation for a year with an accumulated error of less than 0.1 second). In a quartz clock or watch the alternating current from the oscillating circuit containing such a crystal is amplified and the frequency subdivided until it is suitable to drive a synchronous motor, which in turn drives a gear train to operate hands. Alternatively it is used to activate a digital display.

quasars A class of astronomical objects that appear on optical photographs as starlike but have large *redshifts quite unlike those of stars. They were first observed in 1961 when it was found that strong radio emission was emanating from many of these starlike bodies. Over 600 such objects are now known and their redshifts can be as high as 3.53. The redshifts are characteristic of galaxies flying outwards from the centre of the universe at enormous speeds as a result of the *expansion of the universe. This *cosmological redshift* is the explanation of the high observed redshifts of quasars favoured by most astronomers. Some, however, maintain that the redshift could be a local Doppler effect, characteristic of movement relative to the earth and sun of nearby objects in the Galaxy, or a gravitational effect. If the redshifts are cosmological, quasars are the most distant objects in the universe, some being up to 10^{10} light-years away. The origin of quasars is unknown but what we observe now are the emissions made by these objects 10^{10} years ago, when the galaxies are thought to have been forming. It has been suggested that the quasars may have been connected in some way with the birth of the galaxies.

The name *quasar* is a contraction of quasistellar object (*QSO*) or quasistellar galaxy (*QSG*). Quasars that are also radio sources are sometimes called quasistellar radio sources (*QSS*).

quenching 1. The rapid cooling of a metal by immersing it in a bath of liquid in order to improve its properties. Steels are quenched to make them harder but some nonferrous metals are quenched for other reasons (copper, for example, is made softer by quenching). **2.** The process of inhibiting a continuous discharge in a *Geiger counter so that the incidence of further ionizing radiation can cause a new discharge. This is achieved by introducing a quenching vapour, such as ether or a halogen gas, into the tube.

R

racemic mixture (racemate) A mixture of equal quantities of the *d*- and *l*-forms of an optically active compound. Racemic mixtures are denoted by the prefix *dl* (e.g. *dl*-lactic acid). A racemic mixture shows no *optical activity.

rad *See* radiation units.

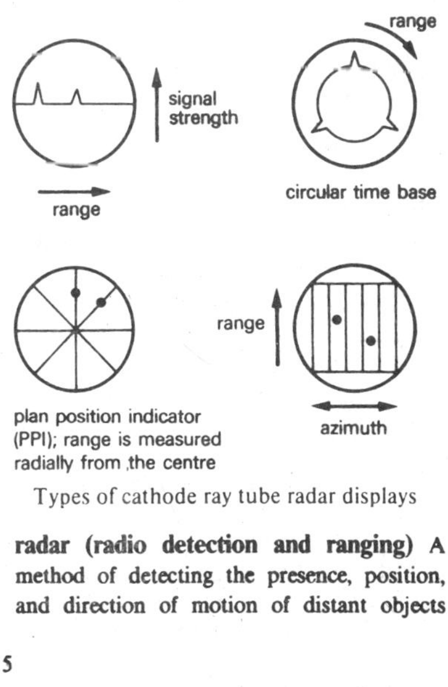

Types of cathode ray tube radar displays

radar (radio detection and ranging) A method of detecting the presence, position, and direction of motion of distant objects (such as ships and aircraft) by means of their ability to reflect a beam of electromagnetic radiation of centimetric wavelengths. It is also used for navigation and guidance. It consists of a transmitter producing radio-frequency radiation, often pulsed, which is fed to a movable aerial from which it is transmitted as a beam. If the beam is interrupted by a solid object, a part of the energy of the radiation is reflected back to the aerial. Signals received by the aerial are passed to the receiver, where they are amplified and detected. An echo from a reflection of a solid object is indicated by a sudden rise in the detector output. The time taken for a pulse to reach the object and be reflected back (t) enables the distance away (d) of the target to be calculated from the equation $d = ct/2$, where c is the speed of light. In some systems the speed of the object can be measured using the *Doppler effect. The output of the detector is usually displayed on a cathode-ray tube in a variety of different formats (see illustration).

radial velocity *See* line-of-sight velocity.

radian *See* circular measure.

radiance Symbol L_e. The radiant intensity per unit transverse area, in a given direction, of a source of radiation. It is measured in $\mathrm{W\,sr^{-1}\,m^{-2}}$.

radiant energy Energy transmitted as electromagnetic radiation. Radiant energy is the only form of energy that can exist in the absence of matter.

radiant exitance *See* exitance.

radiant flux Symbol Φ_e. The total power emitted, received, or passing in the form of electromagnetic radiation. It is measured in watts.

radiant intensity Symbol I_e. The *radiant flux per unit solid angle emitted by a point source. It is measured in watts per steradian.

radiation 1. Energy travelling in the form of electromagnetic waves or photons. **2.** A stream of particles, especially alpha- or beta-particles from a radioactive source or neutrons from a nuclear reactor.

radiation belts *See* Van Allen belts.

radiation damage Harmful changes that occur to inanimate materials and living organisms as a result of exposure to energetic electrons, nucleons, fission fragments, or high-energy electromagnetic radiation. In inanimate materials the damage may be caused by electronic excitation, ionization, transmutation, or displacement of atoms. In organisms, these mechanisms can cause changes to cells that alter their genetic structure, interfere with their division, or kill them. In humans, these changes can lead to *radiation sickness*, *radiation burns* (from large doses of radiation), or to long-term damage of several kinds, the most serious of which result in various forms of cancer (especially leukaemia).

radiation pressure The pressure exerted on a surface by electromagnetic radiation. As radiation carries momentum as well as energy it exerts a force when it meets a surface, i.e. the *photons transfer momentum when they strike the surface. The pressure is usually negligible on large bodies, for example, the pressure of radiation from the sun on the surface of the earth is of the order of 10^{-5} pascal, but on small bodies it can have a considerable effect, driving them away from the radiation source.

radiation temperature The surface temperature of a celestial body as calculated by *Stefan's law, assuming that the body behaves as a *black body. The radiation temperature is usually measured over a narrow portion of the electromagnetic spectrum, such as the visible range (which gives the *optical temperature*).

radiation units Units of measurement used to express the *activity of a radionuclide and the *dose of ionizing radiation. The units *curie*, *roentgen*, *rad*, and *rem* are not coherent with SI units but their temporary use with SI units has been approved while the derived SI units *becquerel*, *gray*, and *sievert* become familiar.

The becquerel (Bq), the SI unit of activity, is the activity of a radionuclide decaying at a rate, on average, of one spontaneous nuclear transition per second. Thus 1 Bq = 1 s^{-1}. The former unit, the curie (Ci), is equal to 3.7×10^{10} Bq. The curie was originally chosen to approximate the activity of 1 gram of radium-226.

The gray (Gy), the SI unit of absorbed dose, is the absorbed dose when the energy per unit mass imparted to matter by ionizing radiation is 1 joule per kilogram. The former unit, the rad (rd), is equal to 10^{-2} Gy.

The sievert (Sv), the SI unit of dose equivalent, is the dose equivalent when the absorbed dose of ionizing radiation multiplied by the stipulated dimensionless factors is 1 J kg^{-1}. As different types of radiation cause different effects in biological tissue a weighted absorbed dose, called the *dose equivalent*, is used in which the absorbed dose is modified by multiplying it by dimensionless factors stipulated by the International Commission on Radiological Protection. The former unit of dose equivalent, the rem (originally an acronym for *r*oentgen *e*quivalent *m*an), is equal to 10^{-2} Sv.

In SI units, exposure to ionizing radiation is expressed in coulombs per kilogram, the quantity of X- or gamma-radiation that produces ion pairs carrying 1 coulomb of charge of either sign in 1 kilogram of pure dry air. The former unit, the roentgen (R), is equal to 2.58×10^{-4} C kg^{-1}.

radiative collision A collision between charged particles in which part of the kinetic energy is radiated in the form of photons.

radical A root of a number or a quantity. The symbol $\sqrt{}$ is called the *radical sign*.

radio A means of transmitting information in which the transmission medium consists of electromagnetic radiation. Information is transmitted by means of the *modulation of a *carrier wave in a transmitter; the modulated carrier wave is fed to a transmitting aerial from which it is broadcast through the atmosphere or through space. A receiving aerial forms part of a *resonant circuit, which can be tuned to the frequency of the carrier wave, enabling the receiver that it feeds selectively to amplify and then to demodulate the transmitted signal. A replica of the original information is thus produced by the receiver. *See also* radio transmission.

radioactive age The age of an archaeological or geological specimen as determined by a process that depends on a radioactive decay. *See* carbon dating; fission-track dating; potassium–argon dating; rubidium–strontium dating; uranium–lead dating.

radioactive series A series of radioactive nuclides in which each member of the series is formed by the decay of the nuclide before it. The series ends with a stable nuclide. Three radioactive series occur naturally, those headed by thorium–232 (*thorium series*), uranium–235 (*actinium series*), and uranium–238 (*uranium series*). All three series end with an isotope of lead. The neptunium series starts with the artificial isotope plutonium–241, which decays to neptunium–237, and ends with thallium–81.

radioactive tracing *See* labelling.

radioactive waste (nuclear waste) Any waste material that contains *radionuclides. These wastes are produced in the mining of radioactive ores, the normal running of nuclear power stations and other reactors, the manufacture of nuclear weapons, and in research laboratories. Because high-level radioactive wastes can be extremely dangerous to all living matter and because they may contain radionuclides having half-lives of many thousands of years, their disposal has to be controlled with great stringency.

The first stage in disposal involves extensive reprocessing to retrieve usable material and reduce the activity of the waste. Subsequent final disposal, usually of solid material in strong metal canisters, involves burial of the material. The two currently favoured categories of sites for burial are deep (600 metres) mined cavities in stable geological formations on land and in red clay some 50–100 metres below the seabed away from tectonic plate margins and where the currents at the bottom of the sea are weak.

radioactivity The spontaneous disintegration of certain atomic nuclei accompanied by the emission of alpha-particles (helium nuclei), beta-particles (electrons), or gamma-radiation (short-wave electromagnetic waves).

Natural radioactivity is the result of the spontaneous disintegration of naturally occurring radioisotopes; these radioisotopes can be arranged in three *radioactive series. The rate of disintegration is uninfluenced by chemical changes or any normal changes in their environment. However, radioactivity can be induced in many nuclides by bombarding them with neutrons or other particles. *See also* half-life; ionizing radiation; radiation units.

radio astronomy The study of the radio-frequency radiation emitted by celestial bodies. This branch of astronomy began in 1932 when a US engineer, Karl Jansky (1905–40), first detected radio waves from outside the earth's atmosphere. *See* radio source; radio telescope; radio window.

radiobiology The branch of biology concerned with the effects of radioactive substances on living organisms and the use of radioactive tracers to study metabolic processes (*see* labelling).

radiocarbon dating *See* carbon dating.

radiochemistry The branch of chemistry concerned with radioactive compounds and with ionization. It includes the study of compounds of radioactive elements and the preparation and use of compounds containing radioactive atoms. *See* labelling; radiolysis.

radio frequencies The range of frequencies, between about 3 kilohertz and 300 gigahertz, over which electromagnetic radiation is used in radio transmission. It is subdivided into eight equal bands, known as very low frequency, low frequency, medium frequency, high frequency, very high frequency, and extremely high frequency.

radio galaxies A *radio source outside the Galaxy that has been identified with an optically visible galaxy. These radio galaxies are distinguished from normal galaxies by having a radio power output some 10^6 times greater (i.e. up to 10^{38} watts rather than 10^{32} W). The source of the radio-frequency energy is unknown.

radiography The process or technique of producing images of an opaque object on photographic film or on a fluorescent screen by means of radiation (either particles or electromagnetic waves of short wavelength, such as X-rays and gamma-rays). The photograph produced is called a *radiograph*. The process is widely used in diagnostic *radiology, using X-rays, and in flaw detection in industrial products, using high-energy X-rays, gamma-radiations, neutron beams, and (more recently) beams of charged particles.

radio interferometry *See* radio telescope.

radioisotope (radioactive isotope) An isotope of an element that is radioactive. *See* labelling.

radiolocation The location of distant objects by means of *radar.

radiology The study and use of X-rays, radioactive materials, and other ionizing radiations for medical purposes, especially for diagnosis (*diagnostic radiology*) and the treatment of cancer and allied diseases (*radiotherapy*).

radiolysis The use of ionizing radiation to produce chemical reactions. The radiation used includes alpha particles, electrons, neutrons, X-rays, and gamma rays from radioactive materials or from accelerators. Energy transfer produces ions and excited species, which undergo further reaction. A particular feature of radiolysis is the formation of short-lived solvated electrons in water and other polar solvents.

radiometric dating (radioactive dating) *See* dating techniques; radioactive age.

radionuclide (radioactive nuclide) A *nuclide that is radioactive.

radio-opaque *See* opacity.

radiosonde A meteorological instrument that measures temperature, pressure, humidity, and winds in the upper atmosphere. It consists of a package of instruments and a radio transmitter attached to a balloon. The data is relayed back to earth by the transmitter. The position of the balloon can be found by radar and from its changes in position the wind velocities can be calculated.

radio source An astronomical object that has been observed with a *radio telescope to emit radio-frequency electromagnetic radiation. Radio sources within the Galaxy include Jupiter, the sun, pulsars, and background radiation arising from *synchrotron radiation. Sources outside the Galaxy include spiral galaxies, *radio galaxies, and *quasars. Radio sources were formerly known as *radio stars*.

radio star *See* radio source.

radio telescope An instrument for detecting and measuring electromagnetic radiation of radio frequencies that have passed through the *radio window in the earth's atmosphere and reached the surface of the earth. There are a great diversity of *radio sources within the universe and radio telescopes are required to detect both continuous emissions and specific spectral lines. They therefore require the highest possible angular resolution so that the details of radio sources can be studied and they should be able to pick up weak signals. The simplest radio telescope consists of a paraboloidal steerable-dish aerial together with ancillary amplifiers. The paraboloidal reflecting dish surface reflects the incoming signal to the principal focus of the reflector. At this point the radio-frequency signals are amplified up to 1000 times and converted to a lower, intermediate, frequency before transmission by cable to the control building. Here the intermediate frequency is amplified again and passed to the detector and display unit. As the radio waves arriving from the surface of the reflector at the focus must be in phase, the surface of the dish must be very accurately constructed; for example, a 100-metre-diameter dish must be accurate to the nearest millimetre, when receiving radiation of 1 cm wavelength. To overcome the problem of constructing large dishes to such a high accuracy, the technique of *radio interferometry* has been developed. In this technique an array of small aerials connected by cable is used to simulate a large dish aerial. In earth-rotation *aperture synthesis*, the individual positions and displacements of an array of only a few such small aerials can be made to simulate an enormous dish aerial as the earth rotates. All but the smallest steerable dishes are constructed from metal mesh so that wind can pass through them. A few very large fixed dishes have been built into the earth's surface.

radiotherapy. *See* radiology.

radio transmission The transmission of radio waves from a transmitting aerial to a receiving aerial. The radiation may take several paths (see illustration). The sum of the line-of-sight ground wave, the reflected ground wave, and the surface wave is called the *ground wave* (or *tropospheric wave*). *Sky waves* (or *ionospheric waves*) are reflected by the *ionosphere and enable long-distance transmissions to be made. The ionization of atoms and molecules in the ionosphere is caused largely by solar ultraviolet and X-radiation and therefore conditions differ between night and day. Ionization in the lower E-region of the ionosphere falls off at night in the absence of sunlight and ions and electrons tend to recombine. However, in the less dense (higher) F-region there are fewer collisions between ions and electrons and therefore fewer recombinations at night. The F-region is therefore a more effective reflector at night.

The UHF and VHF waves used in television broadcasting penetrate the ionosphere with little reflection. Therefore TV broadcasts can only be made over long distances by means of artificial satellites. *See also* radio.

radiotransparent Transparent to radiation, especially to X-rays and gamma-rays.

radio window A region of the electromagnetic spectrum in the radio-frequency band within which radio waves can be transmitted through the earth's atmosphere without significant reflection or attenuation by constituents of the atmosphere. It extends from about 10 megahertz to 100 gigahertz and enables radiation in this range from celestial radio sources to be picked up by *radio telescopes on the earth's surface. Below 100 MHz incoming radio waves are reflected by the *ionosphere and those above 100 GHz are increasingly affected by molecular absorption.

radium Symbol Ra. A radioactive metallic element; a.n. 88; r.a.m. 226.0254; r.d. ~5;

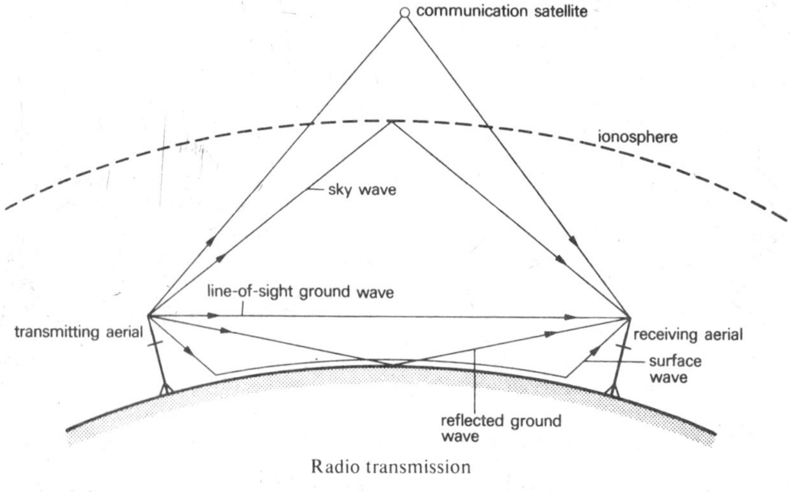

Radio transmission

m.p. 700°C; b.p. 1140°C. It occurs in uranium ores (e.g. pitchblende). The most stable isotope is radium-226 (half-life 1602 years), which decays to radon. It is used as a radioactive source in research and, to some extent, in radiotherapy. The element was isolated from pitchblende in 1898 by Marie and Pierre Curie.

radius of curvature *See* centre of curvature.

radius of gyration Symbol k. The square root of the ratio of the *moment of inertia of a rigid body about an axis to the body's mass, i.e. $k^2 = I/m$, where I is the body's moment of inertia and m is its mass. If a rigid body has a moment of inertia I about an axis and mass m it behaves as if all its mass is rotating at a distance k from the axis.

radius vector *See* polar coordinates.

radon Symbol Rn. A colourless radioactive gaseous element; a.n. 86; r.a.m. 222; d. 9.73 g dm^{-3}; m.p. −71°C; b.p. −61.8°C. At least 20 isotopes are known, the most stable being radon-222 (half-life 3.8 days). It is formed by decay of radium-226 and undergoes alpha decay. It is used in radiotherapy. It was first isolated by Ramsey and Gray in 1908.

rainbow An optical phenomenon that appears as an arc of the colours of the spectrum across the sky when falling water droplets are illuminated from behind by sunlight. The colours are produced by the refraction and internal reflection of the sunlight by the water drops. Two bows may be visible, the inner ring being known as the primary bow and the outer, in which the colours are reversed, as the secondary bow (see illustration).

r.a.m. *See* relative atomic mass.

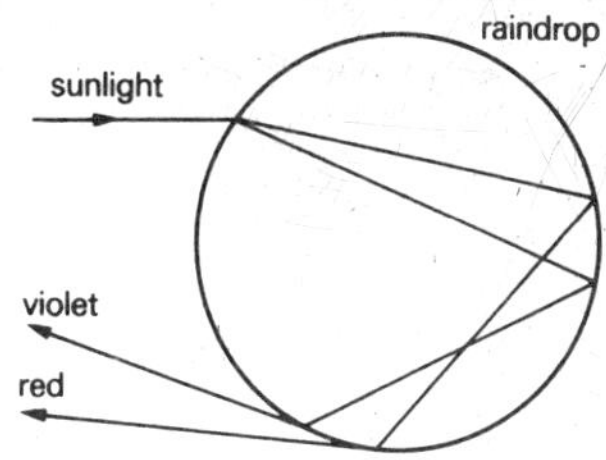

Primary rainbow (one internal reflection)

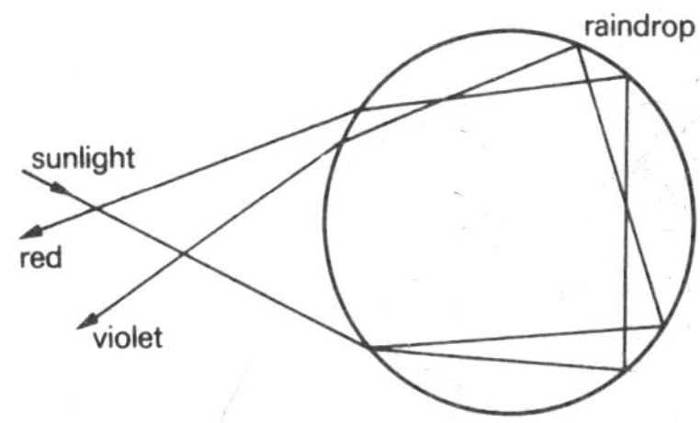

Secondary rainbow (two internal reflections)

Raman effect A type of inelastic *scattering of electromagnetic radiation in which light suffers a change in frequency and a change in phase as it passes through a material medium. The intensity of Raman scattering is about one-thousandth of that in Rayleigh scattering in liquids; for this reason it was not discovered until 1928. However, it was not until the development of the laser that the effect was put to use.

In *Raman spectroscopy* light from a laser is passed through a substance and the scattering is analysed spectroscopically. The new frequencies in the *Raman spectrum* of monochromatic light scattered by a substance are characteristic of the substance. This enables Raman spectroscopy to be used as a means of determining molecular structure and as a tool in chemical analysis. The effect was discovered by the Indian scientist Sir C. V. Raman (1888–1970).

ramjet *See* jet propulsion.

Ramsden eyepiece An eyepiece for optical instruments consisting of two identical plano-convex lenses with their convex faces pointing towards each other. They are separated by a distance of two thirds of the focal length of either lens.

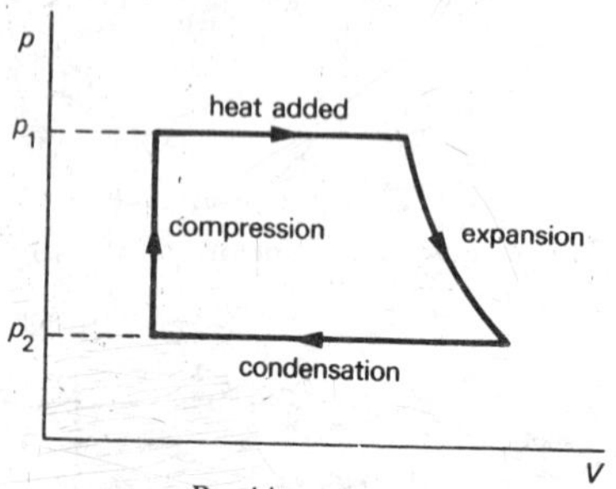

Rankine cycle

Rankine cycle A cycle of operations in a heat engine. The Rankine cycle more closely approximates to the cycle of a real steam engine that does the *Carnot cycle. It therefore predicts a lower ideal thermal efficiency than the Carnot cycle. In the Rankine cycle, heat is added at constant pressure p_1, at which water is converted in a boiler to superheated steam; the steam expands at constant entropy to a pressure p_2 in the cylinder; heat is rejected at constant pressure p_2 in a condenser; the water so formed is compressed at constant entropy to pressure p_1 by a feed pump. The cycle was devised by W. J. M. Rankine (1820–70).

Rankine temperature scale An absolute temperature scale based on the Fahrenheit scale. Absolute zero on this scale, 0°R, is equivalent to −459.67°F and the melting point of ice (32°F) is therefore (459.67 + 32 = 491.67)°R. The scale was devised by W. J. M. Rankine (1820–70).

Raoult's law The partial vapour pressure of a solvent is proportional to its mole fraction. If p is the vapour pressure of the solvent (with a substance dissolved in it) and X the mole fraction of solvent (number of moles of solvent divided by total number of moles) then $p = p_0X$, where p_0 is the vapour pressure of the pure solvent. A solution that obeys Raoult's law is said to be an *ideal solution*. In general the law holds only for dilute solutions, although some mixtures of liquids obey it over a whole range of concentrations. Such solutions are *perfect solutions* and occur when the intermolecular forces between molecules of the pure substances are similar to the forces between molecules of one and molecules of the other. Deviations in Raoult's law for mixtures of liquids cause the formation of *azeotropes. The law was discovered by the French chemist François Raoult (1830–1901).

rarefaction A reduction in the pressure of a fluid and therefore of its density.

raster The pattern of scanning lines on the screen of the cathode-ray tube in a television receiver or other device that provides a visual display.

rationalized units A system of units in which the defining equations have been made to conform to the geometry of the system in a logical way. Thus equations that involve circular symmetry contain the factor 2π, while those involving spherical symmetry contain the factor 4π. *SI units are rationalized; c.g.s. units are unrationalized.

rational number Any number that can be expressed as the ratio of two integers. For example, 0.3333... is rational because it can be written as 1/3. $\sqrt{2}$, however, is *irrational.

ray A narrow beam of radiation or an idealized representation of such a beam on a *ray diagram*, which can be used to indicate the positions of the object and image in a system of lenses or mirrors.

Rayleigh criterion *See* resolving power.

Rayleigh scattering *See* scattering of electromagnetic radiation.

reactance Symbol X. A property of a circuit containing inductance or capacitance that together with any resistance makes up its *impedance. The impedance Z is given by $Z^2 = R^2 + X^2$, where R is the resistance. For a pure capacitance C, the reactance is given by $X_C = 1/2\pi fC$, where f is the frequency of the *alternating current; for a pure inductance L, $X_L = 2\pi fL$. If the resistance, inductance, and capacitance are in series the impedance $Z = \sqrt{[R^2 + (X_L - X_C)^2]}$. Reactance is measured in ohms.

reaction A force that is equal in magnitude but opposite in direction to some other force, in accordance with Newton's third law of motion. Thus if a mass of 10 kg stands on a horizontal surface it exerts a force of $10g$ N on the surface, where g is the acceleration of free fall. The reaction is the force of $10g$ N that the surface exerts on the mass in order to counterbalance the weight and keep the mass stationary.

reaction propulsion *See* jet propulsion.

reactor 1. *See* nuclear reactor. **2.** Any device, such as an inductor or capacitor, that introduces *reactance into an electrical circuit.

real gas A gas that does not have the properties assigned to an *ideal gas. Its molecules have a finite size and there are forces between them (*see* equation of state).

real image *See* image.

real-is-positive convention (real-positive convention) A convention used in optical formulae relating to lenses and mirrors. In this convention, distances from optical components to real objects, images, and foci are taken as positive, whereas distances to virtual points are taken as negative.

Réaumur temperature scale A temperature scale in which the melting point of ice is taken as 0°R and the boiling point of water as 80°R. It was devised by René Antoine Réaumur (1683–1757).

recombination process The process in which a neutral atom or molecule is formed by the combination of a positive ion and a negative ion or electron; i.e. a process of the type:

$$A^+ + B^- \rightarrow AB$$

or

$$A^+ + e^- \rightarrow A$$

In recombination, the neutral species formed is usually in an excited state, from which it can decay with emission of light or other electromagnetic radiation.

rectification 1. (in physics) The process of obtaining a direct current from an alternating electrical supply. *See* rectifier. **2.** (in chemistry) The process of purifying a liquid by distillation.

rectifier An electrical device that allows more current to flow in one direction than the other, thus enabling alternating e.m.f.s to drive only direct current. The device most commonly used for rectification is the semiconductor *diode. In *half-wave rectification*, achieved with one diode, a pulsating current is produced. In *full-wave rectification* two diodes are used, one pair conducting during the first half cycle and the other conducting during the second half (see illustration). The full-wave rectified signal can be smoothed using a capacitor or an inductor.

red giant A *giant star thought to be in the later stages of *stellar evolution. It has a surface temperature in the range 2000–3000 K and a diameter 10–100 times that of the sun. *See also* Hertzsprung–Russell diagram.

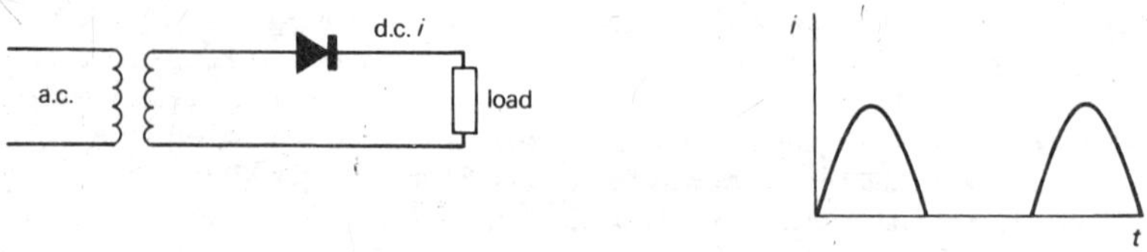

Half-wave rectification

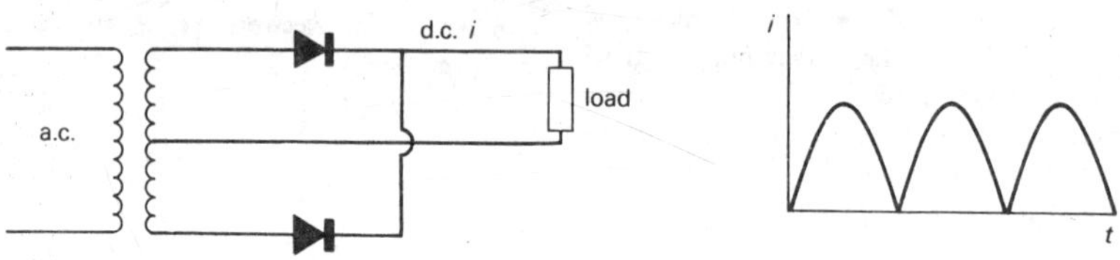

Full-wave rectification

redshift 1. (Doppler redshift) A displacement of the lines in the spectra of certain galaxies towards the red end of the visible spectrum (i.e. towards longer wavelengths). It is usually interpreted as a *Doppler effect resulting from the recession of the galaxies along the line of sight (*see* expansion of the universe). The redshift is usually expressed as $\Delta\lambda/\lambda$, where $\Delta\lambda$ is the shift in wavelength of radiation of wavelength λ. For relatively low velocities of recession this is equivalent to v/c, where v is the relative velocity of recession and c is the speed of light. If very high velocities of recession are involved, a relativistic version of v/c is required (*see* relativistic velocity). The redshift of spectral lines occurs in all regions of the electromagnetic spectrum; ultraviolet can be shifted into the visible region and visible radiation can be shifted into the infrared region. **2. (gravitational** *or* **Einstein redshift)** A similar displacement of spectral lines towards the red caused not by a Doppler effect but by a high gravitational field. This type of redshift was predicted by Einstein and some astronomers believe that this is the cause of the large redshifts of *quasars, which can be as high as 3.35.

reflectance The ratio of the radiant flux reflected by a surface to that falling on it. This quantity is also known as the *radiant reflectance*. The radiant reflectance measured for a specified wavelength of the incident radiant flux is called the *spectral reflectance*.

reflecting telescope (reflector) *See* telescope.

reflection The return of all or part of a beam of particles or waves when it encounters the boundary between two media. The *laws of reflection* state: (1) that the incident ray, the reflected ray, and the normal to the reflecting interface at the point of incidence are all in the same plane; (2) that the angle of incidence equals the angle of reflection. *See also* mirror; reflectance; total internal reflection.

refracting telescope (refractor) *See* telescope.

refraction The change of direction suffered by wavefront as it passes obliquely from one medium to another in which its speed of propagation is altered. The phenomenon occurs with all types of waves, but is most familiar with light waves. In optics the di-

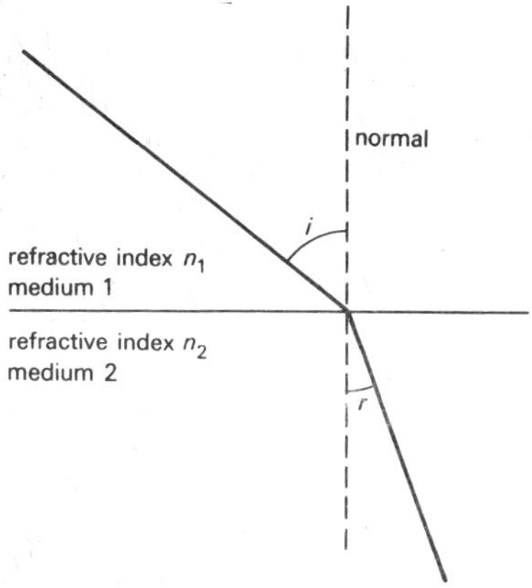

Refraction

rection is changed in accordance with *Snell's law*, i.e. $n_1 \sin i = n_2 \sin r$, where i and r are the angles made by the incident beam of radiation and the refracted beam to the normal (an imaginary line perpendicular to the interface between the two media); n_1 and n_2 are the *refractive indices of the two media. This law is also known as one of the *laws of refraction*. The other law of refraction is that the incident ray, the refracted ray, and the normal at the point of incidence lie in the same plane. The change of direction results from a change in the speed of propagation and the consequent change in wavelength (see illustration).

refractive index (refractive constant) Symbol n. The *absolute refractive index* of a medium is the ratio of the speed of electromagnetic radiation in free space to the speed of the radiation in that medium. As the refractive index varies with wavelength, the wavelength should be specified. It is usually given for yellow light (sodium D-lines; wavelength 589.3 nm). The *relative refractive index* is the ratio of the speed of light in one medium to that in an adjacent medium. *See also* refraction.

refractivity A measure of the extent to which a medium will deviate a ray of light entering its surface. In some contexts it is equal to $(n - 1)$, where n is the *refractive index.

refractometer Any of various instruments for measuring the *refractive index of a substance or medium. An example is the *Pulfrich refractometer*, which is a glass block with a polished top, with a small cell on top of the block for liquid samples. A telescope rotating on a vertical circular scale is used to find the angle (α) between the top of the block and the direction in which the limiting ray (from incident light parallel to the top face) leaves the side of the block. If the refractive index of the block (n_g) is known, that of the liquid can be calculated using $n = \sqrt{(n_g{}^2 - \sin^2\alpha)}$.

refractory 1. Having a high melting point. Metal oxides, carbides, and silicides tend to be refractory materials, and are extensively used for lining furnaces, etc. **2.** A refractory material.

refrigerant *See* refrigeration.

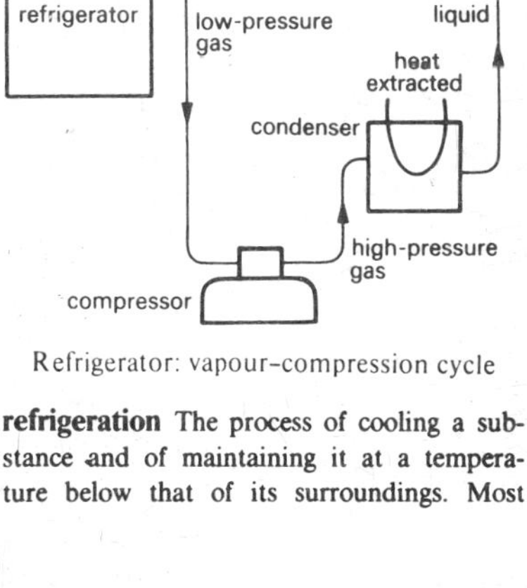

Refrigerator: vapour–compression cycle

refrigeration The process of cooling a substance and of maintaining it at a temperature below that of its surroundings. Most commercial refrigerators use a cycle of operations equivalent to those of a *heat pump. In the *vapour-compression cycle* a volatile liquid (the *refrigerant*) passes to an evaporator where it boils, extracting heat from the refrigerator chamber as it does so. The vapour is then compressed by a compressor, where its temperature and pressure are raised. The compressed vapour then passes to a condenser, where the refrigerant discharges its heat to the surroundings. The liquid so formed then passes to a storage vessel and finally to an expansion valve, through which the high-pressure liquid expands to a low-pressure before entering the evaporator again. The refrigerant used is ammonia, sulphur dioxide, or a haloalkane.

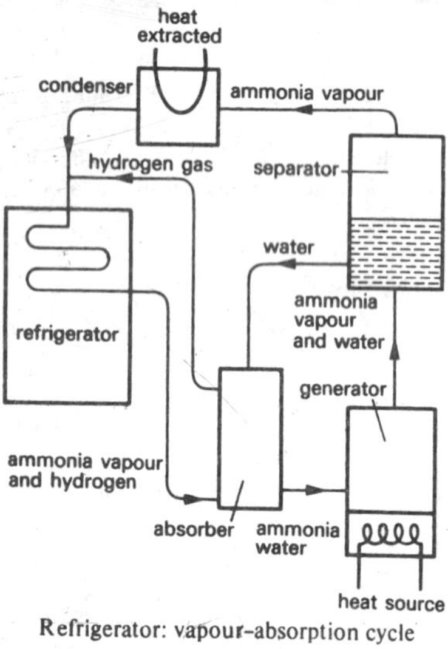

Refrigerator: vapour-absorption cycle

In the *vapour-absorption cycle*, there are no moving parts and energy is supplied as heat, either an electric heater or a gas burner. The refrigerant is usually ammonia, which is liberated from a water solution and moved through the evaporator by a stream of hydrogen gas under pressure. Heat is applied to the generator, raising the ammonia and water vapour to the separator; the ammonia separates from the water and passes to the condenser, where it cools and liquefies, giving off its latent heat to the surroundings. The liquid ammonia is then mixed with hydrogen gas, which carries it through the evaporator and helps in the process of evaporation. The hydrogen and ammonia vapour then enter the absorber, where the water returned from the separator dissolves the ammonia before returning to the generator.

Small-scale refrigeration is sometimes achieved by means of the *Seebeck effect and the *Peltier effect at junctions of *n*- and *p*-type semiconductors.

Regnault's method A technique for measuring gas density by evacuating and weighing a glass bulb of known volume, admitting gas at known pressure, and reweighing. The determination must be carried out at constant known temperature and the result corrected to standard temperature and pressure. The method is named after the French chemist Henri Victor Regnault (1810–78).

relative aperture *See* aperture.

relative atomic mass (atomic weight; r.a.m.) Symbol A_r. The ratio of the average mass per atom of the naturally occurring form of an element to 1/12 of the mass of a carbon-12 atom.

relative density (r.d.) The ratio of the *density of a substance to the density of some reference substance. For liquids or solids it is the ratio of the density (usually at 20°C) to the density of water (at its maximum density). This quantity was formerly called *specific gravity*. Sometimes relative densities of gases are used; for example, relative to dry air, both gases being at s.t.p.

relative humidity *See* humidity.

relative molecular mass (molecular weight) Symbol M_r. The ratio of the average mass per molecule of the naturally occurring form of an element or compound to 1/12 of the mass of a carbon-12 atom. It is equal to the sum of the relative atomic masses of all the atoms that comprise a molecule.

relative permeability *See* permeability.

relative permittivity *See* permittivity.

relativistic mass The mass of a moving body as measured by a stationary observer in the same frame of reference as the body. According to the special theory of *relativity the mass m of a body moving at speed v is given by $m = m_0/\sqrt{(1 - v^2/c^2)}$, where m_0 is its *rest mass and c is the speed of light. The relativistic mass therefore only differs significantly from the rest mass if its speed is a substantial fraction of the speed of light. If $v = c/2$, for example, the relativistic mass is 15% greater than the rest mass.

relativistic mechanics An extension of Newtonian mechanics that takes into account the theory of *relativity.

relativistic quantum theory *See* quantum theory.

relativistic speed A speed that is sufficiently large to make the mass of a body significantly greater than its *rest mass. It is usually expressed as a proportion of the speed of light. *See* relativity; relativistic mass.

relativity One of several theories concerning motion, designed to account for departures from *Newtonian mechanics that occur with very high-speed relative motion. The theory implied is now usually one of two proposed by Albert Einstein (1879–1955). The *special theory* was proposed in 1905 and referred to inertial (nonaccelerated) frames of reference. It assumes that physical laws are identical in all frames of reference and that the speed of light *in vacuo*, c, is constant throughout the universe and is independent of the speed of the observer. The theory develops a system of mathematics to reconcile these apparently conflicting statements. A conclusion of the theory is that the mass of a body, m, increases with its velocity, v, according to the relationship $m = m_0/\sqrt{(1 - v^2/c^2)}$, where m_0 is the *rest mass of the body. Einstein also concluded that if a body loses energy L, its mass diminishes by L/c^2. Einstein generalized this conclusion to the far-reaching postulate that the mass of a body is a measure of its energy content, according to the equation $m = E/c^2$ (or $E = mc^2$ in its more familiar form).

In the *general theory* of 1915, Einstein extended his earlier work to include accelerated systems, which led to his analysis of gravitation. He interpreted the universe in terms of a four-dimensional space–time continuum in which the presence of mass curves space in such a way that the gravitational field is created. The small differences between Newton's interpretation of gravitation and Einstein's interpretation have provided a means of testing the two theories. For example, the motion of the planet Mercury, thought to be anomalous in terms of Newtonian mechanics, can be explained by relativity. Moreover, Einstein's prediction that light rays passing close to the sun would be bent by its gravitational field has also been confirmed, at least qualitatively, by experiments performed during solar eclipses. The general theory is one of several current theories of gravitation; most of these have a good deal in common with it.

relaxation oscillator *See* oscillator.

relay An electrical or electronic device in which a variation in the current in one circuit controls the current in a second circuit. These devices are used in an enormous number of different applications in which electrical control is required. The simplest type is the electromechanical relay in which the first circuit energizes an electromagnet, which operates a switch in a second circuit. The *thyratron gas-filled relay found many uses in the past but has now been largely replaced by the *thyristor solid-state relay.

reluctance Symbol *R*. The ratio of the magnetomotive force to the total magnetic flux in a magnetic circuit or component. It is measured in henries per metre.

reluctivity The reciprocal of magnetic *permeability.

rem *See* radiation units.

remanence (retentivity) The magnetic flux density remaining in a ferromagnetic substance when the saturating field is reduced to zero. *See* hysteresis.

remote sensing The gathering and recording of information concerning the earth's surface by techniques that do not involve actual contact with the object or area under study. These techniques include photography (e.g. aerial photography), multispectral imagery, infrared imagery, and radar. Remote sensing is generally carried out from aircraft and, increasingly, satellites. The techniques are used, for example, in cartography (map making).

resistance Symbol *R*. The ratio of the potential difference across an electrical component to the current passing through it. It is thus a measure of the component's opposition to the flow of electric charge. In general, the resistance of a metallic conductor increases with temperature, whereas the resistance of a *semiconductor decreases with temperature.

resistance thermometer (resistance pyrometer) A *thermometer that relies on the increase of electrical resistance of a metal wire with rising temperature, according to the approximate relationship $R = R_0(1 + aT + bT^2)$, where R is the resistance of the wire at temperature T and R_0 is the resistance of the wire at a reference temperature, usually 0°C; a and b are constants characteristic of the metal of the wire. The metal most frequently used is platinum and the platinum resistance coil is usually incorporated into one arm of a *Wheatstone bridge. The effect of the temperature change on the leads carrying current to the platinum coil is compensated by including a pair of dummy leads within the casing carrying the coil. *See also* thermistor.

resistivity Symbol ρ. A measure of a material's ability to oppose the flow of an electric current. The resistivity of a material is given by RA/l, where R is the resistance of a uniform specimen of the material, having a length l and a cross-sectional area A. It is usually given at 0°C or 20°C and is measured in ohm metres. It was formerly known as *specific resistance*.

resistor A component in an electrical or electronic circuit that is present because of its electrical resistance. For electronic purposes many resistors are either wire-wound or consist of carbon particles in a ceramic binder. The ceramic coating carries a number or colour code indicating the value of the resistance. Some resistors can be varied manually by means of a sliding contact; others are markedly dependent on temperature or illumination.

resolution The separation of a vector quantity into two components, which are usually at right angles to each other. Thus, a force F acting on a body in a vertical plane at an angle θ to the horizontal can be resolved into a horizontal component $F\cos\theta$ and a vertical component $F\sin\theta$, both in the same plane as the original force.

resolving power A measure of the ability of an optical instrument to form separable images of close objects or to separate close wavelengths of radiation. The *chromatic resolving power* for any spectroscopic instrument is equal to $\lambda/\delta\lambda$, where $\delta\lambda$ is the difference in wavelength of two equally strong spectral lines that can barely be separated by the instrument and λ is the average wavelength of these two lines. For a telescope forming images of stars the *angular resolving power* is the smallest angular separation of the images; the *linear resolving power* is the linear separation of the images in the focal plane. In a telescope forming images of two stars, as a result of diffraction by the lens aperture each image consists of a bright central blob surrounded by light and dark rings. According to the *Rayleigh criterion* for resolution, the central ring of one image should fall on the first dark ring of the other. The angular resolving power in radians is then $1.22\lambda/d$, where d is the diameter of the objective lens in centimetres and λ is the wavelength of the light (usually taken as 560 nanometres). For microscopes, the resolving power is usually taken as the minimum distance between two points that can be separated. In both cases, the smaller the resolving power, the better the resolution; to avoid this apparent paradox the resolving power is now sometimes taken as the reciprocals of the quantities stated above.

resonance 1. An oscillation of a system at its natural frequency of vibration, as determined by the physical parameters of the system. It has the characteristic that large amplitude vibrations will ultimately result from low-power driving of the system. Resonance can occur in atoms and molecules, mechanical systems, and electrical circuits (*see* resonant circuit; resonant cavity). **2.** A very short-lived *elementary particle that can be regarded as an excited state of a more stable particle. Resonances decay by the strong interaction (*see* fundamental interactions) in 10^{-24} second.

resonant cavity (cavity resonator) A closed space within a conductor in which an electromagnetic field can be made to oscillate at frequencies above those at which a *resonant circuit will operate. The resonant frequency of the oscillation will depend on the dimensions and the shape of the cavity. The device is used to produce microwaves (*see* klystron; magnetron).

resonant circuit A reactive circuit (*see* reactance) so arranged that it is capable of *resonance. In a *series resonant circuit* a resistor, inductor, and capacitor are arranged in series. Resonance occurs when the *impedance is a minimum and the current amplitude therefore a maximum. In a *parallel resonant circuit* the inductance and capacitance are in parallel and resonance (with minimal current amplitude) occurs at maximum impedance. The frequency at which resonance occurs is called the *resonant frequency*. In a series resonant circuit the impedance Z is $R + \mathrm{i}[\omega L - 1/\omega C]$, where $\omega = 2\pi f$ and f is the frequency, R is the resistance, L is the inductance, and C is the capacitance. At resonance, Z is a minimum and $\omega L = 1/\omega C$, i.e. the circuit behaves as if it is purely resistive. In the parallel circuit, resonance occurs when $R^2 + \omega^2L^2 = L/C$, which in most cases also approximates to $\omega L = 1/\omega C$. Resonant circuits are widely used in *radio to select one carrier-wave frequency in preference to others.

rest energy The *rest mass of a body expressed in energy terms according to the relationship $E_0 = m_0c^2$, where m_0 is the rest mass of the body and c is the speed of light.

restitution coefficient Symbol e. A measure of the elasticity of colliding bodies. For two spheres moving in the same straight line, $e = (v_2 - v_1)/(u_1 - u_2)$, where u_1 and u_2 are the velocities of bodies 1 and 2 before collision ($u_1 > u_2$) and v_1 and v_2 are the velocities of 1 and 2 after impact ($v_2 > v_1$). If the collision is perfectly elastic $e = 1$ and the kinetic energy is conserved; for an inelastic collision $e < 1$.

rest mass The mass of a body at rest when measured by an observer who is at rest in the same frame of reference. *Compare* relativistic mass.

resultant A *vector quantity that has the same effect as two or more other vector quantities of the same kind. *See* parallelogram of vectors.

retardation (deceleration) The rate of reduction of speed, velocity, or rate of change.

retardation plate A transparent plate of a birefringent material, such as quartz, cut parallel to the optic axis. Light falling on the plate at 90° to the optic axis is split into an ordinary ray and an extraordinary ray (*see* double refraction), which travel through the plate at different speeds. By cutting the plate to different thicknesses a specific phase difference can be introduced between the transmitted rays. In the *half-wave plate* a phase difference of π radians, equivalent to a path difference of half a wavelength, is introduced. In the *quarter-wave plate* the waves are out of step by one quarter of a wavelength.

retina The light-sensitive membrane that lines the interior of the eye. The retina consists of two layers. The inner layer contains nerve cells, blood vessels, and two types of light-sensitive cells (*rods and *cones). The outer layer is pigmented, which prevents the back reflection of light and consequent decrease in visual acuity. Light passing through the lens stimulates individual rods and cones, which generates nerve impulses that are transmitted through the optic nerve to the brain, where the visual image is formed.

retrograde motion 1. The apparent motion of a planet from east to west as seen from the earth against the background of the stars. **2.** The clockwise rotation of a planet, as seen from its north pole. *Compare* direct motion.

retrorocket A small rocket motor that produces thrust in the opposite direction to a *rocket's main motor or motors in order to decelerate it.

reverberation time The time taken for the energy density of a sound to fall to the threshold of audibility from a value 10^6 times as great; i.e. a fall of 60 decibels. It is an important characteristic of an auditorium. The optimum value is proportional to the linear dimensions of the auditorium.

reverberatory furnace A metallurgical furnace in which the charge to be heated is kept separate from the fuel. It consists of a shallow hearth on which the charge is heated by flames that pass over it and by radiation reflected onto it from a low roof.

reverse osmosis A method of obtaining pure water from water containing a salt, as in desalination. Pure water and the salt water are separated by a semipermeable membrane and the pressure of the salt water is raised above the osmotic pressure, causing water from the brine to pass through the membrane into the pure water. This process requires a pressure of some 25 atmospheres, which makes it difficult to apply on a large scale.

reversible process Any process in which the variables that define the state of the system can be made to change in such a way that they pass through the same values in the reverse order when the process is reversed. It is also a condition of a reversible process that any exchanges of energy, work, or matter with the surroundings should be reversed in direction and order when the process is reversed. Any process that does not comply with these conditions when it is reversed is said to be an *irreversible process*. All natural processes are irreversible, although some processes can be made to approach closely to a reversible process.

Reynolds number Symbol (*Re*). A dimensionless number used in fluid dynamics to determine the type of flow of a fluid through a pipe, to design prototypes from small-scale models, etc. It is the ratio $v\rho l/\eta$, where v is the flow velocity, ρ is the fluid density, l is a characteristic linear dimension, such as the diameter of a pipe, and η is the fluid viscosity. In a pipe, laminar flow usually occurs if $(Re) < 2000$ and turbulent flow is established if $(Re) > 3000$.

rhe A unit of fluidity equal to the reciprocal of the *poise.

rheology The study of the deformation and flow of matter.

rheopexy The process by which certain thixotropic substances set more rapidly when they are stirred, shaken, or tapped. Gypsum in water is such a *rheopectic substance*.

rheostat A variable *resistor, the value of which can be changed without interrupting the current flow. In the common wire-wound rheostat, a sliding contact moves along the length of the coil of wire.

Richardson (Richardson–Dushman) equation *See* thermionic emission.

Richter scale A logarithmic scale devised in 1935 by C. F. Richter (1900–) to compare the magnitude of earthquakes. The scale ranges from 0 to 10 and the Richter scale value is related to the logarithm of the amplitude of the ground motion divided by the period of the dominant wave, subject to certain corrections. On this scale a value of 2 can just be felt as a tremor and damage to buildings occurs for values in excess of 6. The largest shock recorded had a magnitude of 8.9.

rigidity modulus *See* elastic modulus.

RMS value *See* root-mean-square value.

Rochelle salt Potassium sodium tartrate tetrahydrate, $KNaC_4H_4O_6.4H_2O$. A colourless crystalline salt used for its piezoelectric properties.

Rochon prism An optical device consisting of two quartz prisms; the first, cut parallel to the optic axis, receives the light; the second, with the optic axis at right angles, transmits the ordinary ray without deviation but the extraordinary ray is deflected and can be absorbed by a screen. The device can be used to produce plane-polarized light and it can also be used with ultraviolet radiation.

rock crystal *See* quartz.

rocket A space vehicle or projectile that is forced through space or the atmosphere by *jet propulsion and that carries its own propellants and oxidizers. It is therefore independent of the earth's atmosphere for lift, thrust, or oxygen and is the only known vehicle for travel outside the earth's atmosphere. Rocket motors (or rocket engines) are currently driven by solid or liquid chemical propellants, which burn in an oxidizer carried within the rocket. Typical liquid bipropellant combinations include hydrazine fuel with dinitrogen tetroxide oxidizer and kerosine fuel with nitric acid oxidizer. Experimental rocket motors have also been tested using ionized gases and plasmas to provide thrust (*see also* ion engine). The measure of a rocket motor's performance is its *specific impulse.

rod A type of light-sensitive receptor cell present in the retinas of vertebrates. Rods contain the pigment rhodopsin and are essential for vision in dim light. *Compare* cone.

roentgen The former unit of dose equivalent (*see* radiation units). It is named after the discoverer of X-rays, W. K. Roentgen (1845–1923).

rolling friction *Friction between a rolling wheel and the plane surface on which it is rotating. As a result of any small distortions of the two surfaces, the ideal line contact does not apply and there is a frictional force with a component, F_r, that opposes the motion. This component is proportional to the normal force, N, according to the relationship $F_r = N\mu_r$, where μ_r is called the *coefficient of rolling friction.*

root-mean-square value (RMS value) 1. (in statistics) A typical value of a number (n) of values of a quantity ($x_1, x_2, x_3 \ldots$) equal to the square root of the sum of the squares of the values divided by n, i.e.

$$\text{RMS value} = [\sqrt{(x_1{}^2 + x_2{}^2 + x_3{}^2 \ldots)/n}]$$

2. (in physics) A typical value of a continuously varying quantity, such as an alternating electric current, obtained similarly from many samples taken at regular time intervals during a cycle. Theoretically this can be shown to be the *effective value*, i.e. the value of the equivalent direct current that would produce the same power dissipation in a given resistor. For a sinusoidal current this is equal to $I_m/\sqrt{2}$, where I_m is the maximum value of the current.

Rose's metal An alloy of low melting point (about 100°C) consisting of 50% bismuth, 25–28% lead, and 22–25% tin.

rotary converter A device for converting direct current to alternating current or one d.c. voltage to another. It consists of an electric motor coupled to a generator.

rotational motion The laws relating to the rotation of a body about an axis are analogous to those describing linear motion. The *angular displacement* (θ) of a body is the angle in radians through which a point or line has been rotated in a specified sense about a specified axis. The *angular velocity* (ω) is the angular displacement divided by the time, i.e. $\omega = \mathrm{d}\theta/\mathrm{d}t$, and the *angular acceleration* (α) is the rate of increase of angular velocity, i.e. $\alpha = \mathrm{d}\omega/\mathrm{d}t = \mathrm{d}^2\theta/\mathrm{d}t^2$.

The equations of linear motion have analogous rotational equivalents, e.g.:

$$\omega_2 = \omega_1 + \alpha t$$
$$\theta = \omega_1 t + \alpha t^2/2$$
$$\omega_2{}^2 = \omega_1{}^2 + 2\theta$$

The counterpart of Newton's second law of motion is $T = I\alpha$, where T is the *torque causing the angular acceleration and I is the *moment of inertia of the rotating body.

rotor The rotating part of an electric motor, electric generator, turbine, etc. *Compare* stator.

rubidium–strontium dating A method of dating geological specimens based on the decay of the radioisotope rubidium-87 into the stable isotope strontium-87. Natural rubidium contains 27.85% of rubidium-87, which has a half-life of 4.7×10^{11} years. The ratio ^{87}Rb/^{87}Sr in a specimen gives an estimate of its age (up to several thousand million years).

Rydberg constant Symbol R. A constant that occurs in the formulae for atomic spectra and is related to the binding energy between an electron and an atomic nucleus. It is connected to other constants by the relationship $R = \mu_0{}^2 m e^4 c^3/8h^3$, where μ_0 is the magnetic constant (*see* permeability), m

and e are the mass and charge of an electron, c is the speed of light, and h is the *Planck constant. It has the value $1.097 \times 10^7 \text{ m}^{-1}$.

Rydberg spectrum An absorption spectrum of a gas in the ultraviolet region, consisting of a series of lines that become closer together towards shorter wavelengths, merging into a continuous absorption region. The absorption lines correspond to electronic transitions to successively higher energy levels. The onset of the continuum corresponds to photoionization of the atom or molecule, and can thus be used to determine the ionization pot⌃.tial.

S

sampling The selection of small groups of entities to represent a large number of entities in *statistics. In *random sampling* each individual of a population has an equal chance of being selected as part of the sample. In *stratified random sampling*, the population is divided into strata, each of which is randomly sampled and the samples from the different strata are pooled. In *systematic sampling*, individuals are chosen at fixed intervals; for example, every hundredth article on a production line. In *sampling with replacement*, each individual chosen is replaced before the next selection is made.

satellite 1. (*or* **natural satellite**) A relatively small natural body that orbits a planet. For example, the earth's only natural satellite is the moon. **2.** (*or* **artificial satellite**) A man-made spacecraft that orbits the earth, moon, sun, or a planet. Artificial satellites are used for a variety of purposes. *Communication satellites* are used for relaying telephone, radio, and television signals round the curved surface of the earth (*see* synchronous orbit). They are of two types: *passive satellites* reflect signals from one point on the earth's surface to another; *active satellites* are able to amplify and retransmit the signals that they pick up. *Astronomical satellites* are equipped to gather and transmit to earth astronomical information from space, including conditions in the earth's atmosphere, which is of great value in weather forecasting.

saturated 1. (of a ferromagnetic material) Unable to be magnetized more strongly as all the domains are orientated in the direction of the field. **2.** (of a solution) Containing the maximum equilibrium amount of solute at a given temperature. In a saturated solution the dissolved substance is in equilibrium with undissolved substance; i.e. the rate at which solute particles leave the solution is exactly balanced by the rate at which they dissolve. A solution containing less than the equilibrium amount is said to be *unsaturated*. One containing more than the equilibrium amount is *supersaturated*. Supersaturated solutions can be made by slowly cooling a saturated solution. Such solutions are metastable; if a small crystal seed is added the excess solute crystallizes out of solution. **3.** (of a vapour) *See* vapour pressure.

saturation 1. *See* colour. **2.** *See* supersaturation.

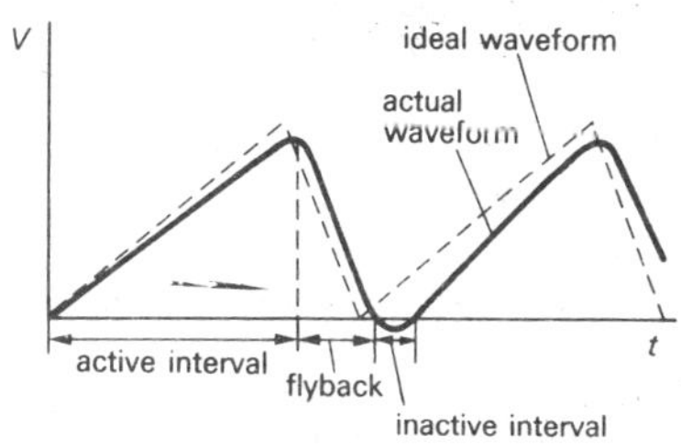

Sawtooth waveform

sawtooth waveform A waveform in which the variable increases uniformly with time for a fixed period, drops sharply to its initial value, and then repeats the sequence periodically. The illustration shows the ideal waveform and the waveform generated by practical electrical circuits. Sawtooth generators are frequently used to provide a time base for electronic circuits, as in the *cathode-ray oscilloscope.

scalar quantity A quantity in which direction is either not applicable (as in temperature) or not specified (as in speed). *Compare* vector quantity.

scalene Denoting a triangle having three unequal sides.

scaler (scaling circuit) An electronic counting circuit that provides an output when it has been activated by a prescribed number of input pulses. A *decade scaler* produces an output pulse when it has received ten or a multiple of ten input pulses; a *binary scaler* produces its output after two input pulses.

scanning The process of repeatedly crossing a surface or volume with a beam, aerial, or moving detector in order to bring about some change to the surface or volume, to measure [illegible] or to detect some object. The [illegible] creen of a television picture tube is scanned by an electron beam in order to produce the picture; an area of the sky may be scanned by the movable dish aerial of a radio telescope in order to detect celestial bodies, etc.

scanning electron microscope *See* electron microscope.

scattering of electromagnetic radiation The process in which electromagnetic radiation is deflected by particles in the matter through which it passes. In *elastic scattering* the photons of the radiation are reflected; i.e. they bounce off the atoms and molecules with unchanged energy and momentum. In this type of scattering, known as *Rayleigh scattering* (after Lord Rayleigh; 1842–1919), there is a change of phase but no frequency change. In *inelastic scattering*, there is interchange of energy between the photons and the particles. Consequently, the scattered photons have a different wavelength as well as a different phase. Examples of inelastic scattering are the *Raman effect and the *Compton effect. *See also* Tyndall effect.

Schlieren photography A technique that enables density differences in a moving fluid to be photographed. In the turbulent flow of a fluid, for example, short-lived localized differences in density create differences of refractive index, which show up on photographs taken by short flashes of light as streaks (German: Schliere). Schlieren photography is used in wind-tunnel studies to show the density gradients created by turbulence and the shock waves around a stationary model.

Schmidt camera *See* telescope.

Schottky defect *See* defect.

Schrödinger equation An equation used in wave mechanics (*see* quantum mechanics) for the wave function of a particle. The time-independent Schrödinger equation is

$$\nabla^2\psi + 8\pi^2 m(E - U)\psi/h^2 = 0$$

where ψ is the wave function; ∇^2 the Laplace operator, h the Planck constant, m the particle's mass, E its total energy, and U its potential energy.

Schwartzchild radius A critical radius of a body of given mass that must be exceeded if light is to escape from that body. It is equal to $2GM/c^2$, where G is the gravitational constant, c is the speed of light, and M is the mass of the body. If the body collapses to such an extent that its radius is less than the Schwartzchild radius the escape velocity becomes equal to the speed of light and the object becomes a *black hole.

The Schwartzchild radius is then the radius of the hole's event horizon.

scintillation counter A type of particle or radiation counter that makes use of the flash of light (scintillation) emitted by an excited atom falling back to its ground state after having been excited by a passing photon or particle. The scintillating medium is usually either solid or liquid and is used in connection with a *photomultiplier, which produces a pulse of current for each scintillation. The pulses are counted with a *scaler. In certain cases, a pulse-height analyser can be used to give an energy spectrum of the incident radiation.

scintillator *See* phosphor.

sclerometer A device for measuring the hardness of a material by determining the pressure on a standard point that is required to scratch it or by determining the height to which a standard ball will rebound from it when dropped from a fixed height. The rebound type is sometimes called a *scleroscope*.

scotopic vision The type of vision that occurs when the *rods in the eye are the principal receptors, i.e. when the level of illumination is low. With scotopic vision colours cannot be identified. *Compare* photopic vision.

screen grid A wire grid in a tetrode or pentode *thermionic valve, placed between the anode and the *control grid to reduce the grid–anode capacitance. *See also* suppressor grid.

screw A simple *machine effectively consisting of an inclined plane wrapped around a cylinder. The mechanical advantage of a screw is $2\pi r/p$, where r is the radius of the thread and p is the *pitch*, i.e. the distance between adjacent threads of the screw measured parallel to its axis.

search coil A small coil in which a current can be induced to detect and measure a magnetic field. It is used in conjunction with a *fluxmeter.

Searle's bar An apparatus for determining the thermal conductivity of a bar of material. The bar is lagged and one end is heated while the other end is cooled, by steam and cold water respectively. At two points d apart along the length of the bar the temperature is measured using a thermometer or thermocouple. The conductivity can then be calculated from the measured temperature gradient.

second 1. Symbol s. The SI unit of time equal to the duration of 9 192 631 770 periods of the radiation corresponding to the transition between two hyperfine levels of the ground state of the caesium-133 atom. **2.** Symbol ″. A unit of angle equal to 1/3600 of a degree or 1/60 of a minute.

secondary cell A *voltaic cell in which the chemical reaction producing the e.m.f. is reversible and the cell can therefore be charged by passing a current through it. *See* accumulator. *Compare* primary cell.

secondary colour Any colour that can be obtained by mixing two *primary colours. For example, if beams of red light and green light are made to overlap, the secondary colour, yellow, will be formed. Secondary colours of light are sometimes referred to as the pigmentary primary colours. For example, if transparent yellow and magenta pigments are overlapped in white light, red will be observed. In this case the red is a pigmentary secondary although it is a primary colour of light.

secondary emission The emission of electrons from a surface as a result of the impact of other charged particles, especially as a result of bombardment with (primary) electrons. As the number of secondary electrons can exceed the number of primary electrons, the process is important in *photomultipliers. *See also* Auger effect.

secondary winding The winding on the output side of a *transformer or *induction coil. *Compare* primary winding.

secular magnetic variation *See* geomagnetism.

sedimentation The settling of the solid particles through a liquid either to produce a concentrated slurry from a dilute suspension or to clarify a liquid containing solid particles. Usually this relies on the force of gravity, but if the particles are too small or the difference in density between the solid and liquid phases is too small, a *centrifuge may be used. In the simplest case the rate of sedimentation is determined by *Stokes's law, but in practice the predicted rate is rarely reached. Measurement of the rate of sedimentation in an *ultracentrifuge can be used to estimate the size of macromolecules.

Seebeck effect (thermoelectric effect) The generation of an e.m.f. in a circuit containing two different metals or semiconductors, when the junctions between the two are maintained at different temperatures. The magnitude of the e.m.f. depends on the nature of the metals and the difference in temperature. The Seebeck effect is the basis of the *thermocouple. It was named after T. J. Seebeck (1770–1831), who actually found that a magnetic field surrounded a circuit consisting of two metal conductors only if the junctions between the metals were maintained at different temperatures. He wrongly assumed that the conductors were magnetized directly by the temperature difference. *Compare* Peltier effect.

Seger cones (pyrometric cones) A series of cones used to indicate the temperature inside a furnace or kiln. The cones are made from different mixtures of clay, limestone, feldspars, etc., and each one softens at a different temperature. The drooping of the vertex is an indication that the known softening temperature has been reached and thus the furnace temperature can be estimated.

seismograph An instrument that records ground oscillations, e.g. those caused by earthquakes, volcanic activity, and explosions. Most modern seismographs are based on the inertia of a delicately suspended mass and depend on the measurement of the displacement between the mass and a point fixed to the earth. Others measure the relative displacement between two points on earth. The record made by a seismograph is known as a *seismogram*.

seismology The branch of geology concerned with the study of earthquakes.

selenium cell Either of two types of *photoelectric cell; one type relies on the photoconductive effect, the other on the photovoltaic effect (*see* photoelectric effect). In the photoconductive selenium cell an external e.m.f. must be applied; as the selenium changes its resistance on exposure to light, the current produced is a measure of the light energy falling on the selenium. In the photovoltaic selenium cell, the e.m.f. is generated within the cell. In this type of cell, a thin film of vitreous or metallic selenium is applied to a metal surface, a transparent film of another metal, usually gold or platinum, being placed over the selenium. Both types of cell are used as light meters in photography.

selenology The branch of astronomy concerned with the scientific study of the *moon.

self-exciting generator A type of electrical generator in which the magnets are excited by current from the generator output.

self inductance *See* inductance.

semiconductor A crystalline solid, such as silicon or germanium, with an electrical conductivity (typically 10^5-10^{-7} siemens per metre) intermediate between that of a conductor (up to 10^9 S m^{-1}) and an insulator (as low as 10^{-15} S m^{-1}). As the atoms in a crystalline solid are close together, the orbitals of their electrons overlap and their individual *energy levels are spread out into *energy bands. Conduction occurs in semiconductors as the result of a net movement, under the influence of an electric field, of electrons in the conduction band and empty states, called *holes*, in the valence band. A hole behaves as if it was an electron with a positive charge. Electrons and holes are known as the *charge carriers* in a semiconductor. The type of charge carrier that predominates in a particular region or material is called the *majority carrier* and that with the lower concentration is the *minority carrier.* An *intrinsic semiconductor* is one in which the concentration of charge carriers is a characteristic of the material itself; electrons jump to the conduction band from the valence band as a result of thermal excitation, each electron that makes the jump leaving behind a hole in the valence band. Therefore, in an intrinsic semiconductor the charge carriers are equally divided between electrons and holes. In *extrinsic semiconductors* the type of conduction that predominates depends on the number and valence of the impurity atoms present. Germanium and silicon atoms have a valence of four. If impurity atoms with a valence of five, such as arsenic, antimony, or phosphorus, are added to the lattice, there will be an extra electron per atom available for conduction, i.e. one that is not required to pair with the four valence electrons of the germanium or silicon. Thus extrinsic semiconductors doped with atoms of valence five give rise to crystals with electrons as majority carriers, the so-called *n-type conductors*. Similarly, if the impurity atoms have a valence of three, such as boron, aluminium, indium, or gallium, one hole per atom is created by an unsatisfied bond. The majority carriers are therefore holes, i.e. *p-type conductors*.

Semiconductor devices have virtually replaced thermionic devices, because they are several orders of magnitude smaller, cheaper in energy consumption, and more reliable. The basic structure for electronic semiconductor devices is the *semiconductor diode* (*see also* transistor). This consists of a silicon crystal doped in such a way that half is *p*-type and half is *n*-type. At the junction between the two halves there is a depletion layer in which electrons from the *n*-type have filled holes from the *p*-type. This sets up a potential barrier, which tends to keep the remaining electrons in the *n*-region and the remaining holes in the *p*-region. However, if the *p*-region is biased with a positive potential, the height of the barrier is reduced; the diode is said to be forward biased, because the majority holes in the *p*-region can then flow to the *n*-region and majority electrons in the *n*-region flow to the *p*-region. When forward biased there is a good current flow across the barrier. On the other hand if the *p*-region is negatively biased, the height of the potential barrier is increased and there is only a small leakage current of minority electrons from the *p*-region able to flow to the *n*-region. Thus the $p-n$ junction acts as an efficient rectifier, for which purpose it is widely used.

semiconductor diode *See* diode; semiconductor.

semipermeable membrane A membrane that is permeable to molecules of the solvent but not the solute in *osmosis. Semipermeable membranes can be made by supporting a film of material (e.g. cellulose) on a wire gauze or porous pot.

series A sequence of terms each of which can be written in a form that is an algebraic function of its position in the series. For example, the *exponential series $1 + x + x^2/2! + x^3/3!$ has an nth term $x^n/n!$. The sum of all the terms from $n = 0$ to $n = \infty$ is written:

$$\sum_{n=0}^{\infty} x^n/n!$$

This series has an infinite number of terms and is therefore called an *infinite series*. A *finite series* has a fixed number of terms. *See also* convergent series.

series circuit A circuit in which the circuit elements are arranged in sequence so that the same current flows through each of them in turn. For resistances in series, the total resistance is the sum of the individual resistances. For capacitors in series, the total capacitance, C, is given by $1/C = 1/C_1 + 1/C_2 + 1/C_3 \ldots.$

series-wound machine *See* shunt.

sets Groups of objects or elements that have at least one characteristic in common. For example, the set X may consist of all the elements x_1, x_2, x_3, etc. This is written $\{x_1, x_2, x_3, \ldots\} = X$. A specific element in a set is characterized by $x_1 \in X$, meaning x_1, is a member of set X. A *subset* of set X, say M, would be written $M \subset X$, i.e. M is contained in X. If x_3 is a member of both subsets M and N, then $x_3 \in (M \cap N)$, i.e. x_3 belongs to the *intersection* of M and N. $M \cup N$ means the *union* of M and N. For example, if M consists of {1, 4, 8} and N consists of {2, 3, 4}then $M \cup N = \{1, 2, 3, 4, 8\}$. In the diagram, the rectangle represents the universal set E, circles represent sets or subsets. These diagrams are called *Venn diagrams*, after John Venn (1834–1923), who invented them.

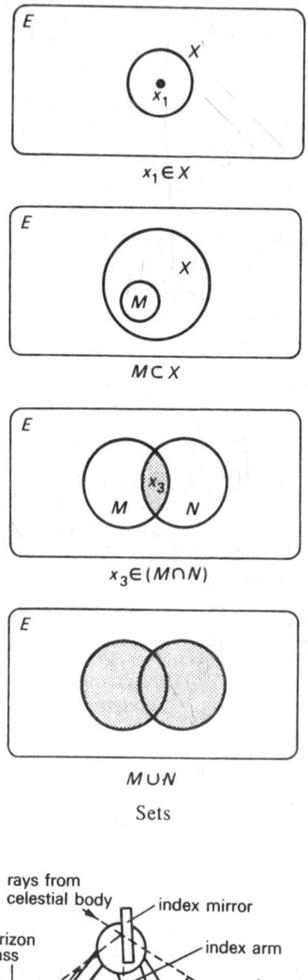

Sets

sextant An instrument used in navigation to measure the altitude of a celestial body. Originally it had an arc of 60° (one sixth of a circle, hence its name) but modern instruments have various angles. The sextant uses two mirrors: the horizon glass, in which only the lower half is silvered, and the in-

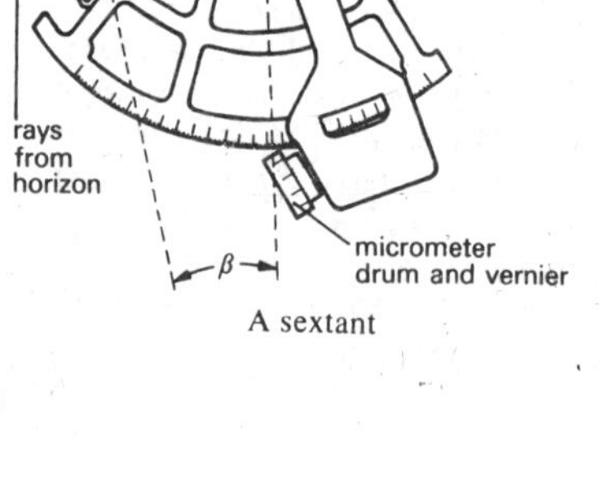

A sextant

dex mirror, which can be rotated about an axis perpendicular to the plane of the instrument. An arm attached to the index glass sweeps round the calibrated arc, from which angles are read. The instrument is aimed at the horizon and the index mirror rotated until the celestial object can also be seen through the telescope. After careful adjustment to make the image of the celestial body just touch the horizon, the angle is read off the graduated scale.

shadow An area of darkness formed on a surface when an object intercepts the light falling on the surface from a source. In the case of a point source the shadow has a sharply defined outline. If the source has an appreciable size the shadow has two distinct regions; one of full-shadow, called the *umbra*, the other of half-shadow, called the *penumbra* (see illustration).

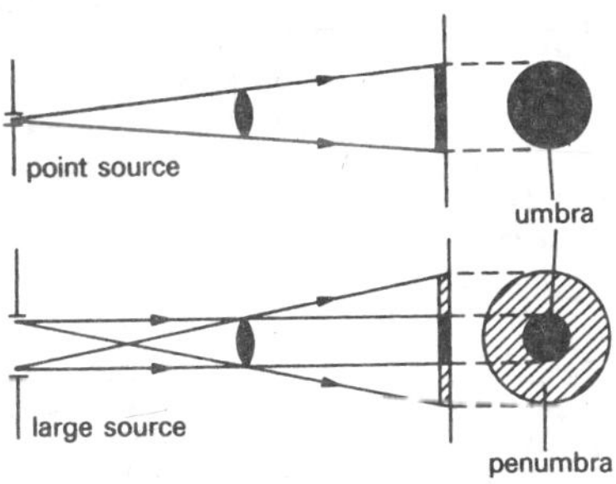

Shadows

shearing force A force that acts parallel to a plane rather than perpendicularly, as with a tensile or compressive force. A *shear stress* requires a combination of four forces acting over (most simply) four sides of a plane and produces two equal and opposite couples. It is measured as the ratio of one shearing force to the area over which it acts, $F/(ab)$ in the diagram. The shear strain is the angular deformation, θ, in circular measure. The *shear modulus* is the ratio of the shear

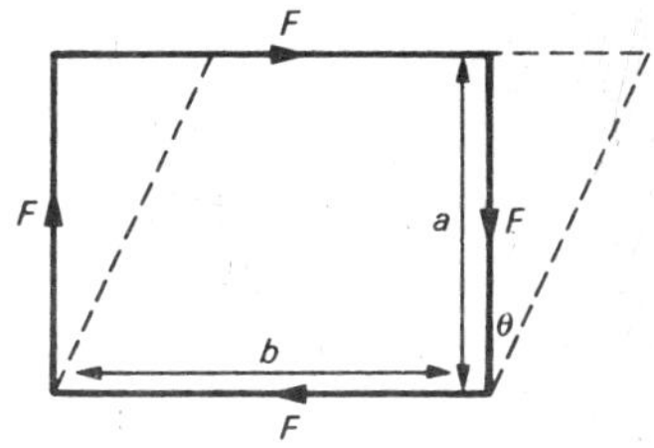

Shearing force

stress to the shear strain (*see also* elastic modulus).

shear modulus *See* elastic modulus; shearing force.

shell model 1. *See* atom. **2.** A model of the atomic nucleus in which nucleons are assumed to move under the influence of a central field in shells that are analogous to atomic electron shells. The model provides a good explanation of the stability of nuclei that have *magic numbers.

sherardizing The process of coating iron or steel with a zinc corrosion-resistant layer by heating the iron or steel in contact with zinc dust to a temperature slightly below the melting point of zinc. At a temperature of about 371°C the two metals amalgamate to form internal layers of zinc–iron alloys and an external layer of pure zinc. The process was invented by Sherard Cowper-Coles (d. 1935).

shielding 1. A barrier surrounding a region to exclude it from the influence of an energy field. For example, to protect a region from an electric field an earthed barrier is required; to protect it from a magnetic field a shield of high magnetic permeability is needed. **2.** A barrier used to surround a source of harmful or unwanted radiations. For example, the core of a *nuclear reactor is surrounded by a cement or lead shield to absorb neutrons and other dangerous radiation.

SHM *See* simple harmonic motion.

shock wave A very narrow region of high pressure and temperature formed in a fluid when the fluid flows supersonically over a stationary object or a projectile flying supersonically passes through a stationary fluid. A shock wave may also be generated by violent disturbances in a fluid, such as a lightning stroke or a bomb blast.

short-sightedness *See* myopia.

shower *See* cosmic radiation.

shunt An electrical resistor or other element connected in parallel with some other circuit or device, to take part of the current passing through it. For example, a shunt is used across the terminals of a galvanometer to increase the current that can pass through the system. A *shunt-wound* electric generator or motor is one in which the field winding is in parallel with the armature circuit. In a *series-wound* electrical machine the field coils and the armature circuit are in series.

sideband The band of frequencies above or below the frequency of the carrier wave in a telecommunications system within which the frequency components of the wave produced by *modulation fall. For example, if a carrier wave of frequency f is modulated by a signal of frequency x, the *upper sideband* will have a frequency $f + x$ and the *lower sideband* a frequency $f - x$.

sidereal day *See* day.

sidereal period The time taken for a planet or satellite to complete one revolution of its orbit measured with reference to the background of the stars. *See also* day; synodic period; year.

siemens Symbol S. The SI unit of electrical conductance equal to the conductance of a circuit or element that has a resistance of 1 ohm. $1\ S = 10^{-1}\ \Omega$. The unit was formerly called the mho or reciprocal ohm. It is named after Sir William Siemens (1823–83).

sievert The SI unit of dose equivalent (*see* radiation units).

signal The variable parameter that contains information and by which information is transmitted in an electronic system or circuit. The signal is created by the *signal generator*, often a voltage source in which the amplitude, frequency, and waveform can be varied.

sign convention A set of rules determined by convention for giving plus or minus signs to distances in the formulae involving lenses and mirrors. The *real-is-positive is the convention now usually adopted. The *New Cartesian convention* is now not widely used. In this convention distances to the right of the pole are treated as positive, those to the left as negative. This system has the advantage of conforming to the sign convention used with Cartesian coordinates in mathematics and is therefore preferred by some for the more complicated calculations.

significant figures The number of digits used in a number to specify its accuracy. The number 6.532 is a value taken to be accurate to four significant figures. The number 7320 is accurate only to three significant figures. Similarly 0.0732 is also only accurate to three significant figures. In these cases the zeros only indicate the order of magnitude of the number, whereas 7.065 is accurate to four significant figures as the zero in this case is significant in expressing the value of the number.

silicon chip A single crystal of a semiconducting silicon material, typically having micrometre dimensions, fabricated in such a way that it can perform a large number of independent electronic functions (*see* integrated circuit).

simple harmonic motion (SHM) A form of periodic motion in which a point or body oscillates along a line about a central point in such a way that it ranges an equal distance on either side of the central point and that its acceleration towards the central point is always proportional to its distance from it. One way of visualizing SHM is to imagine a point rotating around a circle of radius r at a constant angular velocity ω. If the distance from the centre of the circle to the projection of this point on a vertical diameter is y at time t, this projection of the point will move about the centre of the circle with simple harmonic motion. A graph of y against t will be a sine wave, whose equation is $y = r\sin\omega t$ (see diagram). *See also* pendulum.

sine wave (sinusoidal wave) Any waveform that has an equation in which one variable is proportional to the sine of the other. Such a waveform can be generated by an oscillator that executes *simple harmonic motion.

sintering The process of heating and compacting a powdered material at a temperature below its melting point in order to weld the particles together into a single rigid shape. Materials commonly sintered include metals and alloys, glass, and ceramic oxides. Sintered magnetic materials, cooled in a magnetic field, make especially retentive permanent magnets.

sinusoidal oscillator *See* oscillator.

sinusoidal wave *See* sine wave.

siphon An inverted U-tube with one limb longer than the other. Liquid will be transferred from a reservoir at the base of the shorter limb to the end of the longer limb, provided that the U-tube is filled with liquid. The device is useful for emptying an inaccessible container, such as a car's petrol tank.

The pressure (p_1) on the liquid at the base of the short limb (length h_1) is $p - h_1k$, where p is the atmospheric pressure and k is a constant equal to the product of the density of the liquid and the acceleration of free fall. The pressure (p_2) on the liquid at the base of the long limb (length h_2) is $p - h_2k$. Thus for fluid flow to occur through the tube, from short limb to long limb, $p_1 > p_2$, and for this to occur $h_2 > h_1$. Thus if the limbs are of equal length no flow will occur; it will only occur if the limb dipping into the reservoir is shorter than the delivering limb.

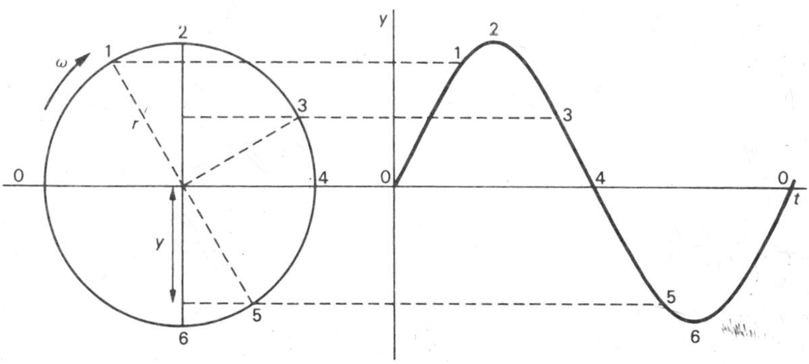

Simple harmonic motion

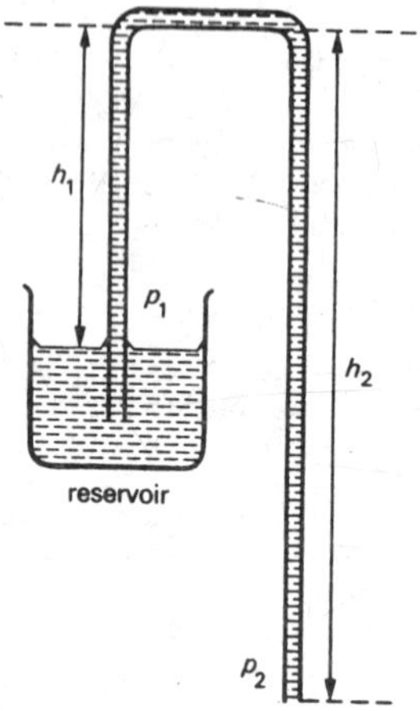

Siphon

SI units Système International d'Unités: the international system of units now recommended for all scientific purposes. A coherent and rationalized system of units derived from the *m.k.s. units, SI units have now replaced *c.g.s. units and *Imperial units. The system has seven *base units* and two *supplementary units* (see Appendix), all other units being derived from these nine units. There are 18 derived units with special names. Each unit has an agreed symbol (a capital letter or an initial capital letter if it is named after a scientist, otherwise the symbol consists of one or two lower-case letters). Decimal multiples of the units are indicated by a set of prefixes; whenever possible a prefix representing 10 raised to a power that is a multiple of three should be used.

skip distance The minimum distance from the transmitter of a radio wave at which reception is possible by means of a sky wave (*see* radio transmission). If a radio wave strikes the ionosphere at a small angle of incidence the wave passes through it and is not reflected. There is therefore a minimum angle of incidence at which reflection occurs for a given frequency. This leads to a region around a transmitter in which sky waves cannot be received. As the frequency of the transmission increases the minimum angle of incidence at which ionospheric reflection occurs becomes greater. Above about 4 megahertz there may be a region of several hundred kilometres around a transmitter, which is within the skip distance and in which ground waves are too attenuated to be effectively received. In this region no reception is possible.

sky wave *See* radio transmission.

slow neutron A neutron with a kinetic energy of less than 10^2 eV (10^{-17} joule). *See also* fast neutron; thermalization.

slug An f.p.s. unit of mass equal to the mass that will acquire an acceleration of 1 ft sec^{-2} when acted on by a force of one pound-force.

smectic *See* liquid crystal.

smelting The process of separating a metal from its ore by heating the ore to a high temperature in a suitable furnace in the presence of a reducing agent, such as carbon, and a fluxing agent, such as limestone. Iron ore is smelted in this way so that the metal melts and, being denser than the molten slag, sinks below the slag, enabling it to be removed from the furnace separately.

smoke A fine suspension of solid particles in a gas.

Snell's law *See* refraction.

sodium-vapour lamp A form of *electric lighting that gives a yellow light as a result of the luminous discharge obtained by the passage of a stream of electrons between tungsten electrodes in a tube containing sodium vapour. To facilitate starting, the tube also contains some neon; for this reason, until the lamp is warm the neon emits a characteristic pink glow. As the sodium vaporizes, the yellow light predominates. Sodium-vapour lamps are widely used as street lights because of their high luminous efficiency and because the yellow light is less absorbed than white light by fog and mist. Low-pressure sodium lamps emit a characteristic yellow light; in high-pressure lamps the atoms are sufficiently close to each other to interact and broaden the spectral lines into the orange and green regions.

soft iron A form of iron that contains little carbon, has high relative permeability, is easily magnetized and demagnetized, and has a small hysteresis loss. Soft iron and other *soft ferromagnetic materials*, such as silicon steel, are used in making parts exposed to rapid changes of magnetic flux, such as the cores of electromagnets, motors, generators, and transformers.

By comparison, *hard ferromagnetic materials*, such as cobalt steel and various alloys of nickel, aluminium, and cobalt, have low relative permeability, are difficult to magnetize, and have a high hysteresis loss. They are used in making permanent magnets.

soft radiation Ionizing radiation of low penetrating power, usually used with reference to X-rays of long wavelength. *Compare* hard radiation.

software *See* computer.

sol A *colloid in which small solid particles are dispersed in a liquid continuous phase.

solar cell An electric cell that converts the sun's electromagnetic energy into a usable electric current. Most solar cells consist of a single-crystal silicon $p-n$ junction. When photons of light energy from the sun fall on or near the *semiconductor junction the electron–hole pairs created are forced by the electric field at the junction to separate so that the holes pass to the p-region and the electrons pass to the n-region. This displacement of free charge creates an electric current when a load is connected across the terminals of the device (see illustration). Individual silicon solar cells cannot be made with a surface area much in excess of 4000 mm^2 and the maximum power delivered by such a cell is approximately 0.6 W at about 0.5 V in full sun. The efficiency of such devices is about 15%. For practical use, therefore, solar cells have to be assembled in arrays. Panels of solar cells have been the exclusive source of power for satellites and space capsules. Their use on earth has been largely limited by their high cost, a reduction in the cost by a factor of 10 being required to make them competitive with other energy sources at present.

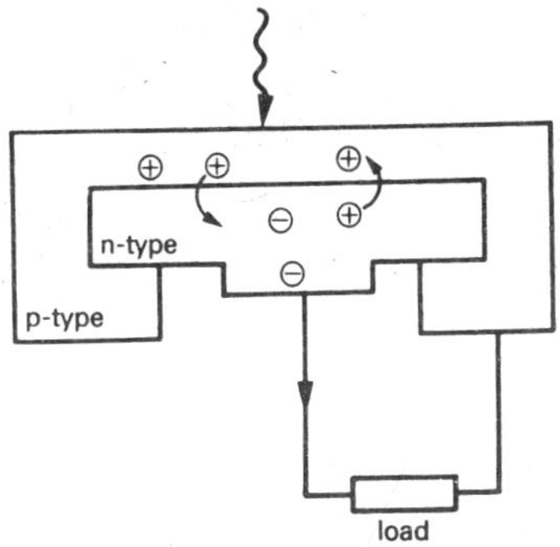

Silicon p–n junction

solar constant The rate at which solar energy is received per unit area at the outer limit of the earth's atmosphere at the mean distance between the earth and the sun. The value is 1.353 kW m^{-2}.

solar day *See* day.

solar energy The electromagnetic energy radiated from the sun. The tiny proportion (about 5×10^{-10} of the total) that falls on the earth is indicated by the *solar constant. The total quantity of solar energy falling on the earth in one year is about 4×10^{18} J, whereas the total annual energy consumption of the earth's inhabitants is only some 3×10^{14} J. The sun, therefore, could provide all the energy needed. The direct ways of making use of solar energy can be divided into thermal methods (*see* solar heating) and nonthermal methods (*see* solar cell).

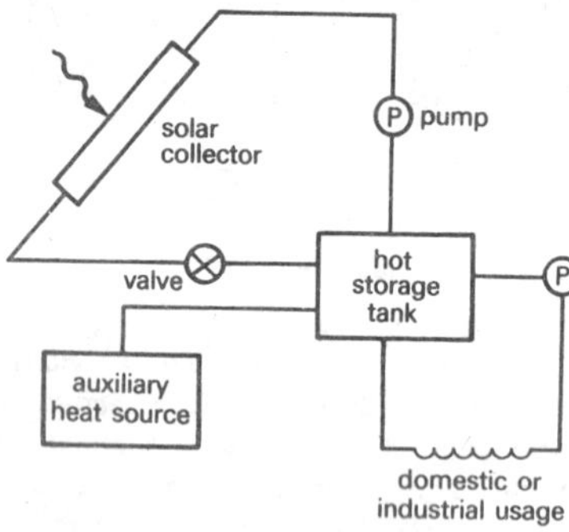

Typical solar heating system

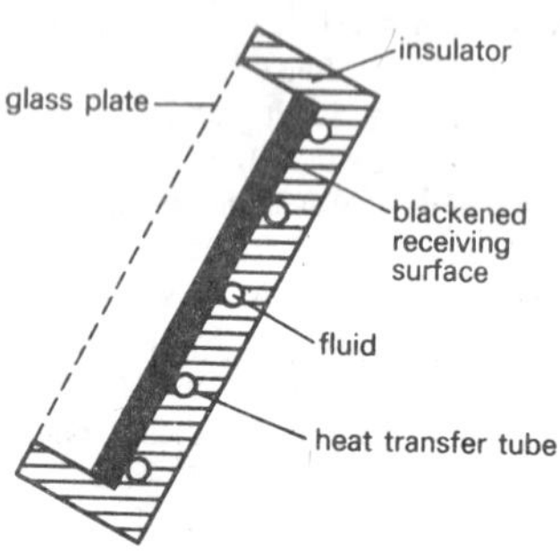

Flat plate solar collector

solar heating A form of domestic or industrial heating that relies on the direct use of solar energy. The basic form of *solar heater* is a thermal device in which a fluid is heated by the sun's rays in a collector (see illustration) and pumped or allowed to flow round a circuit that provides some form of heat storage and some form of auxiliary heat source for use when the sun is not shining. More complicated systems are combined heating-and-cooling devices, providing heat in the winter and air-conditioning in the summer. The simplest form of solar collector is the flat-plate collector, in which a blackened receiving surface is covered by one or more glass plates that acts like a greenhouse (*see* greenhouse effect) and traps the maximum amount of solar energy Tubes attached to the receiving surface carry air, water, or some other fluid to which the absorbed heat is transferred. The whole panel is insulated at the back and can thus form part of the roof of a building. More sophisticated collectors focus the sun's rays using reflectors. *See also* solar cell.

solar parallax The angle subtended by the earth's equatorial radius at the centre of the sun at the mean distance between the earth and the sun (i.e. at 1 astronomical unit). It has the value 8.794 148 arc seconds.

solar prominence A cloud of gas that forms temporarily in the upper *chromosphere or inner *corona of the sun. It has a lower temperature but higher density than its surroundings and is observed as a bright projection.

solar system The sun, the nine major planets (Mercury, Venus, Earth, Mars, Jupiter, Saturn, Uranus, Neptune, and Pluto) and their natural satellites, the asteroids, the comets, and meteoroids. Over 99% of the mass of the system is concentrated in the sun. The solar system as a whole moves in an approximately circular orbit about the centre of the Galaxy, taking about 2.2×10^8 years to complete its orbit.

solar wind A continuous outward flow of charged particles, mostly protons and electrons, from the sun's *corona into interplanetary space. The particles are controlled by the sun's magnetic field and are able to escape from the sun's gravitational field because of their high thermal energy. The average velocity of the particles in the vicinity of the earth is about 450 km s^{-1} and their density at this range is about 8×10^6 protons per cubic metre.

solar year *See* year.

solder An alloy used to join metal surfaces. A *soft solder* melts at a temperature in the range 200–300°C and consists of a tin–lead alloy. The tin content varies between 80% for the lower end of the melting range and 31% for the higher end. *Hard solders* contain substantial quantities of silver in the alloy. *Brazing solders* are usually alloys of copper and zinc, which melt at over 800°C.

solenoid A coil of wire wound on a cylindrical former in which the length of the former is greater than its diameter. When a current is passed through the coil a magnetic field is produced inside the coil parallel to its axis. This field can be made to operate a plunger inside the former so that the solenoid can be used to operate a circuit breaker, valve, or other electromechanical device.

solid A state of matter in which there is a three-dimensional regularity of structure, resulting from the proximity of the component atoms, ions, or molecules and the strength of the forces between them. True solids are crystalline (*see also* amorphous). If a crystalline solid is heated, the kinetic energy of the components increases. At a specific temperature, called the *melting point*, the forces between the components become unable to contain them within the crystal structure. At this temperature, the lattice breaks down and the solid becomes a liquid.

solid angle Symbol Ω. The three-dimensional 'angle' formed by the vertex of a cone. When this vertex is the centre of a sphere of radius r and the base of the cone cuts out an area s on the surface of the sphere, the solid angle in *steradians is defined as s/r^2.

solid solution A crystalline material that is a mixture of two or more components, with ions, atoms, or molecules of one component replacing some of the ions, atoms, or molecules of the other component in its normal crystal lattice. Solid solutions are found in certain alloys. For example, gold and copper form solid solutions in which some of the copper atoms in the lattice are replaced by gold atoms. In general, the gold atoms are distributed at random, and a range of gold–copper compositions is possible. At a certain composition, the gold and copper atoms can each form regular individual lattices (referred to as *superlattices*). Mixed crystals of double salts (such as alums) are also examples of solid solutions. Compounds can form solid solutions if they are isomorphous (*see* isomorphism).

solid-state detector *See* junction detector.

solid-state physics The study of the physical properties of solids, with special emphasis on the electrical properties of semiconducting materials in relation to their electronic structure. *Solid-state devices* are electronic components consisting entirely of solids (e.g. semiconductors, transistors, etc.) without heating elements, as in thermionic valves.

Recently the term *condensed-matter physics* has been introduced to include the study of crystalline solids, amorphous solids, and liquids.

solstice 1. Either of the two points on the *ecliptic midway between the *equinoxes, at which the sun is at its greatest angular distance north (*summer solstice*) or south (*winter solstice*) of the celestial equator. **2.** The time at which the sun reaches either of these points. The summer solstice occurs on June 21 and the winter solstice on December 21 in the northern hemisphere; the dates are reversed in the southern hemisphere.

solubility The quantity of solute that dissolves in a given quantity of solvent to form a saturated solution. Solubility is measured in kilograms per metre cubed, moles per kilogram of solvent, etc. The solubility of a substance in a given solvent depends on the temperature. Generally, for a solid in a liquid, solubility increases with temperature; for a gas, solubility decreases. *See also* concentration.

solubility product Symbol K_s. The product of the concentrations of ions in a saturated solution. For instance, if a compound A_xB_y is in equilibrium with its solution

$$A_xB_y(s) \rightleftharpoons xA^+(aq) + yB^-(aq)$$

the equilibrium constant is

$$K_c = [A^+]^x[B^-]^y/[A_xB_y]$$

Since the concentration of the undissolved solid can be put equal to 1, the solubility product is given by

$$K_s = [A^+]^x[B^-]^y$$

The expression is only true for sparingly soluble salts. If the product of ionic concentrations in a solution exceeds the solubility product, then precipitation occurs.

solute The substance dissolved in a solvent in forming a *solution.

solution A homogeneous mixture of a liquid (the *solvent) with a gas or solid (the *solute*). In a solution, the molecules of the solute are discrete and mixed with the molecules of solvent. There is usually some interaction between the solvent and solute molecules (*see* solvation). Two liquids that can mix on the molecular level are said to be *miscible*. In this case, the solvent is the major component and the solute the minor component. *See also* solid solution.

solvent A liquid that dissolves another substance or substances to form a *solution. *Polar solvents* are compounds such as water and liquid ammonia, which have dipole moments and consequently high dielectric constants. These solvents are capable of dissolving ionic compounds or covalent compounds that ionize (*see* solvation). *Nonpolar solvents* are compounds such as ethoxyethane and benzene, which do not have permanent dipole moments. These do not dissolve ionic compounds but will dissolve nonpolar covalent compounds.

sonar *See* echo.

sonic boom A strong *shock wave generated by an aircraft when it is flying in the earth's atmosphere at supersonic speeds. This shock wave is radiated from the aircraft and where it intercepts the surface of the earth a loud booming sound is heard. The loudness depends on the speed and altitude of the aircraft and is lower in level flight than when the aircraft is undertaking a manoeuvre. The maximum increase of pressure in the shock wave during a trans-oceanic flight of a commercial supersonic transport (SST) is 120 Pa, equivalent to 136 decibels.

sonometer A device consisting essentially of a hollow sounding box with two bridges attached to its top. The string, fixed to the box at one end, is stretched between the two bridges so that the free end can run over a pulley and support a measured load. When the string is plucked the frequency of the note can be matched with that of another sound source, such as a tuning fork. It can be used to verify that the frequency (f) of a stretched string is given by $f = (1/2l)\sqrt{(T/m)}$, where l is the length of the string, m is its mass per unit length, and T is its tension.

Originally called the *monochord*, the sonometer was widely used as a tuning aid, but is now used only in teaching laboratories.

sorption *Absorption of a gas by a solid.

sorption pump A type of vacuum pump in which gas is removed from a system by absorption on a solid (e.g. activated charcoal or a zeolite) at low temperature.

sound A vibration in an elastic medium at a frequency and intensity that is capable of being heard by the human ear. The frequency of sounds lie in the range 20–20 000 Hz, but the ability to hear sounds in the upper part of the frequency range declines with age (*see also* pitch). Vibrations that have a lower frequency than sound are called *infrasounds* and those with a higher frequency are called *ultrasounds*.

Sound is propagated through an elastic fluid as a longitudinal *sound wave*, in which a region of high pressure travels through the fluid at the *speed of sound in that medium. At a frequency of about 10 kilohertz the maximum excess pressure of a sound wave in air lies between 10^{-4} Pa and 10^{3} Pa. Sound travels through solids as either longitudinal or transverse waves.

source The electrode in a field-effect *transistor from which electrons or holes enter the interelectrode space.

space 1. A property of the universe that enables physical phenomena to be extended into three mutually perpendicular directions. In Newtonian physics, space, time, and matter are treated as quite separate entities. In Einsteinian physics, space and time are combined into a four-dimensional continuum (*see* space–time) and in the general theory of *relativity matter is regarded as having an effect on space, causing it to curve. **2.** (*or* **outer space**) The part of the universe that lies outside the earth's atmosphere.

space probe An unmanned spacecraft that investigates features within the solar system. A *planetary probe* examines the conditions on or in the vicinity of one or more planets and a *lunar probe* is designed to obtain information about the moon. Probes are propelled by rocket motors and once out of the earth's gravitational field use their propulsion systems for course changes. Many are powered by panels of *solar cells, for both internal operation and radio communications.

space-reflection symmetry *See* parity.

space–time (space–time continuum) A geometry that includes the three dimensions and a *fourth dimension* of time. In Newtonian physics, space and time are considered as separate entities and whether or not events are simultaneous is a matter that is regarded as obvious to any competent observer. In Einstein's concept of the physical universe, based on a system of geometry devised by H. Minkowski (1864–1909), space and time are regarded as entwined, so that two observers in relative motion could disagree regarding the simultaneity of distant events. In Minkowski's geometry, an event is identified by a *world point* in a four-dimensional continuum.

spallation A type of nuclear reaction in which the interacting nuclei disintegrate into a large number of protons, neutrons, and other light particles, rather than exchanging nucleons between them.

spark *See* electric spark.

spark chamber A device for detecting charged particles. It consists of a chamber, filled with helium and neon at atmospheric pressure, in which a stack of 20 to 100 plates are placed; the plates are connected alternately to the positive and negative terminals of a source of high potential (10 000 V or more). An incoming particle creates ion pairs in its track, causing the gas to become conducting and sparks to jump between the plates. The light from the sparks is focused to obtain stereoscopic photographs of the particles' tracks. It can also function as a counter (called a *spark counter*) when connected to suitable counting circuits. Some versions use crossed sets of parallel wires rather than plates; simple patterns may have a single wire near a plate, in the open atmosphere.

spark counter *See* spark chamber.

special theory of relativity *See* relativity.

specific 1. Denoting that an extensive physical quantity so described is expressed per unit mass. For example, the *specific latent heat* of a body is its latent heat per unit mass. When the extensive physical quantity is denoted by a capital letter (e.g. L for latent heat), the specific quantity is denoted by the corresponding lower-case letter (e.g. l for specific latent heat). **2.** In some older physical quantities the adjective 'specific' was added for other reasons (e.g. specific gravity, specific resistance). These names are now no longer used.

specific activity *See* activity.

specific charge The ratio of the charge of an *elementary particle or other charged body to its mass.

specific gravity *See* relative density; specific.

specific heat capacity *See* heat capacity.

specific humidity *See* humidity.

specific impulse A measure of the thrust available from a rocket propellant. It is the ratio of the thrust produced to the fuel consumption.

specific latent heat *See* latent heat.

specific resistance *See* resistivity; specific.

specific surface The surface area of a particular substance per unit mass, expressed in $m^2\ kg^{-1}$. It provides a measure of the surface area available for a process, such as adsorption, for a given mass of a powder or porous substance.

specific volume The volume of a substance per unit mass. The reciprocal of density, it has the units $m^3\ kg^{-1}$.

spectral class (spectral type) A form of classification used for stars, based on their spectra. The *Harvard classification*, introduced in 1890 and modified in the 1920s, is based on the seven star types known as O, B, A, F, G, K, M; see table below.

O hottest blue stars; ionized helium lines dominant

B hot blue stars; neutral helium lines dominant, no ionized helium

A blue blue-white stars; hydrogen lines dominant

F white stars; metallic lines strengthen, hydrogen lines weaken

G yellow stars; ionized calcium lines dominant

K orange-red stars; neutral metallic lines dominant, some molecular bands

M coolest red stars; molecular bands dominant

spectrograph *See* spectroscope.

spectrometer Any of various instruments for producing a spectrum and measuring the wavelengths, energies, etc., involved. A simple type, for visible radiation, is a spectroscope equipped with a calibrated scale allowing wavelengths to be read off or calculated. In the X-ray to infrared region of the electromagnetic spectrum, the spectrum is produced by dispersing the radiation with a prism or diffraction grating (or crystal, in the case of hard X-rays). Some form of photoelectric detector is used, and the spectrum can be obtained as a graphical plot, which shows how the intensity of the radiation varies with wavelength. Such instruments are also called *spectrophotometers.* Spectrometers also exist for investigating the gamma-ray region and the microwave and radio-wave regions of the spectrum (*see* electron-spin resonance; nuclear magnetic resonance). Instruments for obtaining spectra of particle beams are also called spectrometers (*see* spectrum; photoelectron spectroscopy).

spectrophotometer *See* spectrometer.

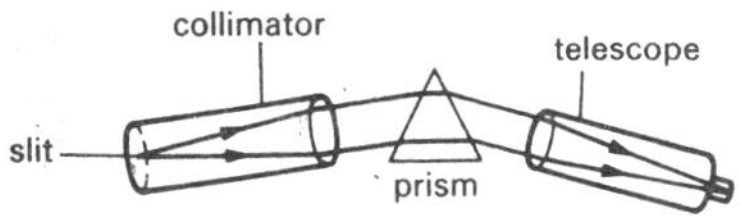

Spectroscope

spectroscope An optical instrument that produces a *spectrum for visual observation. The first such instrument was made by R. W. Bunsen; in its simplest form it consists of a hollow tube with a slit at one end by which the light enters and a collimating lens at the other end to produce a parallel beam, a prism to disperse the light, and a telescope for viewing the spectrum (see illustration). In the *spectrograph*, the spectroscope is provided with a camera to record the spectrum. For a broad range of spectroscopic work, from the ultraviolet to the infrared, a diffraction grating is used instead of a prism. *See also* spectrometer.

spectroscopic binary *See* binary stars.

spectroscopy The study of methods of producing and analysing *spectra using *spectroscopes, *spectrometers, spectrographs, and spectrophotometers. The interpretations of the spectra so produced can be used for chemical analysis, examining atomic and molecular energy levels and molecular structures, and for determining the composition and motions of celestial bodies (*see* redshift).

spectrum **1.** A distribution of entities or properties arrayed in order of increasing or decreasing magnitude. For example, a beam of ions passed through a mass spectrograph, in which they are deflected according to their charge-to-mass ratios, will have a range of masses called a *mass spectrum*. A *sound spectrum* is the distribution of energy over a range of frequencies of a particular source. **2.** A range of electromagnetic energies arrayed in order of increasing or decreasing wavelength or frequency (*see* electromagnetic spectrum). The *emission spectrum* of a body or substance is the characteristic range of radiations it emits when it is heated, bombarded by electrons or ions, or absorbs photons. The *absorption spectrum* of a substance is produced by examining, through the substance and through a spectroscope, a continuous spectrum of radiation. The energies removed from the continuous spectrum by the absorbing medium show up as black lines or bands; with a substance capable of emitting a spectrum these are in exactly the same positions in the spectrum as the emission lines and bands would occur in the emission spectrum.

Emission and absorption spectra may show a *continuous spectrum*, a *line spectrum*, or a *band spectrum*. A continuous spectrum contains an unbroken sequence of frequencies over a relatively wide range; it is produced by incandescent solids, liquids, and compressed gases. Line spectra are discontinuous lines produced by excited atoms and ions as they fall back to a lower energy level. Band spectra (closely grouped bands of lines) are characteristic of molecular gases or chemical compounds. *See also* spectroscopy.

speculum An alloy of copper and tin formerly used in reflecting *telescopes to make the main mirror as it could be cast, ground, and polished to make a highly reflective surface. It has now been largely replaced by silvered glass for this purpose.

speed The ratio of a distance covered by a body to the time taken. Speed is a *scalar quantity, i.e. no direction is given. Velocity is a *vector quantity, i.e. both the rate of travel and the direction are specified.

speed of light Symbol c. The speed at which electromagnetic radiation travels. The speed of light in a vacuum is $2.997\,924\,58 \times 10^8$ m s^{-1}. When light passes through any material medium its speed is reduced (*see* refractive index). The speed of light in a vacuum is the highest speed attainable in the universe (*see* relativity; Cerenkov radiation). It is a universal constant and is independent of the speed of the observer. Since October 1983 it has formed the basis of the definition of the *metre.

speed of sound Symbol c. The speed at which sound waves are propagated through a material medium. In air at 20°C sound travels at 344 m s^{-1}, in water at 20°C it travels at 1461 m s^{-1}, and in steel at 20°C at 5000 m s^{-1}. The speed of sound in a medium depends on the medium's modulus of elasticity (E) and its density (ρ) according to the relationship $c = \sqrt{(E/\rho)}$. For a solid E is the Young modulus; for a liquid E is the bulk modulus (*see* elastic modulus); and for a gas $E = \gamma p$, where γ is the ratio of the principal specific *heat capacities and p is the pressure of the gas. For an ideal gas the relationship takes the form $c = \sqrt{(\gamma rT)}$, where r is the gas constant per unit mass and T is the thermodynamic temperature. This equation shows how the speed of sound in a gas is related to its temperature. This relationship can be written $c = c_0\sqrt{(1 + t/273)}$, where c_0 is the speed of sound in a particular gas at 0°C and t is the temperature in °C.

sphere The figure generated when a circle is rotated about a diameter. The volume of a sphere is $4\pi r^3/3$ and its surface area is $4\pi r^2$, where r is its radius. In Cartesian coordinates the equation of a sphere centred at the origin is $x^2 + y^2 + z^2 = r^2$.

spherical aberration *See* aberration.

spherical mirror *See* mirror.

spherical polar coordinates *See* polar coordinates.

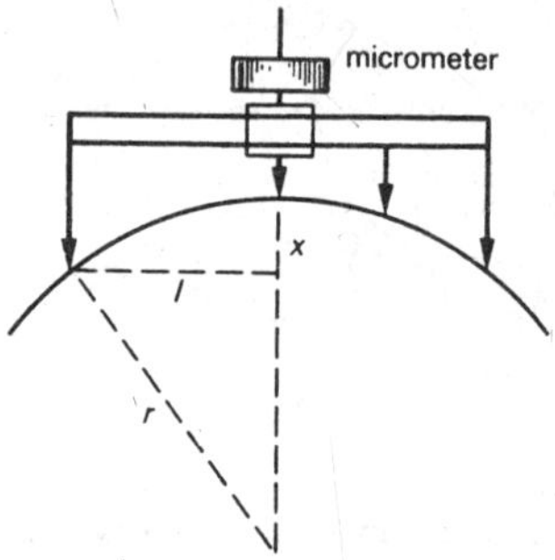

Spherometer

spherometer An instrument for measuring the curvature of a surface. The usual instrument for this purpose consists of a tripod, the pointed legs of which rest on the spherical surface at the corners of an equilateral triangle. In the centre of this triangle is a fourth point, the height of which is adjusted by means of a micrometer screw. If the distance between each leg and the axis through the micrometer screw is l, and the height of the micrometer point above (or below) a flat surface is x, the radius (r) of the sphere is given by $r = (l^2 + x^2)/2x$.

spiegel (spiegeleisen) A form of *pig iron containing 15–30% of manganese and 4–5% of carbon. It is added to steel in a Bessemer converter as a deoxidizing agent and to raise the manganese content of steel.

spin (intrinsic angular momentum) Symbol s. The part of the total angular momentum of a particle, atom, nucleus, etc., that can continue to exist even when the particle is apparently at rest, i.e. when its translational motion is zero and therefore its orbital angular momentum is zero. A molecule, atom, or nucleus in a specified energy level, or a particular elementary particle, has a particular spin, just as it has a particular charge or mass. According to *quantum theory, this is quantized and is restricted to multiples of $h/2\pi$, where h is the *Planck constant. Spin is characterized by a quantum number s. For example, for an electron $s = \pm\frac{1}{2}$, implying a spin of $+h/4\pi$ when it is spinning in one direction and $-h/4\pi$ when it is spinning in the other. Because of their spin, charged particles also have their own intrinsic *magnetic moments and in a magnetic field the spin of the particles lines up at an angle to the direction of the field, precessing around this direction.

spiral galaxy *See* galaxy.

spring balance A simple form of *balance in which a force is measured by the extension it produces in a helical spring. The extension, which is read off a scale, is directly proportional to the force, provided that the spring is not overstretched. The device is often used to measure the weight of a body approximately.

sputtering The process by which some of the atoms of an electrode (usually a cathode) are ejected as a result of bombardment by heavy positive ions. Although the process is generally unwanted, it can be used to produce a clean surface or to deposit a uniform film of a metal on an object in an evacuated enclosure.

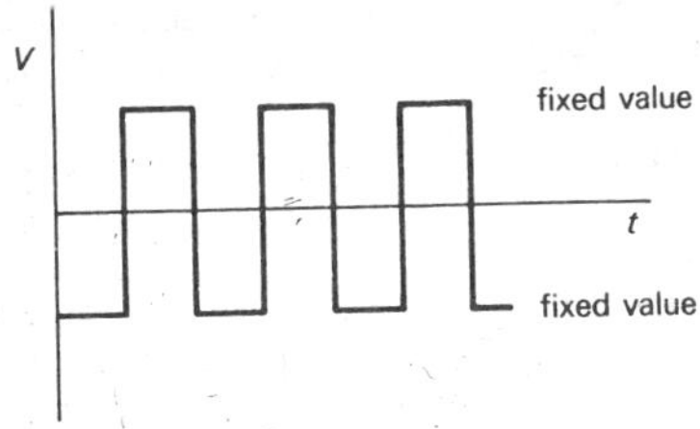

Square wave

square wave A train of rectangular voltage pulses that alternate between two fixed values for equal lengths of time. The time of transition between each fixed value is negligible compared to the duration of the fixed value.

stable equilibrium *See* equilibrium.

stainless steel A form of *steel containing at least 11–12% of chromium, a low percentage of carbon, and often some other elements, notably nickel and molybdenum. Stainless steel does not rust or stain and therefore has a wide variety of uses in industrial, chemical, and domestic environments. A particularly successful alloy is the steel known as 18–8, which contains 18% Cr, 8% Ni, and 0.08% C.

standard cell A *voltaic cell, such as a *Clarke cell, or *Weston cell, used as a standard of e.m.f.

standard deviation A measure of the dispersion of data in statistics. For a set of values $a_1, a_2, a_3, \ldots a_n$, the mean m is given by $(a_1 + a_2 + \ldots + a_n)/n$. The *deviation* of each value is absolute value of the difference from the mean $|m - a_1|$, etc. The standard deviation is the square root of the mean of the squares of these values, i.e.

$$\sqrt{[(|m - a_1|^2 + \ldots |m - a_n|^2)/n]}$$

When the data is continuous the sum is replaced by an integral.

standard electrode An electrode (a half cell) used in measuring electrode potential. *See* hydrogen half cell.

standard electrode potential *See* electrode potential.

standard form A way of writing a number, especially a large or small number, in which only one integer appears before the decimal point, the value being adjusted by multiplying by the appropriate power of 10. For example, 236,214 would be written in the standard form as $2.362\ 14 \times 10^5$; likewise

0.006821047 would be written 6.821 047 × 10^{-3}. Note that in the standard form, commas are not used, the digits are grouped into threes and a space is left between groups.

standard state A state of a system used as a reference value in thermodynamic measurements. Standard states involve a reference value of pressure (usually 1 atmosphere, 101.325 kPa) or concentration (usually 1 M). Thermodynamic functions are designated as 'standard' when they refer to changes in which reactants and products are all in their standard and their normal physical state. For example, the standard molar enthalpy of formation of water at 298 K is the enthalpy change for the reaction

$$H_2(g) + \tfrac{1}{2}O_2(g) \rightarrow H_2O(l)$$

$\Delta H^{\ominus}_{298} = -285.83$ kJ mol^{-1}. Note the superscript $^{\ominus}$ is used to denote standard state and the temperature should be indicated.

standard temperature and pressure *See* s.t.p.

standing wave *See* stationary wave.

star A self-luminous celestial body, such as the *sun, that generates nuclear energy within its core. Stars are not distributed uniformly throughout the universe, but are collected together in vast *galaxies. The age and lifetime of a star are related to its mass (*see* stellar evolution; Hertzsprung–Russell diagram).

stat- A prefix attached to the name of a practical electrical unit to provide a name for a unit in the electrostatic system of units, e.g. statcoulomb, statvolt. *Compare* ab-. In modern practice both absolute and electrostatic units have been replaced by *SI units.

state of matter One of the three physical states in which matter can exist, i.e. *solid, *liquid, or *gas. *Plasma is sometimes regarded as the fourth state of matter.

static electricity The effects produced by electric charges at rest, including the forces between charged bodies (*see* Coulomb's law) and the field they produce (*see* electric field).

statics The branch of mechanics concerned with bodies that are acted upon by balanced forces and couples so that they remain at rest or in unaccelerated motion. *Compare* dynamics.

stationary orbit *See* synchronous orbit.

stationary state A state of a system when it has an energy level permitted by *quantum mechanics. A transition from one stationary state to another can only occur by the emission or absorption of an appropriate quanta of energy in the form of photons.

stationary wave (standing wave) A form of *wave in which the profile of the wave does not move through the medium but remains stationary. This is in contrast to a *travelling* (or *progressive*) *wave*, in which the profile moves through the medium at the speed of the wave. A stationary wave results when a travelling wave is reflected back along its own path. In a stationary wave there are points at which the displacement is zero; these are called *nodes*. Points of maximum displacement are called *antinodes*. The distance between a node and its neighbouring antinode is one quarter of a wavelength. In a stationary wave all the points along the wave have different amplitudes and the points between successive nodes are in phase; in a travelling wave every point vibrates with the same amplitude and the phase of vibration changes for different points along its path.

statistical mechanics The branch of physics in which statistical methods are applied to the microscopic constituents of a system in order to predict its macroscopic properties. The earliest application of this method was Boltzmann's attempt to explain the thermodynamic properties of gases on the basis of the statistical properties of large assemblies of molecules.

In classical statistical mechanics, each particle is regarded as occupying a point in *phase space*, i.e. to have an exact position and momentum at any particular instant. The probability that this point will occupy any small volume of the phase space is taken to be proportional to the volume. The Maxwell–Boltzmann law gives the most probable distribution of the particles in phase space.

With the advent of quantum theory, the exactness of these premises was disturbed (by the Heisenberg uncertainty principle). In the *quantum statistics that evolved as a result, the phase space is divided into cells, each having a volume h^f, where h is the Planck constant and f is the number of degrees of freedom of the particles. This new concept led to Bose–Einstein statistics, and for particles obeying the Pauli exclusion principle, to Fermi–Dirac statistics.

statistics The branch of mathematics concerned with the inferences that can be drawn from numerical data on the basis of probability. A *statistical inference* is a conclusion drawn about a population as a result of an analysis of a representative sample. *See* sampling.

stator The stationary electromagnetic structure of an electric motor or electric generator. *Compare* rotor.

steady-state theory The cosmological theory that the universe has always existed in a steady state, that it had no beginning, will have no end, and has a constant mean density. To compensate for the observed *expansion of the universe this theory postulates that matter is created throughout the universe at a rate of about 10^{-10} nucleon per metre cubed per year as a property of space. Because it has failed to account for the *microwave background radiation or the evidence of evolution in the universe it has lost favour to the *big-bang theory, in spite of its other attractions. It was first proposed by Hermann Bondi (1919–), Thomas Gold (1920–), and Fred Hoyle (1915–) in 1948.

steam distillation A method of distilling liquids that are immiscible with water by bubbling steam through them. It depends on the fact that the vapour pressure (and hence the boiling point) of a mixture of two immiscible liquids is lower than the vapour pressure of either pure liquid.

steam engine A *heat engine in which the thermal energy of steam is converted into mechanical energy. It consists of a cylinder fitted with a piston and valve gear to enable the high-pressure steam to be admitted to the cylinder when the piston is near the top of its stroke. The steam forces the piston to the bottom of its stroke and is then exhausted from the cylinder usually into a condenser. The reciprocating motion of the piston is converted to rotary motion of the flywheel by means of a connecting rod, crosshead, and crank. The steam engine reached its zenith at the end of the 19th century, since when it has been replaced by the steam turbine and the internal-combustion engine. *See also* Rankine cycle.

steam point The temperature at which the maximum vapour pressure of water is equal to the standard atmospheric pressure (101 325 Pa). On the Celsius scale it has the value 100°C.

steel Any of a number of alloys consisting predominantly of iron with varying proportions of carbon (up to 1.7%) and, in some cases, small quantities of other elements (*alloy steels*), such as manganese, silicon, chromium, molybdenum, and nickel. Steels

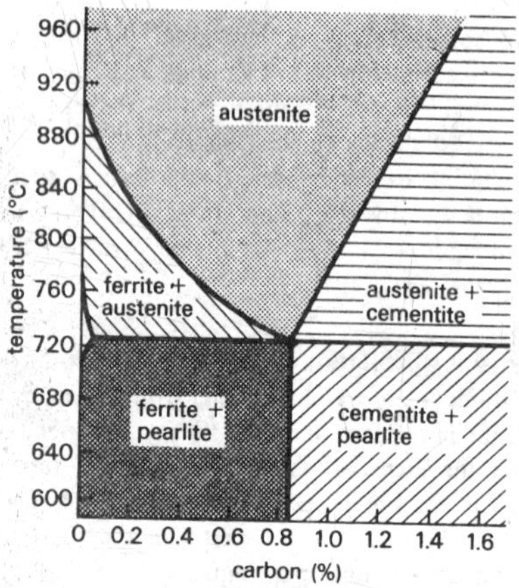

Phase diagram for steel

containing over 11–12% of chromium are known as *stainless steels.

Carbon steels exist in three stable crystalline phases: *ferrite* has a body-centred cubic crystal, *austenite* has a face-centred cubic crystal, and *cementite* has an orthorhombic crystal. *Pearlite* is a mixture of ferrite and cementite arranged in parallel plates. The phase diagram shows how the phases form at different temperatures and compositions.

Steels are made by the *basic-oxygen process (L–D process), which has largely replaced the *Bessemer process and the *open-hearth process, or in electrical furnaces.

Stefan's law (Stefan–Boltzmann law) The total energy radiated per unit surface area of a *black body in unit time is proportional to the fourth power of its thermodynamic temperature. The constant of proportionality, the *Stefan constant* (or *Stefan–Boltzmann constant*) has the value 5.6697×10^{-8} $J\,s^{-1}\,m^{-2}\,K^{-4}$. The law was discovered by Joseph Stefan (1853–93) and theoretically derived by Ludwig Boltzmann (1844–1906).

stellar evolution The changes that occur to a *star during its lifetime, from birth to final extinction. The process is illustrated on, and largely explained by, a *Hertzsprung–Russell diagram. A star is believed to form from a condensation of interstellar matter, which collects either by chance or for unexplained reasons, and grows by attracting other matter towards itself as a result of its gravitational field. This initial cloud of cold contracting matter, called a *prostar*, builds up an internal pressure as a result of its gravitational contraction. The pressure raises the temperature until it reaches 5–10 × 10^6 K, at which temperature the thermonuclear conversion of hydrogen to helium begins. In our *sun, a typical star, hydrogen is converted at a rate of some 10^{11} $kg\,s^{-1}$ with the evolution of some 6×10^{25} $J\,s^{-1}$ of energy. It is estimated that the sun contains sufficient hydrogen to burn at this rate for 10^{10} years and that it still has half its life to live as a main-sequence star (*see* Hertzsprung–Russell diagram). Eventually, however, this period of stability comes to an end, because the thermonuclear energy generated in the interior is no longer sufficient to counterbalance the gravitational contraction. The core, which is now mostly helium, collapses until a sufficiently high temperature is reached in a shell of unburnt hydrogen round the core to start a new phase of thermonuclear reaction. This burning of the shell causes the star's outer envelope to expand and cool, the temperature drop changes the colour from white to red and the star becomes a *red giant* or a *supergiant* if the original star was very large. The core now contracts, reaching a temperature of 10^8 K, and the helium in the core acts as the thermonuclear energy source. This reaction produces carbon, but a star of low mass relatively soon runs out of helium and the core collapses into a *white dwarf*, while the outer regions drift away into space. Larger stars (several times larger than the sun) have sufficient helium for the process to continue so that heavier elements, up to iron, are formed. But iron is the heaviest element that can be formed with the production of energy and when the helium has all been consumed there is a catastrophic collapse of the whole massive structure, resulting in a *supernova explosion. The current theory suggests that thereafter the burnt out remnants become *neutron stars and *black holes.

steradian Symbol sr. The supplementary *SI unit of solid angle equal to the *solid angle that encloses a surface on a sphere equal to the square of the radius of the sphere.

stere A unit of volume equal to 1 m^3. It is not now used for scientific purposes.

stimulated emission *See* laser.

stochastic process Any process in which there is a random variable.

stokes Symbol St. A c.g.s. unit of kinematic viscosity equal to the ratio of the viscosity of a fluid in poises to its density in grams per cubic centimetre. 1 stokes = 10^{-4} m^2 s^{-1}. It is named after Sir George Stokes (1819–1903).

Stokes's law A law that predicts the frictional force F on a spherical ball falling through a viscous medium under gravity. According to this law $F = 6\pi r\eta v$, where r is the radius of the ball, v is its velocity, and η is the viscosity of the medium. The sphere accelerates until it reaches a steady terminal speed, when F is equal to the gravitational force on the sphere, less any upthrust. The law was discovered by Sir George Stokes (1819–1903).

stop A circular aperture that limits the effective size of a lens in an optical system. It may be adjustable, as the iris diaphragm in a camera, or have a fixed diameter, as the disc used in some telescopes.

stopping power A measure of the ability of matter to reduce the kinetic energy of a particle passing through it. The *linear stopping power*, $-\mathrm{d}E/\mathrm{d}x$, is the energy loss of a particle per unit distance. The *mass stopping power*, $\mathrm{d}E/\rho\mathrm{d}x$, is the linear stopping power divided by the density (ρ) of the substance. The *atomic stopping power*, $\mathrm{d}E/n\mathrm{d}x$ = $(A/\rho N)\mathrm{d}E/\mathrm{d}x$, is energy loss per atom per unit area perpendicular to the particle's motion, i.e. n is the number of atoms in unit volume of the substance, N is the Avogadro number, and A is the relative atomic mass of the substance. The relative stopping power is the ratio of the stopping power of a substance to that of a standard substance, usually aluminium, oxygen, or air.

storage ring A large evacuated toroidal ring forming a part of some particle accelerators. The rings are designed like *synchrotrons, except that they do not accelerate the particles circling within them but supply just sufficient energy to make up for losses (mainly *synchrotron radiation). The storage rings are usually built tangentially to the associated accelerator so that particles can be transferred accurately between them. At *CERN in Geneva, two interlaced storage rings are used, containing protons rotating in opposite directions. At the intersections very high collision energies (up to 1700 GeV) can be achieved.

s.t.p. Standard temperature and pressure, formerly known as *N.T.P.* (normal temperature and pressure). The standard conditions used as a basis for calculations involving quantities that vary with temperature and pressure. These conditions are used when comparing the properties of gases. They are 273.15 K (or 0°C) and 101 325 Pa (or 760.0 mmHg).

strain A measure of the extent to which a body is deformed when it is subjected to a *stress. The *linear strain* or *tensile strain* is the ratio of the change in length to the original length. The *bulk strain* or *volume strain* is the ratio of the change in volume to the original volume. The *shear strain* is the angular distortion in radians of a body subjected to a *shearing force. *See also* elasticity; elastic modulus.

strain gauge A device used to measure a small mechanical deformation in a body (*see* strain). The most widely used devices are metal wires or foils or semiconductor materials, such as a single silicon crystal, which are attached to structural members; when the members are stretched under tensile *stress the resistance of the metal or semiconductor element increases. By making the resistance a component in a *Wheatstone-bridge circuit an estimate of the strain can be made by noting the change in resistance. Other types of strain gauge rely on changes of capacitance or the magnetic induction between two coils, one of which is attached to the stressed member.

strain hardening (work hardening) An increase in the resistance to the further plastic deformation of a body as a result of a rearrangement of its internal structure when it is strained, particularly by repeated stress. *See also* elasticity.

strangeness Symbol *s*. A property of certain elementary particles called hadrons (K-mesons and hyperons) that decay more slowly than would have been expected from the large amount of energy released in the process. These particles were assigned the quantum number *s* to account for this behaviour. For nucleons and other nonstrange particles $s = 0$; for strange particles *s* does not equal zero but has an integral value. In quark theory (*see* elementary particles) hadrons with the property of strangeness contain a strange quark or its antiquark.

stratosphere *See* earth's atmosphere.

streamline flow A type of fluid flow in which no *turbulence occurs and the particles of the fluid follow continuous paths, either at constant velocity or at a velocity that alters in a predictable and regular way (*see also* laminar flow).

stress The force per unit area on a body that tends to cause it to deform (*see* strain). It is a measure of the internal forces in a body between particles of the material of which it consists as they resist separation, compression, or sliding in response to externally applied forces. *Tensile stress* and *compressive stress* are axial forces per unit area applied to a body that tend either to extend it or compress it linearly. *Shear stress* is a tangential force per unit area that tends to shear a body. *See also* elasticity; elastic modulus.

stroboscope A device for making a moving body intermittently visible in order to make it appear stationary. It may consist of a lamp flashing at regular intervals or a shutter that enables it to be seen intermittently. *Stroboscopic photography* is the taking of very short-exposure pictures of moving objects using an electronically controlled flash lamp.

strong interaction *See* fundamental interactions.

subatomic particle *See* elementary particles.

subcritical *See* critical mass; critical reaction; multiplication factor.

sublimate A solid formed by sublimation.

sublimation A direct change of state from solid to gas.

subshell *See* atom.

subsonic speed A speed that is less than *Mach 1.

subtractive process *See* colour.

sun The *star at the centre of the *solar system. A typical main-sequence dwarf star (*see* Hertzsprung–Russell diagram; stellar evolution), the sun is some 149 600 000 km from earth. It has a diameter of about 1 392 000 km and a mass of 1.9×10^{30} kg. Hydrogen and helium are the primary constituents (about 75% hydrogen, 25% helium), with less than 1% of heavier elements. In the central core, some 400 000 km in diameter, hydrogen is converted into helium by thermonuclear reactions, which generate vast quantities of energy. This energy is radiated into space and provides the earth with all the light and heat necessary to have created and maintained life (*see* solar constant). The surface of the sun, the *photosphere, forms the boundary between its opaque interior and its transparent atmosphere. It is here that *sunspots occur. Above the photosphere is the *chromosphere and above this the *corona, which extends tenuously into interplanetary space. *See also* solar wind.

sunspot A dark patch in the sun's *photosphere resulting from a localized fall in temperature to about 4000 K. Most spots have a central very dark umbra surrounded by a lighter penumbra. Sunspots tend to occur in clusters and to last about two weeks. The number of sunspots visible fluctuates over an eleven-year cycle – often called the *sunspot cycle*. The cause is thought to be the presence of intense localized magnetic fields, which suppress the convection currents that bring hot gases to the photosphere.

superconductivity The absence of measurable electrical resistance in certain substances at temperatures close to 0 K. First discovered in 1911 in mercury, superconductivity is now known to occur in some 26 metallic elements and many compounds and alloys. The temperature below which a substance becomes superconducting is called the *transition temperature* (or *critical temperature*), the highest known transition temperature being 23 K for an alloy of niobium, aluminium, and germanium.

The theoretical explanation of the phenomenon was given by J. Bardeen (1908–), L. N. Cooper (1930–), and J. R. Schrieffer (1931–) in 1957 and is known as the *BCS theory*. According to this theory an electron moving through an elastic crystal lattice creates a slight distortion of the lattice as a result of Coulomb forces between the positively charged lattice and the negatively charged electron. If this distortion persists for a finite time it can affect a second passing electron. In 1956 Cooper showed that the effect of this phenomenon is for the current to be carried in superconductors not by individual electrons but by bound pairs of electrons, called *Cooper pairs*. The BCS theory is based on a *wave function in which all the electrons are paired. Because the total momentum of a Cooper pair is unchanged by the interaction between one of its electrons and the lattice, the flow of electrons continues indefinitely.

Superconducting coils in which large currents can circulate indefinitely can be used to create powerful magnetic fields and are used for this purpose in some particle accelerators and in other devices.

supercooling 1. The cooling of a liquid to below its freezing point without a change from the liquid to solid state taking place. In this metastable state the particles of the liquid lose energy but do not fall into place in the lattice of the solid crystal. If the liquid is seeded with a small crystal, crystallization usually takes place and the temperature returns to the freezing point. This is a common occurrence in the atmosphere where water droplets frequently remain unfrozen at temperatures well below 0°C until disturbed, following which they rapidly freeze. The supercooled droplets, for example, rapidly freeze on passing aircraft forming 'icing', which can be a hazard. **2.** The analogous cooling of a vapour to make it supersaturated until a disturbance causes condensation to occur, as in the Wilson *cloud chamber.

supercritical *See* critical mass; critical reaction; multiplication factor.

superdense theory *See* big-bang theory.

superficial expansivity *See* expansivity.

superfluidity The property of liquid helium at very low temperatures that enables it to flow without friction. Both helium isotopes possess this property, but ^{4}He becomes superfluid at 2.172 K, whereas ^{3}He does not become superfluid until a temperature of 0.00093 K is reached. There is a basic connection between superfluidity and *superconductivity, so that sometimes a superconductor is called a charged superfluid.

supergiant The largest and most luminous type of star. They are formed from the most massive stars and are therefore very rare. They lie above the giants on the *Hertzsprung–Russell diagram. *See also* stellar evolution.

superheterodyne receiver A widely used type of radio receiver in which the incoming radio-frequency signal is mixed with an internally generated signal from a local oscillator. The output of the mixer has a carrier frequency equal to the difference between the transmitted carrier frequency and the locally generated frequency, still retains the transmitted modulation, and is called the *intermediate frequency* (IF). The IF signal is amplified before being passed to the audio-frequency amplifier. This system enables the IF signal to be amplified with less distortion, greater gain, better selectivity, and easier elimination of noise than can be achieved by amplifying the radio-frequency signal.

super high frequency (SHF) A radio frequency in the range 3–30 gigahertz.

superlattice *See* solid solution.

supernova An explosive brightening of a star in which the energy radiated by it increases by a factor of 10^{10}. It takes several years to fade and while it lasts dominates the whole galaxy in which it lies. It is estimated that there could be a supernova explosion in the Milky Way every 30 years, although only six have actually been observed in the last 1000 years. A supernova explosion occurs when a star has burnt up all its available helium and the core collapses catastrophically (*see* stellar evolution). *Compare* nova.

superplasticity The ability of some metals and alloys to stretch uniformly by several thousand percent at high temperatures, unlike normal alloys, which fail after being stretched 100% or less. Since 1962, when this property was discovered in an alloy of zinc and aluminium (22%), many alloys and ceramics have been shown to possess this property. For superplasticity to occur, the metal grain must be small and rounded and the alloy must have a slow rate of deformation.

supersaturated solution *See* saturated.

supersaturation 1. The state of the atmosphere in which the relative humidity is over 100%. This occurs in pure air where no condensation nuclei are available. Supersaturation is usually prevented in the atmosphere by the abundance of condensation nuclei (e.g. dust, sea salt, and smoke particles). **2.** The state of any vapour whose pressure exceeds that at which condensation usually occurs (at the prevailing temperature).

supersonic *See* Mach number.

supplementary units *See* SI units.

suppressor grid A wire grid in a pentode *thermionic valve placed between the *screen grid and the anode to prevent electrons produced by *secondary emission from the anode from reaching the screen grid.

surd A quantity that cannot be expressed as a *rational number. It consists of the root of an arithmetic member (e.g. $\sqrt{3}$), which cannot be exactly determined, or the sum or difference of such roots.

surface tension Symbol γ. The property of a liquid that makes it behave as if its surface is enclosed in an elastic skin. The property results from intermolecular forces: a molecule in the interior of a liquid experiences a force of attraction from other molecules equally from all sides, whereas a molecule at the surface is only attracted by molecules below it in the liquid. The surface tension is defined as the force acting over the surface per unit length of surface perpendicular to the force. It is measured in newtons per metre. It can equally be defined as the energy required to increase the surface area by one square metre, i.e. it can be measured in joules per metre squared (which is equivalent to $N\,m^{-1}$).

The property of surface tension is responsible for the formation of liquid drops, soap bubbles, and meniscuses, as well as the rise of liquids in a capillary tube (*capillarity*), the absorption of liquids by porous substances, and the ability of liquids to wet a surface.

surfactant (surface active agent) A substance, such as a *detergent, added to a liquid to increase its spreading or wetting properties by reducing its *surface tension.

susceptance Symbol *B*. The reciprocal of the *reactance of a circuit and thus the imaginary part of its *admittance. It is measured in siemens.

susceptibility 1. (*or* **magnetic susceptibility**) Symbol χ_m. The dimensionless quantity describing the contribution made by a substance when subjected to a magnetic field to the total magnetic flux density present. It is equal to $\mu_r - 1$, where μ_r is the relative *permeability of the material. Diamagnetic materials have a low negative susceptibility, paramagnetic materials have a low positive susceptibility, and ferromagnetic materials have a high positive value. **2.** (*or* **electric susceptibility**) Symbol χ_e. The dimensionless quantity referring to a *dielectric equal to $P/\varepsilon_0 E$, where P is the electric polarization, E is the electric intensity producing it, and ε_0 is the electric constant. The electric susceptibility is also equal to $\varepsilon_r - 1$, where ε_r is the relative *permittivity of the dielectric.

suspension A mixture in which small solid or liquid particles are suspended in a liquid or gas.

synchrocyclotron A form of *cyclotron in which the frequency of the accelerating potential is synchronized with the increasing period of revolution of a group of the accelerated particles, resulting from their relativistic increase in mass as they reach *relativistic speeds. The accelerator is used with protons, deuterons, and alpha-particles.

synchronous motor *See* electric motor.

synchronous orbit (geosynchronous orbit) An orbit of the earth made by an artificial *satellite with a period exactly equal to the earth's period of rotation on its axis, i.e. 23 hours 56 minutes 4.1 seconds. If the orbit is inclined to the equatorial plane the satellite will appear from the earth to trace out a figure-of-eight track once every 24 hours. If the orbit lies in the equatorial plane and is circular, the satellite will appear to be stationary. This is called a *stationary orbit* (or *geostationary orbit*) and it occurs at an altitude of 35 900 km. Most communication satellites are in stationary orbits, with three or more spaced round the orbit to give worldwide coverage.

synchronous rotation The rotation of a natural satellite in which the period of rotation is equal to its orbital period. The moon, for example, is in synchronous rotation about the earth.

synchrotron A particle accelerator used to impart energy to electrons and protons in order to carry out experiments in particle physics and in some cases to make use of the *synchrotron radiation produced. The particles are accelerated in closed orbits (often circular) by radio-frequency fields. Magnets are spaced round the orbit to bend the trajectory of the particles and separate focusing magnets are used to keep the particles in a narrow beam. The radio-frequency accelerating cavities are interspersed between the magnets. The motion of the particles is automatically synchronized with the rising magnetic field, as the field strength has to increase as the particle energy increases; the frequency of the accelerating field also has to increase synchronously.

synchrotron radiation (magnetobremsstrahlung) Electromagnetic radiation that is emitted by charged particles moving at relativistic speeds in circular orbits in a magnetic field. The rate of emission is inversely proportional to the product of the radius of curvature of the orbit and the fourth power of the mass of the particles. For this reason, synchrotron radiation is not a problem in the design of proton *synchrotrons but it is significant in electron synchrotrons. The greater the circumference of a synchrotron, the less important is the loss of energy by synchrotron radiation. In *storage rings, synchrotron radiation is the principal cause of energy loss.

However, since the 1950s it has been realized that synchrotron radiation is itself a very useful tool and many accelerator laboratories have research projects making use of the radiation on a secondary basis to the main high-energy research. The radiation used for these purposes is primarily in the ultraviolet and X-ray frequencies.

Much of the microwave radiation from celestial radio sources outside the Galaxy is believed to originate from electrons moving in circular paths in celestial magnetic fields; it is also called synchrotron radiation as it is analogous to the radiation occurring in a synchrotron.

synodic month (lunation) The interval between new *moons. It is equal to 29 days, 12 hours, and 44 minutes.

synodic period The mean time taken by any object in the solar system to move between successive returns to the same position, relative to the sun as seen from the earth. Since a planet is best observed at opposition the synodic period of a planet, S, is easier to measure than its *sidereal period, P. For inferior planets $1/S = 1/P - 1/E$; for superior planets $1/S = 1/E - 1/P$, where E is the sidereal period of the earth.

Système International d'Unités *See* SI units.

syzygy The situation that occurs when the sun, the moon (or a planet), and the earth are in a straight line. This occurs when the moon (or planet) is at *conjunction or *opposition.

T

tachometer An instrument for measuring angular speed, especially the number of revolutions made by a rotating shaft in unit time. Various types of instrument are used, including mechanical, electrical, and electronic devices. The widely used electrical-generator tachometer consists of a small generator in which the output voltage is a measure of the rate of rotation of the shaft that drives it.

tachyon A hypothetical particle that has a speed in excess of the *speed of light. According to electromagnetic theory, a charged particle travelling through a medium at a speed in excess of the speed of light in that medium emits *Cerenkov radiation. A charged tachyon would emit Cerenkov radiation even in a vacuum. No such particle has yet been detected. According to the special theory of *relativity, it is impossible to accelerate a particle up to the speed of light because its energy E, given by $E = mc^2/\sqrt{(1 - v^2/c^2)}$, would have to become infinite. The theory, however, does not forbid the existence of particles with $v > c$ (where c is the speed of light). In such cases the expression in the brackets becomes negative and the energy would be imaginary.

tandem generator A type of particle generator, essentially consisting of a *Van de Graaff generator that maintains one electrode at a high positive potential; this electrode is placed between two earthed electrodes. Negative ions are accelerated from earth potential to the positively charged electrode, where surplus electrons are stripped from the ions to produce positive ions, which are accelerated again from the positive electrode back to earth. Thus the ions are accelerated twice over by a single high potential. This tandem arrangement enables energies up to 30 MeV to be achieved.

tangent 1. A line that touches a curve or a plane that touches a surface. **2.** *See* trigonometric functions.

tangent galvanometer A type of galvanometer, now rarely used, in which a small magnetic needle is pivoted horizontally at the centre of a vertical coil that is adjusted to be parallel to the horizontal component of the earth's magnetic field. When a current I is passed through the coil, the needle is deflected so that it makes an angle θ with its equilibrium position parallel to the earth's field. The value of I is given by $I = (2Hr\tan\theta)/n$, where H is the strength of the earth's horizontal component of magnetizing force, r is the radius of the coil, and n is the number of turns in the coil. Although not now used for measuring current, the instrument provides a means of measuring the earth's magnetizing force.

tau particle *See* elementary particles; lepton.

Taylor series The infinite power series of derivatives into which a function $f(x)$ can be expanded, for a fixed value of the variable $x = a$:

$$f(x) = f(a) + f'(a)(x - a) + f''(a)(x - a)^2/2! + \ldots$$

When $a = 0$, the series formed is known as *Maclaurin's series*:

$$f(x) = f(0) + f'(0)x + f''(0)x^2/2! + \ldots$$

The series was discovered by Brook Taylor (1685–1731) and the special case was named after Colin Maclaurin (1698–1746).

tektite A small black, greenish, or yellowish glassy object found in groups on the earth's surface and consisting of a silicaceous material unrelated to the geological formations in which it is found. Tektites are believed to have formed on earth as a result of the impact of meteorites.

telecommunications The study and application of means of transmitting information, either by wires or by electromagnetic radiation.

telescope An instrument that collects radiation from a distant object in order to produce an image of it or enable the radiation to be analysed. An *optical telescope* uses visible radiation (*see also* radio telescope). Optical *astronomical telescopes* fall into two main classes: *refracting telescopes* and *reflecting telescopes*. Refracting telescopes use a converging lens to collect the light and the resulting image is magnified by the eyepiece, a lens of short focal length. This type of instrument was first constructed in 1608 by Hans Lippershey (1587–1619) in Holland and developed in the following year as an astronomical instrument by Galileo, who used a diverging lens as eyepiece (see illustration (a)). The *Galilean telescope* was later improved by Johannes Kepler (1571–1630), who substituted a converging eyepiece lens, and this form of the instrument is still in use for small astronomical telescopes. For terrestrial telescopes an additional lens is usually inserted in order to provide an upright image (see illustration (b)). However the Galilean design provides an upright image originally, and so is still used for the compact telescopes of opera glasses.

The first reflecting telescope was produced by Newton in 1668. This used a concave mirror to collect and focus the light and a small secondary mirror at an angle of 45° to the main beam to reflect the light into the magnifying eyepiece. The *Gregorian telescope*, designed by James Gregory (1638–75), and the *Cassegrainian telescope*, designed by N. Cassegrain (*fl.* 1670s), use

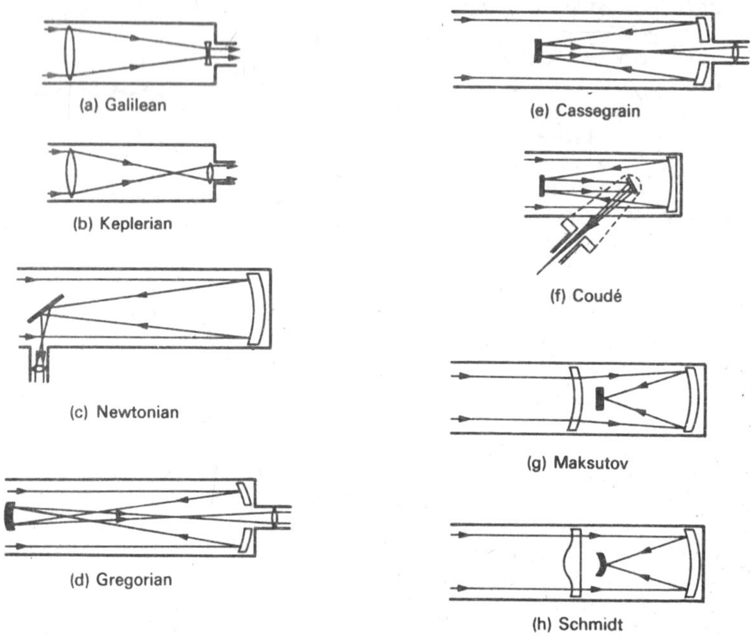

Telescopes

different secondary optical systems (see illustrations (d) and (e)). The *coudé telescope* (French: angled) is sometimes used with larger instruments as it increases their focal lengths (see illustration (f)).

Catadioptic telescopes use both lenses and mirrors. The most widely used astronomical instruments in this class are the *Maksutov telescope* and the *Schmidt camera*. The Maksutov instrument has a thick spheroidal meniscus lens to introduce spherical aberration to the image, which is corrected by the spheroidal principal mirror (see illustration (g)). The Schmidt camera has a correcting plate of complicated aspherical shape close to the centre of curvature of the spheroidal principal mirror. This instrument is used primarily for photographic purposes (see illustration (h)).

television The transmission and reception of moving images by means of radio waves or cable. The scene to be transmitted is focused onto a photoelectric screen in the television *camera. This screen is scanned by an electron beam. The camera produces an electric current, the instantaneous magnitude of which is proportional to the brightness of the portion of the screen being scanned. In Europe the screen is scanned by 625 lines and 25 such frames are produced every second. In the USA 525 lines and 30 frames per second are used. The picture signal so produced is used to modulate a VHF or UHF carrier wave and is transmitted with an independent sound signal, but with colour information (if any) incorporated into the brief gaps between the picture lines. The signals received by the receiving aerial are demodulated in the receiver; the demodulated picture signal controls the electron beam in a cathode-ray tube, on the screen of which the picture is reconstructed. *See also* colour television.

television tube *See* cathode rays.

temperament The way in which the intervals between notes on keyboard instruments are distributed throughout the scale to ensure that music in all keys sounds in tune. The problem can be illustrated by a piano keyboard. Taking a low C and a high C seven octaves above, the interval should be $2^7 = 128$. However, in passing through the cycle of 12 keys, each using as its fundamental the fifth of its predecessor, the interval between Cs becomes $(3/2)^{12} = 129.75$. The difference between 129.75 and 128 is known as the *comma of Pythagoras*. The *equal-temperament scale*, which has been in use since the time of J. S. Bach, distributes the comma of Pythagoras equally between the 12 intervals of the scale over seven octaves. Thus each fifth becomes $(128)^{1/12} = 1.4983$. All forms of temperament involve a measure of compromise; this system is now regarded as the best.

temperature The property of a body or region of space that determines whether or not there will be a net flow of heat into it or out of it from a neighbouring body or region and in which direction (if any) the heat will flow. If there is no heat flow the bodies or regions are said to be in *thermodynamic equilibrium* and at the same temperature. If there is a flow of heat, the direction of the flow is from the body or region of higher temperature. Broadly, there are two methods of quantifying this property. The empirical method is to take two or more reproducible temperature-dependent events and assign *fixed points* on a scale of values to these events. For example, the Celsius temperature scale uses the freezing point and boiling point of water as the two fixed points, assigns the values 0 and 100 to them, respectively, and divides the scale between them into 100 degrees. This method is serviceable for many practical purposes (*see* temperature scales), but lacking a theoretical basis it is awkward to use in many scientific contexts. In the 19th century, Lord Kelvin proposed a thermodynamic method to specify temperature, based on the measurement of the quantity of heat flowing between bodies at different temperatures. This concept relies on an absolute scale of temperature with an *absolute zero of temperature, at which no body can give up heat. He also used Sadi Carnot's concept of an ideal frictionless perfectly efficient heat engine (*see* Carnot cycle). This Carnot engine takes in a quantity of heat q_1 at a temperature T_1, and exhausts heat q_2 at T_2, so that $T_1/T_2 = q_1/q_2$. If T_2 has a value fixed by definition, a Carnot engine can be run between this fixed temperature and any unknown temperature T_1, enabling T_1 to be calculated by measuring the values of q_1 and q_2. This concept remains the basis for defining *thermodynamic temperature*, quite independently of the nature of the working substance. The unit in which thermodynamic temperature is expressed is the *kelvin. In practice thermodynamic temperatures cannot be measured directly; they are usually inferred from measurements with a gas thermometer containing a nearly ideal gas. This is possible because another aspect of thermodynamic temperature is its relationship to the *internal energy of a given amount of substance. This can be shown most simply in the case of an ideal monatomic gas, in which the internal energy per mole (U) is equal to the total kinetic energy of translation of the atoms in one mole of the gas (a monatomic gas has no rotational or vibrational energy). According to *kinetic theory, the thermodynamic temperature of such a gas is given by $T = 2U/3R$, where R is the universal *gas constant.

temperature coefficient A coefficient that determines the rate of change of some physical property with change in temperature. For example, the dependence of the resistance (R) of a material on the temperature T, is given by $R = R_0 + \alpha T + \beta T^2$, where R_0 is the resistance at absolute zero and α and β are constants. If β is negligible, then α is the *temperature coefficient of resistance.*

temperature scales A number of empirical scales of *temperature have been in use: the *Celsius scale is widely used for many purposes and in certain countries the *Fahrenheit scale is still used. These scales both rely on the use of *fixed points,* such as the freezing point and the boiling point of water, and the division of the *fundamental interval* between these two points into units of temperature (100 degrees in the case of the Celsius scale and 180 degrees in the Fahrenheit scale).

However, for scientific purposes the scale in use is the *International Practical Temperature Scale (1968),* which is designed to conform as closely as possible to thermodynamic temperature and is expressed in the unit of thermodynamic temperature, the *kelvin. The eleven fixed points of the scale are given in the table, with the instruments specified for interpolating between them. Above the freezing point of gold, a radiation pyrometer is used, based on Planck's law of radiation. The scale is expected to be refined in the late 1980s.

tempering The process of increasing the toughness of an alloy, such as steel, by heating it to a predetermined temperature, maintaining it at this temperature for a predetermined time, and cooling it to room temperature at a predetermined rate. In steel, the purpose of the process is to heat the alloy to a temperature that will enable the excess carbide to precipitate out of the supersaturated solid solution of *martensite and then to cool the saturated solution fast enough to prevent further precipitation or grain growth. For this reason steel is quenched rapidly by dipping into cold water.

temporary magnetism Magnetism in a body that is present when the body is in a

	T/K	t/°C
triple point of equilibrium hydrogen	13.81	−259.34
temperature of equilibrium hydrogen when its vapour pressure is 25/76 standard atmosphere	17.042	−256.108
b.p. of equilibrium hydrogen	20.28	−252.87
b.p. of neon	27.102	−246.048
triple point of oxygen	54.361	−218.789
b.p. of oxygen	90.188	−182.962
triple point of water	273.16	0.01
b.p. of water	373.15	100
f.p. of zinc	692.73	419.58
f.p. of silver	1235.08	961.93
f.p. of gold	1337.58	1064.43

Temperature scales

magnetic field but that largely disappears when it is removed from the field.

tensile strength A measure of the resistance that a material offers to tensile *stress. It is defined as the stress, expressed as the force per unit cross-sectional area, required to break it.

tensimeter A form of differential manometer with two sealed bulbs attached to the limbs. It is used to measure the difference in vapour pressure between liquids sealed into the bulbs. If one liquid has a known vapour pressure (often water is used) that of the other can be determined.

tensiometer Any apparatus for measuring *surface tension.

tera- Symbol T. A prefix used in the metric system to denote one million million times. For example, 10^{12} volts = 1 teravolt (TV).

terminal 1. The point at which electrical connection is made to a device or system. 2. A device at which data is put into a *computer or taken from it.

terminal speed The constant speed finally attained by a body moving through a fluid under gravity when there is a zero resultant force acting on it. *See* Stokes's law.

terminator The boundary, on the surface of the moon or a planet, between the sunlit area and the dark area.

terrestrial magnetism *See* geomagnetism.

tertiary colour A colour obtained by mixing two *secondary colours.

tesla Symbol T. The SI unit of magnetic flux density equal to one weber of magnetic flux per square metre, i.e. 1 T = 1 $Wb\,m^{-2}$. It is named after Nikola Tesla (1870–1943), Croatian-born US electrical engineer.

Tesla coil A device for producing a high-frequency high-voltage current. It consists of a *transformer with a high turns ratio, the primary circuit of which includes a spark gap and a fixed capacitor; the secondary circuit is tuned by means of a variable capacitor to resonate with the primary. It was devised by Nikola Tesla (1870–1943). Tesla coils are commonly used to excite luminous discharges in glass vacuum apparatus, in order to detect leaks.

tetragonal *See* crystal system.

tetrahedron A polyhedron with four triangular faces. In a *regular tetrahedron* all four triangles are congruent equilateral triangles. It constitutes a regular triangular *pyramid.

tetrode A *thermionic valve with a *screen grid* placed between the anode and the control grid of a *triode to reduce the capacitance between these two electrodes and so improve the valve's performance as an amplifier or oscillator at high frequencies. The screen grid is maintained at a fixed potential.

theodolite An optical surveying instrument for measuring horizontal and vertical angles. It consists of a sighting telescope, with crosshairs in the eyepiece for focusing on the target, which can be rotated in both the horizontal and vertical planes. It is mounted on a tripod and a spirit level is used to indicate when the instrument is horizontal. The angles are read off graduated circles seen through a second eyepiece in the instrument.

therm A practical unit of energy defined as 10^5 British thermal units. 1 therm is equal to 1.055×10^8 joules.

thermal capacity *See* heat capacity.

thermal conductivity *See* conductivity.

thermal diffusion The diffusion that takes place in a fluid as a result of a temperature gradient. If a column of gas is maintained so that the lower end is cooler than the upper end, the heavier molecules in the gas will tend to remain at the lower-temperature end and the lighter molecules will diffuse to the higher-temperature end. This property has been used in the separation of gaseous isotopes (*see* Clusius column).

thermal equilibrium *See* equilibrium.

thermal expansion *See* expansivity.

thermalization The reduction of the kinetic energy of neutrons in a thermal *nuclear reactor by means of a *moderator; the process of producing thermal neutrons.

thermal neutrons *See* moderator; nuclear reactors; thermalization.

thermal reactor *See* nuclear reactor.

thermionic emission The emission of electrons, usually into a vacuum, from a heated conductor. The emitted current density, J, is given by the *Richardson* (or *Richardson–Dushman*) *equation*, i.e. $J = AT^2\exp(-W/kT)$, where T is the thermodynamic temperature of the emitter, W is its *work function, k is the Boltzmann constant, and A is a constant. Thermionic emission is the basis of the *thermionic valve and the *electron gun in cathode-ray tubes.

thermionics The branch of electronics concerned with the study and design of devices based on the emission of electrons from metal or metal-oxide surfaces as a result of high temperatures. The primary concern of thermionics is the design of *thermionic valves and the electron guns of cathode-ray tubes and other devices.

thermionic valve An electronic valve based on *thermionic emission. In such valves the cathode is either directly heated by passing a current through it or indirectly heated by placing it close to a heated filament. Directly heated cathodes are usually made of tungsten wire, whereas indirectly heated cathodes are usually coated with barium and strontium oxides. Most electronic valves are thermionic vacuum devices, although a few have cold cathodes and some are gas-filled (*see* thyratron). *See* diode; triode; tetrode; pentode.

thermistor A semiconductor electronic device the resistance of which decreases as its temperature increases. It consists of a bead, rod, or disc of various oxides of manganese, nickel, cobalt, copper, iron, or other metals. Thermistors are used as thermometers, often forming one element in a resistance bridge. They are used for this purpose in such applications as bearings, cylinder heads, and transformer cores. They are also used to compensate for the increased resistance of ordinary resistors when hot, and in vacuum gauges, time-delay switches, and voltage regulators.

thermite A stoichiometric powdered mixture of iron(III) oxide and aluminium for the reaction:

$$2Al + Fe_2O_3 \rightarrow Al_2O_3 + 2Fe$$

The reaction is highly exothermic and the increase in temperature is sufficient to melt the iron produced. It has been used for localized welding of steel objects (e.g. railway lines) in the *Thermit process*. Thermite is also used in incendiary bombs.

thermochemistry The branch of physical chemistry concerned with heats of chemical reaction, heats of formation of chemical compounds, etc.

thermocouple A device consisting of two dissimilar metal wires or semiconducting rods welded together at their ends. A thermoelectric e.m.f. is generated in the device when the ends are maintained at different temperatures, the magnitude of the e.m.f. being related to the temperature difference. This enables a thermocouple to be used as a thermometer over a limited temperature range. One of the two junctions, called the *hot* or *measuring junction*, is exposed to the temperature to be measured. The other, the *cold* or *reference junction*, is maintained at a known reference temperature. The e.m.f. generated is measured by a suitable millivoltmeter or potentiometer incorporated into the circuit. *See* Seebeck effect; thermopile.

thermodynamics The study of the laws that govern the conversion of energy from one form to another, the direction in which heat will flow, and the availability of energy to do work. It is based on the concept that in an isolated system anywhere in the universe there is a measurable quantity of energy called the *internal energy (U) of the system. This is the total kinetic and potential energy of the atoms and molecules of the system of all kinds that can be transferred directly as heat; it therefore excludes chemical and nuclear energy. The value of U can only be changed if the system ceases to be isolated. In these circumstances U can change by the transfer of mass to or from the system, the transfer of heat (Q) to or from the system, or by the work (W) being done on or by the system. For an adiabatic ($Q = 0$) system of constant mass, $\Delta U = W$. By convention, W is taken to be positive if work is done on the system and negative if work is done by the system. For nonadiabatic systems of constant mass, $\Delta U = Q + W$. This statement, which is equivalent to the law of conservation of energy, is known as the *first law of thermodynamics*.

All natural processes conform to this law, but not all processes conforming to it can occur in nature. Most natural processes are irreversible, i.e. they will only proceed in one direction (*see* reversible process). The direction that a natural process can take is the subject of the *second law of thermodynamics*, which can be stated in a variety of ways. R. Clausius (1822–88) stated the law in two ways: "heat cannot be transferred from one body to a second body at a higher temperature without producing some other effect" and "the entropy of a closed system increases with time". These statements introduce the thermodynamic concepts of *temperature (T) and *entropy (S), both of which are parameters determining the direction in which an irreversible process can go.

The temperature of a body or system determines whether heat will flow into it or out of it; its entropy is a measure of the unavailability of its energy to do work. Thus T and S determine the relationship between Q and W in the statement of the first law. This is usually presented by stating the second law in the form $\Delta U = T\Delta S - W$.

The second law is concerned with changes in entropy (ΔS). The *third law of thermodynamics* provides an absolute scale of values for entropy by stating that for changes involving only perfect crystalline solids at *absolute zero, the change of the total entropy is zero. This law enables absolute values to be stated for entropies.

One other law is used in thermodynamics. Because it is fundamental to, and assumed by, the other laws of thermodynamics it is usually known as the *zeroth law of thermodynamics*. This states that if two bodies are each in thermal equilibrium with a third body, then all three bodies are in thermal equilibrium with each other. *See also* enthalpy; free energy.

thermodynamic temperature *See* temperature.

thermoelectricity An electric current generated by heat. There are three interrelated *thermoelectric effects*. *See* Peltier effect; Seebeck effect; Thomson effect.

thermograph **1.** A recording thermometer used in meteorology to obtain a continuous record of temperature changes over a period on a graph. **2.** A record so obtained. **3.** A record obtained by the technique of *thermography.

thermography A medical technique that makes use of the infrared radiation from the human skin to detect an area of elevated skin temperature that could be associated with an underlying cancer. The heat radiated from the body varies according to the local blood flow, thus an area of poor circulation produces less radiation. A tumour, on the other hand, has an abnormally increased blood supply and is revealed on the *thermogram* (or *thermograph*) as a 'hot spot'. The technique is used particularly in mammography, the examination of the infrared radiation emitted by human breasts in order to detect breast cancer.

thermoluminescence *Luminescence produced in a solid when its temperature is raised. It arises when free electrons and *holes, trapped in a solid as a result of exposure to ionizing radiation, unite and emit photons of light. The process is made use of in *thermoluminescent dating*, which assumes that the number of electrons and holes trapped in a sample of pottery is related to the length of time that has elapsed since the pottery was fired. By comparing the luminescence produced by heating a piece of pottery of unknown age with the luminescence produced by heating similar materials of known age, a fairly accurate estimate of the age of an object can be made.

thermoluminescent dating *See* thermoluminescence.

thermometer An instrument used for measuring the *temperature of a substance. A number of techniques and forms are used in thermometers depending on such factors as the degree of accuracy required and the range of temperatures to be measured, but they all measure temperature by making use of some property of a substance that varies with temperature. For example, *liquid-in-glass thermometers* depend on the expansion of a liquid, usually mercury or alcohol coloured with dye. These consist of a liquid-filled glass bulb attached to a partially filled capillary tube. In the *bimetallic thermometer* the unequal expansion of two dissimilar metals that have been bonded together into a narrow strip and coiled is used to move a pointer round a dial. The *gas thermometer*, which is more accurate than the liquid-in-glass thermometer, measures the variation in the pressure of a gas kept at constant volume. The *resistance thermometer* is based on the change in resistance of conductors or semiconductors with temperature change. Platinum, nickel, and copper are the metals most commonly used in resistance thermometers. *See also* pyrometry; thermistor; thermocouple.

thermonuclear reaction *See* nuclear fusion; thermonuclear reactor.

thermonuclear reactor (fusion reactor) A reactor in which *nuclear fusion takes place with the controlled release of energy. Although thermonuclear reactors do not yet exist, intense research in many parts of the world is being carried out with a view to achieving such a machine. There are two central problems in the creation of a self-sustaining thermonuclear reactor: heating the reacting nuclides to the enormous *ignition temperature (about 40×10^6 K for a deuterium–tritium reaction) and containing the reacting nuclides for long enough for the fusion energy released to exceed the energy required to achieve the ignition temperature (*see* Lawson criterion). The two methods being explored are *magnetic containment* and *pellet fusion.*

In the closed magnetic-containment device the fusion *plasma is contained in a toroidal-shaped reactor, called a *Tokamak,* in which strong magnetic fields guide the charged plasma particles round the toroid without allowing them to contact the container walls. In open-ended magnetic systems the plasma is trapped between magnetic mirrors (strong magnetic fields) at the two ends of a straight containment vessel.

In pellet fusion the objective is to heat and compress a tiny pellet of the nuclear fuels, by means of a laser or an electron beam, so rapidly that fusion is achieved before the pellet flies apart. Results with this type of equipment have been comparable to those achieved by magnetic confinement.

thermonuclear weapon *See* nuclear weapons.

thermopile A device used to detect and measure the intensity of radiant energy. It consists of a number of *thermocouples connected together in series to achieve greater sensitivity. The hot junctions of the thermocouples are blackened and exposed to the radiation to be detected or measured, while the cold junctions are shielded from the radiation. The thermoelectric e.m.f. generated enables the hot junction excess temperature to be calculated and the radiant intensity to be deduced. They are used in various applications, from a safety device that produces an electric signal if a pilot light blows out to an instrument to measure the heat radiation received from the sun.

thermostat A device that controls the heating or cooling of a substance in order to maintain it at a constant temperature. It consists of a temperature-sensing instrument connected to a switching device. When the temperature reaches a predetermined level the sensor switches the heating or cooling source on or off according to a predetermined program. The sensing thermometer is often a *bimetallic strip that triggers a simple electrical switch. Thermostats are used for space-heating controls, in water heaters and refrigerators, and to maintain the environment of a scientific experiment at a constant temperature.

thixotropy *See* Newtonian fluid.

Thomson effect (Kelvin effect) When an electric current flows through a conductor, the ends of which are maintained at different temperatures, heat is evolved at a rate approximately proportional to the product of the current and the temperature gradient. If either the current or the temperature gradient is reversed heat is absorbed rather than being evolved. It is named after Sir William Thomson (later Lord Kelvin; 1824–1907).

Thomson scattering The scattering of electromagnetic radiation by free charged particles, especially by electrons. The energy lost by the radiation is accounted for by classical theory as a result of the radiation emitted by the electrons when they are accelerated in the transverse electric field of the radiation. It is named after Sir J. J. Thomson (1856–1940).

thorium Symbol Th. A grey radioactive metallic element belonging to the actinoids; a.n. 90; r.a.m. 232.038; r.d. 11.5–11.9 (17°C); m.p. 1740–1760°C; b.p. 4780–4800°C. It occurs in monazite sand in Brazil, India, and USA. The isotopes of thorium have mass numbers from 223 to 234 inclusive; the most stable isotope, thorium-232, has a half-life of 1.39×10^{10} years. It can be used as a nuclear fuel for breeder reactors as thorium-232 captures slow neutrons to breed uranium-233. Thorium dioxide (*thoria*, ThO_2) is used on gas mantles and in special refractories. The element was discovered by J. J. Berzelius in 1829.

thorium series *See* radioactive series.

threshold The minimum value of a parameter or variable that will produce a specified effect.

threshold frequency *See* photoelectric effect.

thrust The propelling force generated by an aircraft engine or rocket. It is usually calculated as the product of the rate of mass discharge and the velocity of the exhaust gases relative to the vehicle.

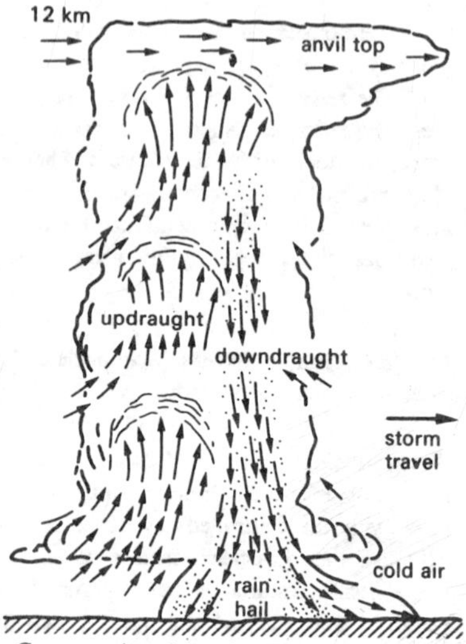

Cross-section through a thunderstorm cell

thunderstorm A convective storm accompanied by *lightning and thunder and a variety of weather conditions, especially heavy rain or hail, high winds, and sudden temperature changes. Thunderstorms originate when intense heating causes a parcel of moist air to rise, leading to instability and the development of cumulonimbus cloud – a towering cloud with a characteristic anvil-shaped top. The exact mechanisms of thunderstorms are not fully understood. They occur most frequently in the tropics but are also common in the mid-latitudes.

thyratron A thermionic valve (usually a triode) that functions as a gas-filled relay. A positive pulse fed to a correctly biased thyratron causes a discharge to start and to continue until the anode voltage has been reduced. It has now been replaced by its solid-state counterpart, the silicon-controlled rectifier.

thyristor A silicon-controlled rectifier whose anode–cathode current is controlled by a signal applied to a third electrode (the gate) in much the same way as in a thyratron valve. It consists usually of a four-layer chip comprising three $p-n$ junctions.

tidal energy *See* tides.

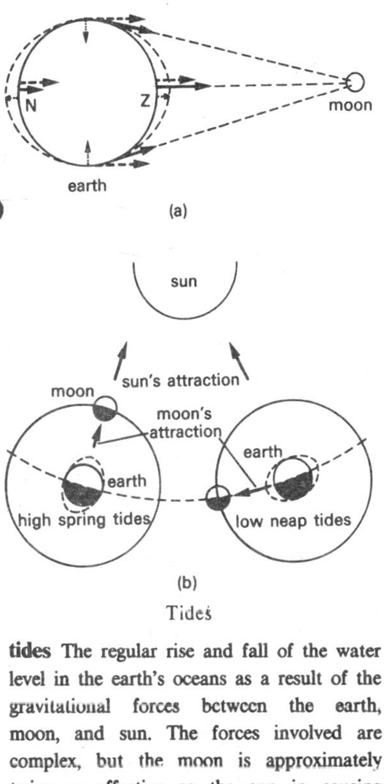

Tides

tides The regular rise and fall of the water level in the earth's oceans as a result of the gravitational forces between the earth, moon, and sun. The forces involved are complex, but the moon is approximately twice as effective as the sun in causing tides. In illustration (a) the resultant gravitational forces between the moon and various points on the earth (solid lines) are shown as the vector sums of the tide-generating forces (broken lines) and a constant force (dotted lines) that is the same at all points on the earth and is equal to the moon's attraction on the earth's centre. The resultant force when the moon is in zenith (Z in the illustration) is greater than that at nadir (N) because Z is closer to the moon than N and the force is proportional to the square of the distance according to *Newton's law of gravitation. Illustration (b) shows how at full and new moon the sun and moon act together to produce the high *spring tides*, while at quarter moon the forces are at right angles to each other causing the low *neap tides*.

The use of *tidal energy*, estimated at some 4×10^{18} J per annum at known tidal sites, dates back to medieval tidal mills. Modern tidal power stations use specially designed turbines, operated by tidal waters, to drive generators.

timbre *See* quality of sound.

time A dimension that enables two otherwise identical events that occur at the same point in space to be distinguished (*see* space–time). The interval between two such events forms the basis of time measurement. For general purposes, the earth's rotation on its axis provides the units of the clock (*see* day) and the earth's orbit round the sun (*see* year) provides the units of the calendar. For scientific purposes, intervals of time are now defined in terms of the frequency of a specified electromagnetic radiation (*see* second). *See also* time dilation.

time dilation (time dilatation) The principle, predicted by Einstein's special theory of *relativity, that intervals of time are not absolute but are relative to the motion of the observers. If two identical clocks are synchronized and placed side by side in an inertial frame of reference they will read the same time for as long as they both remain side by side. However, if one of the clocks has a velocity relative to the other, which remains beside a stationary observer, the travelling clock will show, to that observer, that less time has elapsed than the stationary clock. In general, the interval between two events is shorter by a factor $\sqrt{(1 - v^2/c^2)}$, when measured in a frame of reference travelling at a velocity v relative to another frame of reference; *c* is the speed of light. The principle has been verified in a number of ways; for example, by comparing the lifetimes of fast mesons, which increase with the speed of the particles to an extent predicted by this factor.

time-lapse photography A form of ciné photography used to record a slow process, such as plant growth. A series of single exposures of the object is made on ciné film at predetermined regular intervals. The film produced is then projected at normal speeds and the process appears to be taking place at an extremely high rate.

Titan *See* Saturn.

Tokamak *See* thermonuclear reactor.

tomography The use of X-rays to photograph a selected plane of a human body with other planes eliminated. The *CAT* (*computerized axial tomography*) *scanner* is a ring-shaped X-ray machine that rotates through 180° around the horizontal patient, making numerous X-ray measurements every few degrees. The vast amount of information acquired is built into a three-dimensional image of the tissues under examination by the scanner's own computer. The patient is exposed to a dose of X-rays only some 20% of that used in a normal diagnostic X-ray.

topology The branch of geometry concerned with the properties of solids that are unchanged by continuous deformations, such as twisting or stretching. It is useful in the analysis of complicated electrical networks.

torque (moment of a force or couple) The product of a force and its perpendicular distance from a point about which it causes rotation or *torsion. The unit of torque is the newton metre, a vector product, unlike the joule, also equal to a newton metre, which is a scalar product. A turbine produces a torque on its central rotating shaft. *See also* couple.

torr A unit of pressure, used in high-vacuum technology, defined as 1 mmHg. 1 torr is equal to 133.322 pascals. The unit is named after Evangelista Torricelli (1609–47).

Torricellian vacuum The vacuum formed when a long tube, closed at one end and filled with mercury, is inverted into a mercury reservoir so that the open end of the tube is below the surface of the mercury. The pressure inside the Torricellian vacuum is the vapour pressure of mercury, about 10^{-7} torr.

torsion A twisting deformation produced by a *torque or *couple. A *torsion bar* is a form of spring in which one end of a bar is fixed and a torque is applied to the other end. Torsion bars are used in the suspension systems of motor vehicles.

torsion balance An instrument for measuring very weak forces. It consists of a horizontal rod fixed to the end of a vertical wire or fibre or to the centre of a taut horizontal wire. The forces to be measured are applied to the end or ends of the rod. The turning of the rod may be measured by the displacement of a beam of light reflected from a plane mirror attached to it. The best-known form is that used by Henry Cavendish (1731–1810) and later by Sir Charles Boys (1855–1944) to determine the *gravitational constant; in this form the balance is calibrated by determining the torsional coefficient of the suspension by treating the device as a torsional pendulum.

torus A solid generated by rotating a circle about an external line in its plane, also called an *anchor ring*. It has the shape of the inner tube of a tyre. If *r* is the radius of the rotating circle and *R* the distance between the centre of the circle and the axis of rotation, the volume of the torus is $2\pi^2Rr^2$ and the surface area is $4\pi^2Rr$. In Cartesian coordinates, if the z-axis is the axis of rotation, the equation of the torus is $[\sqrt{(x^2 + y^2)} - R]^2 + z^2 = r^2$.

total internal reflection The total reflection of a beam of light at the interface of one medium and another medium of lower refractive index, when the angle of incidence to the second medium exceeds a specific *critical angle.*

If a beam of light passing through a medium A (say glass) strikes the boundary to a medium B of lower refractive index (say air) with a small angle of incidence i, part will be refracted, with an angle of refraction r, and part will be reflected (see illustration a). If i is increased it will reach a critical angle c, at which $r = 90°$(see illustration b). If i is now increased further, no refraction can occur and all the light energy is reflected by the interface (see illustration c). This total internal reflection occurs when c (given by $n\sin c = 1$) is exceeded (n is the refractive index of A relative to B). The critical angle of optical glass is usually about 40° and total internal reflection is made use of by incorporating prisms in some optical instruments instead of mirrors.

totality The period during a total *eclipse of the sun in which the view of the sun's surface from a point on the earth is totally obscured by the moon. The maximum duration of totality is 7.67 minutes, but it is usually less.

total-radiation pyrometer *See* pyrometry.

tracing (radioactive tracing) *See* labelling.

transcendental number A number that is not algebraic, such as π or e. A *transcendental function* is also nonalgebraic, such as a^x, $\sin x$, or $\log x$.

transducer A device for converting a nonelectrical signal, such as sound, light, heat, etc., into an electrical signal, or vice versa. Thus microphones and loudspeakers are *electroacoustic transducers.* An *active transducer* is one that can itself introduce a power gain and has its own power source. A *passive transducer* has no power source other than the actuating signal and cannot introduce gain.

transformer A device for transferring electrical energy from one alternating-current circuit to another with a change of voltage, current, phase, or impedance. It consists of a primary winding of N_p turns magnetically linked by a ferromagnetic core or by proximity to the secondary winding of N_s turns. The *turns ratio* (N_s/N_p) is approximately equal to V_s/V_p and to I_p/I_s, where V_p and I_p are the voltage and current fed to the primary winding and V_s and I_s are the voltage and current induced in the secondary winding, assuming that there are no

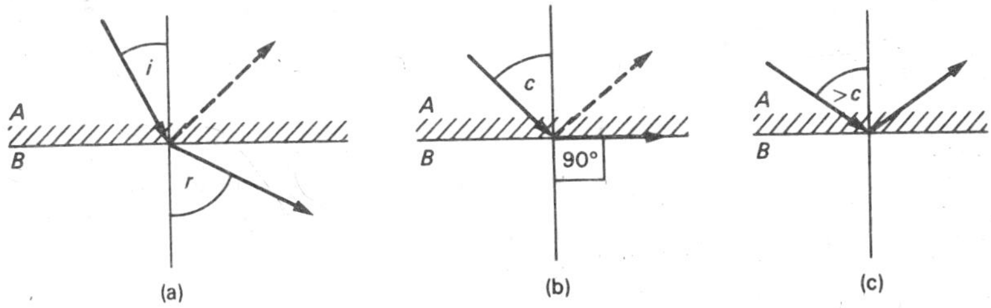

Total internal reflection

power losses in the core. In practice, however, there are *eddy-current and *hysteresis losses in the core, incomplete magnetic linkage between the coils, and heating losses in the coils themselves. By the use of a *laminated core and careful design, transformers with 98% efficiency can be achieved.

transient A brief disturbance or oscillation in a circuit caused by a sudden rise in the current or e.m.f.

transistor A *semiconductor device capable of amplification in addition to rectification. It is the basic unit in radio, television, and computer circuits, having almost completely replaced the *thermionic valve. The *point-contact transistor*, which is now obsolete, was invented in 1948. It consists of a small germanium crystal with two rectifying point contacts attached to it; a third contact, called the *base*, makes a low-resistance nonrectifying (ohmic) connection with the crystal. Current flowing through the device between the point contacts is modulated by the signal fed to the base. This type of transistor was replaced by the *junction transistor*, which was developed in 1949–50. The *field-effect transistor* (*FET*) was a later invention. *Bipolar transistors*, such as the junction transistor, depend on the flow of both majority and minority carriers, whereas in *unipolar transistors*, such as the FET, the current is carried by majority carriers only.

In the bipolar junction transistor, two *p*-type semiconductor regions are separated by a thin *n*-type region, making a $p-n-p$ structure. Alternatively, an $n-p-n$ structure can also be used. In both cases the thin central region is called the base and one outer region of the sandwich is called the *emitter*, the other the *collector*. The emitter–base junction is forward-biased and the collector–base junction is reverse-biased. In the $p-n-p$ transistor, the forward bias causes

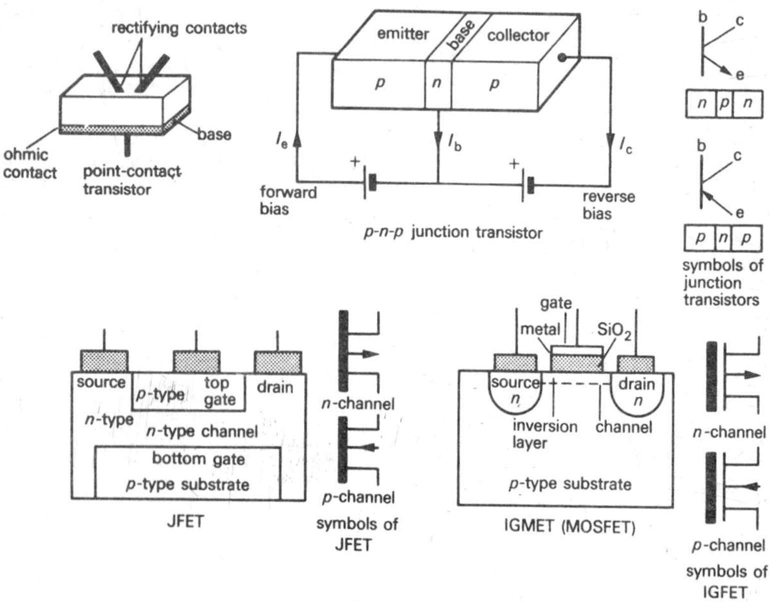

Transistors

holes in the emitter region to flow across the junction into the base; as the base is thin, the majority of holes are swept right across it (helped by the reverse bias), into the collector. The minority of holes that do not flow from the base to the collector combine with electrons in the *n*-type base. This recombination is balanced by a small electron flow in the base circuit. The diagram illustrates the (conventional) current flow using the *common-base* type of connection. If the emitter, base, and collector currents are I_e, I_b, and I_c, respectively, then $I_e = I_b + I_c$ and the current gain is I_c/I_b.

Field-effect transistors are of two kinds, the *junction FET* (*JFET* or *JUGFET*) and the *insulated-gate FET* (*IGFET*; also known as a *MOSFET*, i.e. metal-oxide-semiconductor FET). Both are unipolar devices and in both the current flows through a narrow *channel* between two electrodes (the *gate*) from one region, called the *source*, to another, called the *drain*. The modulating signal is applied to the gate. In the JFET, the channel consists of a semiconductor material of relatively low conductivity sandwiched between two regions of high conductivity of the opposite polarity. When the junctions between these regions are reverse-biased, *depletion layers form, which narrow the channel. At high bias the depletion layers meet and pinch-off the channel completely. Thus the voltage applied to the two gates controls the thickness of the channel and thus its conductivity. JFETs are made with both *n*-type and *p*-type channels.

In the IGFET, a wafer of semiconductor material has two highly doped regions of opposite polarity diffused into it, to form the source and drain regions. An insulating layer of silicon dioxide is formed on the surface between these regions and a metal conductor is evaporated on to the top of this layer to form the gate. When a positive voltage is applied to the gate, electrons move along the surface of the *p*-type substrate below the gate, producing a thin surface of *n*-type material, which forms the channel between the source and drain. This surface layer is called an *inversion layer*, as it has opposite conductivity to that of the substrate. The number of induced electrons is directly proportional to the gate voltage, thus the conductivity of the channel increases with gate voltage. IGFETs are also made with both *p*-type and *n*-type channels.

transition point (transition temperature) **1.** The temperature at which one crystalline form of a substance changes to another form. **2.** The temperature at which a substance changes phase. **3.** The temperature at which a substance becomes superconducting (*see* superconductivity). **4.** The temperature at which some other change, such as a change of magnetic properties (*see also* Curie temperature), takes place.

translation Motion of a body in which all the points in the body follow parallel paths.

translucent Permitting the passage of radiation but not without some scattering or diffusion. For example, frosted glass allows light to pass through it but an object cannot be seen clearly through it because the light rays are scattered by it. *Compare* transparent.

transmission coefficient *See* transmittance.

transmission electron microscope *See* electron microscope.

transmittance (transmission coefficient) The ratio of the energy of some form of radiation transmitted through a surface to the energy falling on it. The reciprocal of the transmittance is the *opacity*.

transmitter **1.** The equipment used to generate and broadcast radio-frequency electromagnetic waves for communication purposes. It consists of a carrier-wave generator, a device for modulating the carrier wave in accordance with the information to be broadcast, amplifiers, and an aerial system. **2.** The part of a telephone system that converts sound into electrical signals.

transparent Permitting the passage of radiation without significant deviation or absorption. *Compare* translucent. A substance may be transparent to radiation of one wavelength but not to radiation of another wavelength. For example, some forms of glass are transparent to light but not to ultraviolet radiation, while other forms of glass may be transparent to all visible radiation except red light. *See also* radiotransparent.

transport number Symbol t. The fraction of the total charge carried by a particular type of ion in the conduction of electricity through electrolytes.

transuranic elements Elements with an atomic number greater than 92, i.e. elements above uranium in the periodic table.

transverse wave *See* wave.

travelling wave *See* wave.

triangle of vectors A triangle constructed so that each of its sides represents one of three coplanar *vectors acting at a point with no resultant. If the triangle is completed, with the sides representing the vectors in both magnitude and direction, so that there are no gaps between the sides, then the vectors are in equilibrium. If the three vectors are forces, the figure is called a *triangle of forces*; if they are velocities, it is a *triangle of velocities*.

triboelectricity *Static electricity produced as a result of friction.

tribology The study of friction, lubrication, and lubricants.

triboluminescence *Luminescence caused by friction; for example, some crystalline substances emit light when they are crushed as a result of static electric charges generated by the friction.

triclinic *See* crystal system.

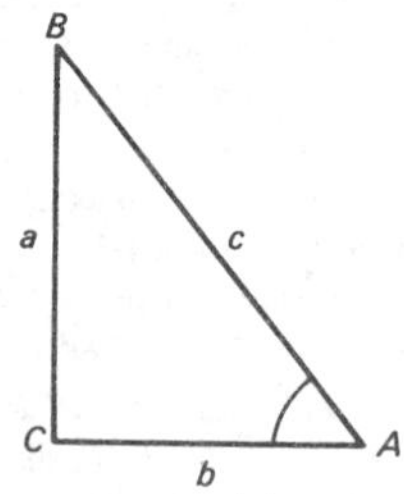

Trigonometric functions

trigonometric functions Functions defined in terms of a right-angled triangle and widely used in the solution of many mathematical problems. They are defined as:

tangent of angle A, written $\tan A = a/b$

sine of angle A, written $\sin A = a/c$

cosine of angle A, written $\cos A = b/c$,

where a is the length of the side opposite the angle A, b is the length of the side opposite the angle B, and c is the hypotenuse of the triangle.

The reciprocal functions are:

cotangent of angle A, written

$$\cot A = 1/\tan A = b/a$$

secant of angle A, written

$$\sec A = 1/\cos A = c/b$$

cosecant of angle A, written

$$\operatorname{cosec} A = 1/\sin A = c/a$$

triode A *thermionic valve with three electrodes. Electrons produced by the heated cathode flow to the anode after passing through the negatively biased *control grid. Small voltage fluctuations superimposed on the grid bias cause large fluctuations in the anode current. The triode was thus the first electronic device capable of amplification. Its role has now been taken over by the transistor.

triple point The temperature and pressure at which the vapour, liquid, and solid phases of a substance are in equilibrium. For water the triple point occurs at 273.16 K and 611.2 Pa. This value forms the basis of the definition of the *kelvin and the thermodynamic *temperature scale.

tritiated compound *See* labelling.

tritium Symbol T. An isotope of hydrogen with mass number 3; i.e. the nucleus contains 2 neutrons and 1 proton. It is radioactive (half-life 12.3 years), undergoing beta decay to helium-3. Tritium is used in *labelling.

triton The nucleus of a tritium atom.

tropical year *See* year.

troposphere *See* earth's atmosphere.

tuning fork A metal two-pronged fork that when struck produces an almost pure tone of a predetermined frequency. It is used for tuning musical instruments and in experiments in acoustics.

tunnel diode (Esaki diode) A semiconductor diode, discovered in 1957 by L. Esaki (1925–), based on the *tunnel effect. It consists of a highly doped $p-n$ semiconductor junction, which short circuits with negative bias and has negative resistance over part of its range when forward biased. Its fast speed of operation makes it a useful device in many electronic fields.

tunnel effect An effect in which electrons are able to tunnel through a narrow *potential barrier that would constitute a forbidden region if the electrons were treated as classical particles. That there is a finite probability of an electron tunnelling from one classically allowed region to another arises as a consequence of *quantum mechanics. The effect is made use of in the *tunnel diode.

turbine A machine in which the energy of a fluid is converted into rotary mechanical

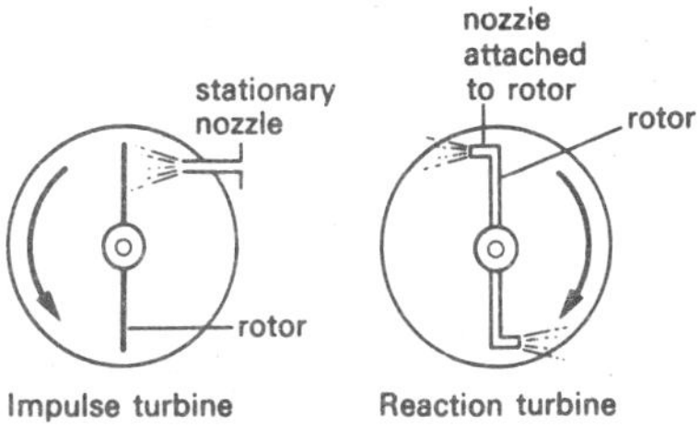

Principle of the turbine

energy. The most widely used turbines are the *steam turbines* and *water turbines* that provide some 95% of the world's electric power (in the form of *turbogenerators) and the *gas turbines* that power all the world's jet-propelled aircraft. In the *impulse turbine* a high-pressure low-velocity fluid is expanded through stationary nozzles, producing low-pressure high-velocity jets, which are directed onto the blades of a rotor. The rotor blades reduce the speed of the jets and thus convert some of the fluid's kinetic energy into rotary mechanical energy of the rotor shaft. In the *reaction turbine* the discharge nozzles are themselves attached to the rotor. The acceleration of the fluid leaving the nozzles produces a force of reaction on the nozzles, causing the rotor to move in the opposite direction to that of the fluid. Many turbines work on a combination of the impulse and reaction principles.

turbogenerator A steam turbine driving an electric generator. This is the normal method of generating electricity in power stations. In a conventional power station the steam is raised by burning a fossil fuel (coal, oil, or natural gas); in a nuclear power station the steam is raised by heat transfer from a nuclear reactor.

turbojet *See* jet propulsion.

turbulence A form of fluid flow in which the particles of the fluid move in irregular paths, resulting in an exchange of momentum from one portion of a fluid to another. Turbulent flow takes over from *laminar flow when high values of the *Reynolds number are reached.

turns ratio *See* transformer.

tweeter A small loudspeaker capable of reproducing sounds of relatively high frequency, i.e. 5 kilohertz upwards. In high-fidelity equipment a tweeter is used in conjunction with a *woofer.

Tyndall effect The scattering of light as it passes through a medium containing small particles. If a polychromatic beam of light is passed through a medium containing particles with diameters less than about one-twentieth of the wavelength of the light, the scattered light appears blue. This accounts for the blue appearance of tobacco smoke. At higher particle diameters, the scattered light remains polychromatic. It is named after John Tyndall (1820–93). *See also* scattering of electromagnetic radiation.

U

ultracentrifuge A high-speed centrifuge used to measure the rate of sedimentation of colloidal particles or to separate macromolecules, such as proteins or nucleic acids, from solutions. Ultracentrifuges are electrically driven and are capable of speeds up to 60 000 rpm.

ultrahigh frequency (UHF) A radio frequency in the range 3×10^9–0.3×10^9 Hz; i.e. having a wavelength in the range 10 cm to 1 m.

ultramicroscope A form of microscope that uses the *Tyndall effect to reveal the presence of particles that cannot be seen with a normal optical microscope. Colloidal particles, smoke particles, etc., are suspended in a liquid or gas in a cell with a black background and illuminated by an intense cone of light that enters the cell from the side and has its apex in the field of view. The particles then produce diffraction-ring systems, appearing as bright specks on the dark background.

ultrasonics The study and use of pressure waves that have a frequency in excess of 20 000 Hz and are therefore inaudible to the human ear. *Ultrasonic generators* make use of the *piezoelectric effect, *ferroelectric materials, or *magnetostriction to act as transducers in converting electrical energy into mechanical energy. Ultrasonics are used in medical diagnosis, particularly in conditions such as pregnancy, in which X-rays could have a harmful effect. Ultrasonic techniques are also used industrially to test for flaws in metals, to clean surfaces, to test the thickness of parts, and to form colloids.

ultraviolet microscope A *microscope that has quartz lenses and slides and uses *ultraviolet radiation as the illumination. The use of shorter wavelengths than the visible range enables the instrument to resolve smaller objects and to provide greater magnification than the normal optical microscope. The final image is either photographed or made visible by means of an *image converter.

ultraviolet radiation (UV) Electromagnetic radiation having wavelengths between that of violet light and long X-rays, i.e. between 400 nanometres and 4 nm. In the range 400–300 nm the radiation is known as the *near ultraviolet*. In the range 300–200 nm it is known as the *far ultraviolet*. Below 200 nm it is known as the *extreme ultraviolet* or the *vacuum ultraviolet*, as absorption by the oxygen in the air makes the use of evacuated apparatus essential. The sun is a strong emitter of UV radiation but only the near UV reaches the surface of the earth as the ozone in the atmosphere absorbs all wavelengths below 290 nm.

Most UV radiation for practical use is produced by various types of *mercury-vapour lamps. Ordinary glass absorbs UV radiation and therefore lenses and prisms for use in the UV are made from quartz.

umbra *See* shadow.

uncertainty principle (Heisenberg uncertainty principle; principle of indeterminism) The principle that it is not possible to know with unlimited accuracy both the position and momentum of a particle. This principle, discovered in 1927 by Werner Heisenberg (1901–76), is usually stated in the form: $\Delta x \Delta p_x > h/2\pi$, where Δx is the uncertainty in the x-coordinate of the particle, Δp_x is the uncertainty in the x-component of the particle's momentum, and h is the *Planck constant. An explanation of the uncertainty is that in order to locate a particle exactly, an observer must be able to bounce off it a photon of radiation; this act of location itself alters the position of the particle in an unpredictable way. To locate the position accurately, photons of short wavelength would have to be used. The high momenta of such photons would cause a large effect on the position. On the other hand, using photons of lower momenta would have less effect on the particle's position, but would be less accurate because of the longer wavelength. The principle has had a profound effect on scientific thought as it appears to upset the classical relationship between cause and effect at the atomic level.

underdamped *See* damping.

uniaxial crystal A double-refracting crystal (*see* double refraction) having only one *optic axis.

unified-field theory A comprehensive theory that would relate the electromagnetic, gravitational, strong, and weak interactions (*see* fundamental interactions) in one set of equations. In its original context the expression referred only to the unification of general *relativity and classical electromagnetic theory. No such theory has yet been found but some progress has been made in the unification of the electromagnetic and weak interactions.

union *See* set theory.

unit A specified measure of a physical quantity, such as length, mass, time, etc., specified multiples of which are used to express magnitudes of that physical quantity. For scientific purposes previous systems of units have now been replaced by *SI units.

unit cell The group of particles (atoms, ions, or molecules) in a crystal that is repeated in three dimensions in the *crystal lattice. *See also* crystal system.

unit magnetic pole *See* magnetic poles.

universal constant *See* fundamental constant.

universal motor *See* electric motor.

universe All the matter, energy, and space that exists. *See* cosmology; heat death of the universe.

unstable equilibrium *See* equilibrium.

upper atmosphere The upper part of the *earth's atmosphere above about 300 km. This is the part of the atmosphere that cannot be reached by balloons.

upthrust *See* Archimedes' principle.

uranium Symbol U. A white radioactive metallic element belonging to the actinoids;

a.n. 92; r.a.m. 238.03; r.d. 19.05 (20°C); m.p. 1132±1°C; b.p. 3818°C. It occurs as uraninite, from which the metal is extracted by an ion-exchange process. Three isotopes are found in nature: uranium–238 (99.28%), uranium–235 (0.71%), and uranium–234 (0.006%). As uranium–235 undergoes *nuclear fission with slow neutrons it is the fuel used in *nuclear reactors and *nuclear weapons; uranium has therefore assumed enormous technical and political importance since their invention. It was discovered by M. H. Klaproth in 1789.

uranium–lead dating A group of methods of *dating certain rocks that depends on the decay of the radioisotopes uranium–238 to lead–206 (half-life 4.5×10^9 years) or the decay of uranium–235 to lead–207 (half-life 7.1×10^8 years). One form of uranium–lead dating depends on measuring the ratio of the amount of helium trapped in the rock to the amount of uranium present (since the decay $^{238}U \rightarrow {}^{206}Pb$ releases eight alpha-particles). Another method of calculating the age of the rocks is to measure the ratio of radiogenic lead (^{206}Pb, ^{207}Pb, and ^{208}Pb) present to nonradiogenic lead (^{204}Pb). These methods give reliable results for ages of the order 10^7–10^9 years.

uranium series *See* radioactive series.

UV *See* ultraviolet radiation.

V

vacancy *See* defect.

vacuum A space in which there is a low pressure of gas, i.e. relatively few atoms or molecules. A *perfect vacuum* would contain no atoms or molecules, but this is unobtainable as all the materials that surround such a space have a finite *vapour pressure. In a *soft* (or *low*) *vacuum* the pressure is reduced to about 10^{-2} pascal, whereas a *hard* (or *high*) vacuum has a pressure of 10^{-2}–10^{-7} pascal. Below 10^{-7} pascal is known as an *ultrahigh vacuum*. *See also* vacuum pump.

vacuum pump A pump used to reduce the gas pressure in a container. The normal laboratory rotary oil-seal pump can maintain a pressure of 10^{-1} Pa. For pressures down to 10^{-7} Pa a *diffusion pump is required. *Ion pumps can achieve a pressure of 10^{-9} Pa and a *cryogenic pump combined with a diffusion pump can reach 10^{-13} Pa.

vacuum tube *See* thermionic valve.

valence band *See* energy bands.

valence electron An electron in one of the outer shells of an atom that takes part in forming chemical bonds.

valency (valence) The combining power of an atom or radical, equal to the number of hydrogen atoms that the atom could combine with or displace in a chemical compound (hydrogen has a valency of 1). It is equal to the ionic charge in ionic compounds; for example, in Na_2S, sodium has a valency of 1 (Na^+) and sulphur a valency of 2 (S^{2-}). In covalent compounds it is equal to the number of bonds formed; in CO_2 both carbon and oxygen have a valency of 2.

valve *See* thermionic valve.

Van Allen belts (radiation belts) Belts that are sources of intense radiation surrounding the earth, consisting of high-energy charged particles trapped in the earth's magnetic field within which they follow roughly helical paths. They were discovered in 1958 by James Van Allen (1914–) as a result of radiation detectors carried by Explorer satellites. The lower belt, extending from 1000 to 5000 km above the equator,

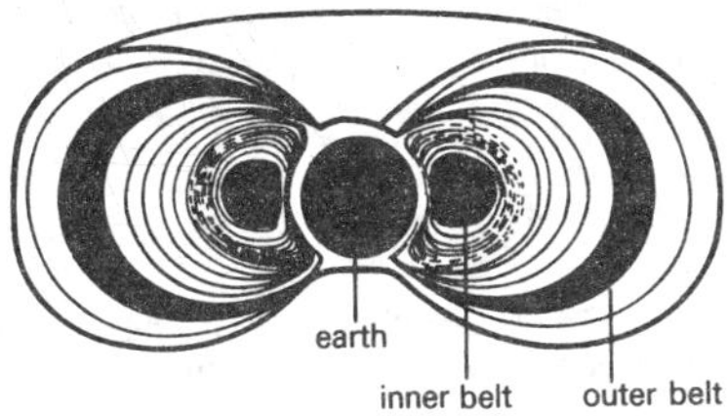

Van Allen belts

contains electrons and protons, while the upper belt, 15 000–25 000 km above the equator, contains mainly electrons (see illustration).

Van de Graaff generator An electrostatic generator used to produce a high voltage, usually in the megavolt range. It consists of a large metal dome-shaped terminal mounted on a hollow insulating support. An endless insulating belt runs through the support from the base to a pulley within the spherical terminal. In the original type, charge is sprayed by point discharge from metal needles, held at a potential of about 10 kV, on to the bottom of the belt. A row of needles near the upper belt pulley removes the charge from the belt and passes it to the outer surface of the spherical terminal. The voltage achieved by the device is proportional to the radius of the spherical terminal. A typical device with a terminal having a radius of 1 m will produce about 1 MV. However, terminals can be made smaller, for a given voltage, by enclosing the apparatus in nitrogen at a pressure of 10–20 atmospheres (1–2 MPa) to reduce sparking. Generators having a positive-ion source are fitted with an evacuated tube through which the particles can be accelerated for research purposes. Machines having an electron source are used for various medical and industrial purposes. The generator was invented by R. J. Van de Graaff (1901–67).

Modern patterns of the generator have a chainlike belt of alternate links of metal and insulator. The metal links are charged by contact with a metal pulley, and discharge to the dome in the same way. This permits much higher current drain that the point discharge.

van der Waals' equation *See* equation of state.

van der Waals' force An attractive force between atoms or molecules, named after J. D. van der Waals (1837–1923). The force accounts for the term a/V^2 in the van der Waals equation (*see* equation of state). These forces are much weaker than those arising from valence bonds and are inversely proportional to the seventh power of the distance between the atoms or molecules. They are the forces responsible for nonideal behaviour of gases and for the lattice energy of molecular crystals. There are three factors causing such forces: (1) dipole–dipole interaction, i.e. electrostatic attractions between two molecules with permanent dipole moments; (2) dipole-induced dipole interactions, in which the dipole of one molecule polarizes a neighbouring molecule; (3) dispersion forces arising because of small instantaneous dipoles in atoms.

van't Hoff factor Symbol *i*. A factor appearing in equations for *colligative properties, equal to the ratio of the number of actual particles present to the number of undissociated particles. It was first suggested by Jacobus van't Hoff (1852–1911).

van't Hoff's isochore An equation for the variation of equilibrium constant with temperature

$$(\mathrm{d}\log_e K)/\mathrm{d}T = \Delta H/RT^2$$

where T is the thermodynamic temperature and ΔH the enthalpy of the reaction.

vapour density The density of a gas or vapour relative to hydrogen, oxygen, or air. Taking hydrogen as the reference substance, the vapour density is the ratio of the mass of a particular volume of a gas to the mass of an equal volume of hydrogen under identical conditions of pressure and temperature. Taking the density of hydrogen as 1, this ratio is equal to half the relative molecular mass of the gas.

vapour pressure The pressure exerted by a vapour. All solids and liquids give off vapours, consisting of atoms or molecules of the substances that have evaporated from the condensed forms. These atoms or molecules exert a vapour pressure. If the substance is in an enclosed space, the vapour pressure will reach a maximum value that depends only on the nature of the substance and the temperature. This maximum value occurs when there is a dynamic equilibrium between the atoms or molecules escaping from the liquid or solid and those that strike the surface of the liquid or solid and return to it. The vapour is then said to be a *saturated vapour* and the pressure it exerts is the *saturated vapour pressure*.

variation *See* geomagnetism.

variometer **1.** A variable inductor consisting of two coils connected in series and able to move relative to each other. It can be used to measure inductance. **2.** Any of several devices for detecting and measuring changes in the geomagnetic elements (*see* geomagnetism).

vector quantity A physical quantity in which both the magnitude and the direction must be stated (*compare* scalar quantity). Force, velocity, and field strength are examples of vector quantities. Note that distance and speed are scalar quantities, whereas displacement and velocity are vector quantities. Vector quantities must be treated by *vector algebra*, for example, the resultant of two vectors may be found by a *parallelogram of vectors. *See also* triangle of vectors.

velocity Symbol v. The rate of displacement of a body. It is the *speed of a body in a specified direction. Velocity is thus a *vector quantity, whereas speed is a scalar quantity.

velocity modulation *See* klystron.

velocity ratio (distance ratio) The ratio of the distance moved by the point of application effort in a simple *machine to the distance moved by the point of application load in the same time.

Venn diagram *See* set theory.

Venturi tube A device for mixing a fine spray of liquid with a gas or measuring a flow rate of a gas. It consists of two tapered sections of pipe joined by a narrow throat. The fluid velocity in the throat is increased and the pressure is therefore reduced. By attaching manometers to the three sections of the tube, the pressure drop can be measured and the flow rate through the throat can be calculated. In a carburettor, the petrol from the float chamber is made into a fine spray by being drawn through a jet into the low pressure in the throat of a Venturi tube, where it mixes with the air being drawn into the engine. The device was invented by G. B. Venturi (1746–1822).

vernier A short auxiliary scale placed beside the main scale on a measuring instrument to enable subdivisions of the main scale to be read accurately. The vernier scale is usually calibrated so that each of its divisions is 0.9 of the main scale divisions. The zero on the vernier scale is set to the observed measurement on the main scale and by noting which division on the vernier scale is exactly in line with a main scale division, the second decimal place of the measurement is obtained (see illustration). The device was invented by Pierre Vernier (1580–1637) in about 1630.

very high frequency (VHF) A radio frequency in the range 3×10^8–0.3×10^8

Hz, i.e. having a wavelength in the range 1–10 m.

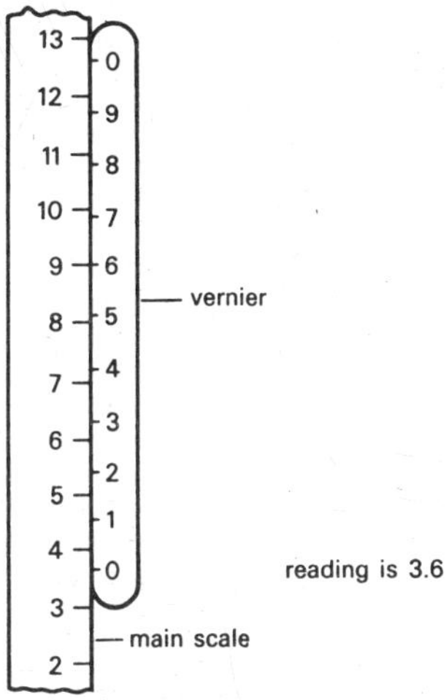

Vernier scale

very low frequency (VLF) A radio frequency in the range 3×10^4–0.3×10^4 Hz, i.e. having a wavelength in the range 10–100 km.

Victor Meyer's method A method of measuring vapour density, devised by Victor Meyer (1848 97). A weighed sample in a small tube is dropped into a heated bulb with a long neck. The sample vaporizes and displaces air, which is collected over water and the volume measured. The vapour density can then be calculated.

virial equation A gas law that attempts to account for the behaviour of real gases, as opposed to an ideal gas. It takes the form $pV = RT + Bp + Cp^2 + Dp^3 + \ldots,$ where B, C, and D are known as *virial coefficients*.

virtual image *See* image.

virtual state The state of the *virtual particles* that are exchanged between two interacting charged particles. These particles, called *photons, are not in the real state, i.e. directly observable; they are constructs to enable the phenomenon to be explained in terms of *quantum mechanics.

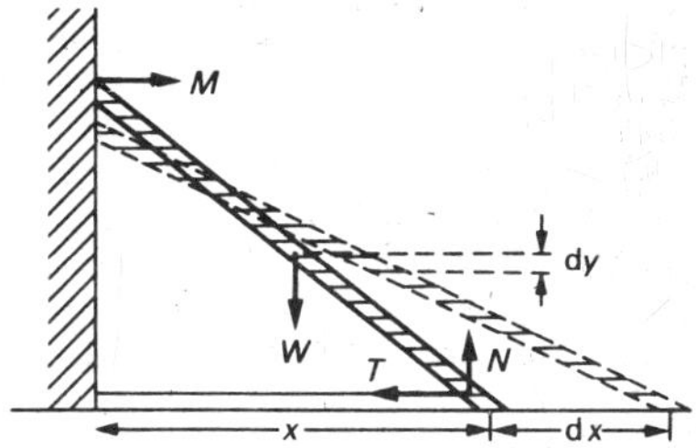

Virtual work

virtual work The imaginary work done when a system is subjected to infinitesimal hypothetical displacements. According to the *principle of virtual work*, the total work done by all the forces acting on a system in equilibrium is zero. This principle can be used to determine the forces acting on a system in equilibrium. For example, the illustration shows a ladder leaning against a wall, with the bottom of the ladder attached to the wall by a horizontal weightless string. The tension, T, in the string can be calculated by assuming that infinitesimal movement dx and dy take place as shown. Then by applying the principle of virtual work, $Tdx + Wdy = 0$. As dx and dy can be calculated from the geometry, T can be found.

viscometer An instrument for measuring the viscosity of a fluid. In the *Ostwald viscometer*, used for liquids, a bulb in a capillary tube is filled with the liquid and the time taken for the meniscus to reach a mark on the capillary, below the bulb, is a measure of the viscosity. The *falling-sphere viscometer*, based on *Stokes' law, enables the speed of fall of a ball falling through a sample of the fluid to be measured. Various other devices are used to measure viscosity.

viscosity A measure of the resistance to flow that a fluid offers when it is subjected to shear stress. For a *Newtonian fluid, the force, F, needed to maintain a velocity gradient, dv/dx, between adjacent planes of a fluid of area A is given by: $F = \eta A(dv/dx)$, where η is a constant, called the coefficient of viscosity. In SI units it has the unit pascal second (in the c.g.s. system it was measured in *poise). Non-Newtonian fluids, such as clays, do not conform to this simple model. *See also* kinematic viscosity.

visible spectrum The *spectrum of electromagnetic radiations to which the human eye is sensitive. *See* colour.

visual binary *See* binary stars.

visual-display unit (VDU) The part of a *computer system or word processor on which text or diagrams are displayed. It consists of a *cathode-ray tube and usually has its own input keyboard attached.

volt Symbol V. The SI unit of electric potential, potential difference, or e.m.f. defined as the difference of potential between two points on a conductor carrying a constant current of one ampere when the power dissipated between the points is one watt. It is named after Alessandro Volta (1745–1827).

voltage Symbol V. An e.m.f. or potential difference expressed in volts.

voltage divider (potential divider; potentiometer) A resistor or a chain of resistors connected in series that can be tapped at one or more points to obtain a known fraction of the total voltage across the whole resistor or chain. In the illustration, V is the total voltage across the divider and v is required voltage, then $v/V = R_2/(R_1 + R_2)$.

voltaic cell (galvanic cell) A device that produces an e.m.f. as a result of chemical reactions that take place within it. These reactions occur at the surfaces of two electrodes, each of which dips into an electrolyte. The first voltaic cell, devised by Alessandro Volta (1745–1827), had electrodes of two different metals dipping into brine. *See* primary cell; secondary cell.

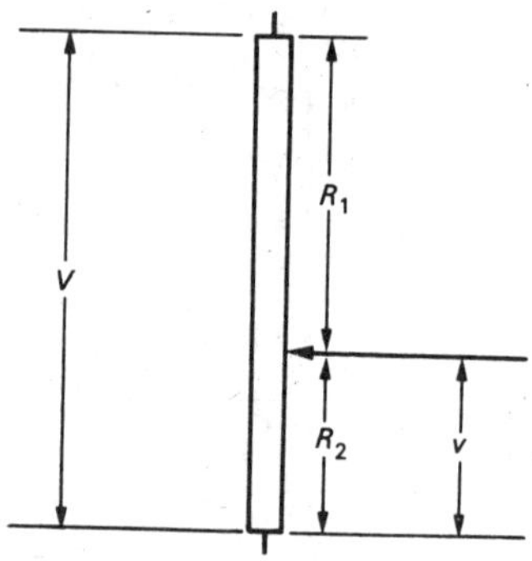

Voltage divider

voltaic pile An early form of battery, devised by Alessandro Volta, consisting of a number of flat *voltaic cells joined in series. The liquid electrolyte was absorbed into paper or leather discs.

voltameter (coulometer) 1. An electrolytic cell formerly used to measure quantity of electric charge. The increase in mass (m) of the cathode of the cell as a result of the deposition on it of a metal from a solution of its salt enables the charge (Q) to be determined from the relationship $Q = m/z$, where z is the electrochemical equivalent of the metal. **2.** Any other type of electrolytic cell used for measurement.

voltmeter An instrument used to measure voltage. *Moving-coil instruments are widely used for this purpose; generally a galvanometer is used in series with a resistor of high values (sometimes called a *multiplier*). To measure an alternating potential difference a rectifier must be included in the circuit. A moving-iron instrument can be used for either d.c. or a.c. without a rectifier. *Cathode-ray oscilloscopes are also used as voltmeters. The electronic *digital voltmeter* displays the value of the voltage in digits. The input is repeatedly sampled by the voltmeter and the instantaneous values are displayed.

volume Symbol *V*. The space occupied by a body or mass of fluid.

W

wall effect Any effect resulting from the nature or presence of the inside wall of a container on the system it encloses.

watt Symbol W. The SI unit of power, defined as a power of one joule per second. In electrical contexts it is equal to the rate of energy transformation by an electric current of one ampere flowing through a conductor the ends of which are maintained at a potential difference of one volt. The unit is named after James Watt (1736–1819).

wattmeter An instrument for measuring the power in watts in an alternating-current electric circuit. In a direct-current circuit, power is usually determined by separate measurements of the voltage and the current.

The *electrodynamic wattmeter* consists of two coils, one fixed (current) coil and one movable (potential) coil. The fixed coil carries the load current, and the movable coil carries a current proportional to the voltage applied to the measured circuit. The deflection of the needle attached to the movable coil indicates the power.

wave A periodic disturbance in a medium or in space. In a *travelling wave* (or *progressive wave*) energy is transferred from one place to another by the vibrations (*see also* stationary wave). In a wave passing over the

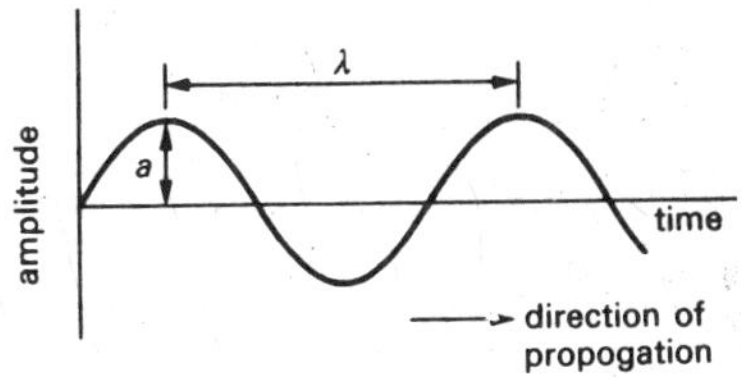

Sine wave

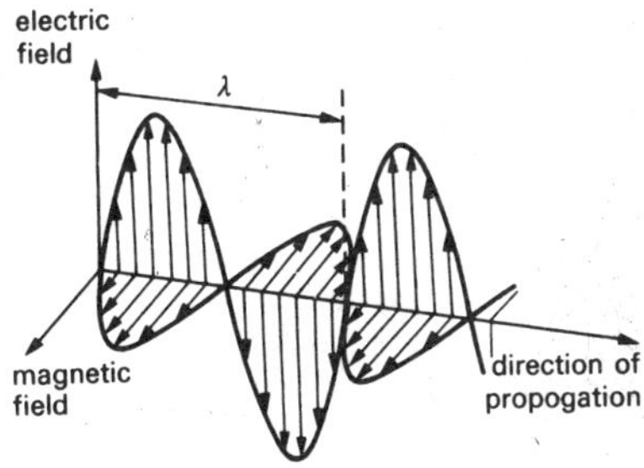

Electromagnetic waves

surface of water, for example, the water rises and falls as the wave passes but the particles of water on average do not move forward with the wave. This is called a *transverse wave* because the disturbances are at right angles to the direction of propagation. The water surface moves up and down while the waves travel across the surface of the water. Electromagnetic waves are also of this kind, with electric and magnetic fields varying in a periodic way at right angles to each other and to the direction of propagation. In sound waves, the air is alternately compressed and rarefied by displacements in the direction of propagation. Such waves are called *longitudinal waves*.

The chief characteristics of a wave are its *speed of propagation*, its *frequency*, its *wavelength*, and its *amplitude*. The speed of propagation is the distance covered by the wave in unit time. The frequency is the number of complete disturbances (cycles) in unit time, usually expressed in *hertz. The wavelength is the distance in metres between successive points of equal phase in a wave.

The amplitude is the maximum difference of the disturbed quantity from its mean value.

Generally, the amplitude (a) is half the peak-to-peak value. There is a simple relationship between the wavelength (λ) and the frequency (f), i.e. $\lambda = c/f$, where c is the speed of propagation. The energy transferred by a progressive *sine wave is proportional to a^2f^2. *See also* simple harmonic motion.

wave equation A partial differential equation of the form:

$$\partial^2u/\partial x^2 + \partial^2u/\partial y^2 + \partial^2u/\partial z^2 = (1/c^2)\partial^2u/\partial t^2$$

It represents the propagation of a wave, where u is the displacement and c the speed of propagation. *See also* Schrödinger's equation.

wave form The shape of a wave or the pattern representing a vibration. It can be illustrated by drawing a graph of the periodically varying quantity against distance for one complete wavelength. *See also* sine wave.

wavefront A line or surface within a two- or three-dimensional medium through which waves are passing, being the locus of all adjacent points at which the disturbances are in phase. At large distances from a small source in a uniform medium, the fronts are small parts of a sphere of very large radius and they can be considered as plane. For example, sunlight reaches the earth with plane wavefronts.

wave function A function $\psi(x,y,z)$ appearing in *Schrödinger's equation in wave mechanics. The wave function is a mathematical expression involving the coordinates of a particle in space. If the Schrödinger equation can be solved for a particle in a given system (e.g. an electron in an atom) then, depending on the boundary conditions, the solution is a set of allowed wave functions (*eigenfunctions*) of the particle, each corresponding to an allowed energy level (*eigenvalue*). The physical significance of the wave function is that the square of its absolute value, $|\psi|^2$, at a point is proportional to the probability of finding the particle in a small element of volume, $dxdydz$, at that point. For an electron in an atom, this gives rise to the idea of atomic and molecular *orbitals.

wave guide A hollow tube through which microwave electromagnetic radiation can be transmitted with relatively little attenuation. They often have a rectangular cross section, but some have a circular cross section. In transverse electric (TE) modes the electric vector of the field has no component in the direction of propagation. In transverse magnetic (TM) modes, the magnetic vector has no such component.

wavelength *See* wave.

wave mechanics *See* quantum mechanics.

wavemeter A device for measuring the wavelength of electromagnetic radiation. For frequencies up to about 100 MHz a wavemeter consists of a tuned circuit with a suitable indicator to establish when resonance occurs. Usually the tuned circuit includes a variable capacitor calibrated to read wavelengths and resonance is indicated by a current-detecting instrument. At higher frequencies a cavity-resonator in a waveguide is often used. The cavity resonator is fitted with a piston, the position of which determines the resonant frequency of the cavity.

wave number Symbol σ. The number of cycles of a wave in unit length. It is the reciprocal of the wavelength (*see* wave).

wave–particle duality The concept that waves carrying energy may have a corpuscular aspect and that particles may have a wave aspect; which of the two models is the more appropriate will depend on the properties the model is seeking to explain. For example, waves of electromagnetic radiation need to be visualized as particles, called *photons, to explain the *photoelectric effect while electrons need to be thought of as de Broglie waves in *electron diffraction. *See also* complementarity; de Broglie wavelength; light.

wave theory *See* light.

weak interaction *See* fundamental interactions.

weber Symbol Wb. The SI unit of magnetic flux equal to the flux that, linking a circuit of one turn, produces in it an e.m.f. of one volt as it is reduced to zero at a uniform rate in one second. It is named after Wilhelm Weber (1804–91).

weight The force by which a body is gravitationally attracted to the earth. *See* mass.

weightlessness A condition of a body when it is an infinite distance from any other body. In practice the appearance of weightlessness occurs in space when the gravitational attraction of the earth on a body in space is equal to the centripetal force required by its orbital motion so that the body is effectively in free fall. Weighlessness can also be simulated for short periods in an aircraft flying a parabolic flight path, so that its occupants are again in free fall.

Weston cell (cadmium cell) A type of primary *voltaic cell, which is used as a standard; it produces a constant e.m.f. of 1.0186 volts at 20°C. The cell is usually made in an H-shaped glass vessel with a mercury anode covered with a paste of cadmium sulphate and mercury(I) sulphate in one leg and a cadmium amalgam cathode covered with cadmium sulphate in the other leg. The electrolyte, which connects the two electrodes by means of the bar of the H, is a saturated solution of cadmium sulphate. In some cells sulphuric acid is added to prevent the hydrolysis of mercury sulphate.

wet-and-dry bulb hygrometer *See* hygrometer.

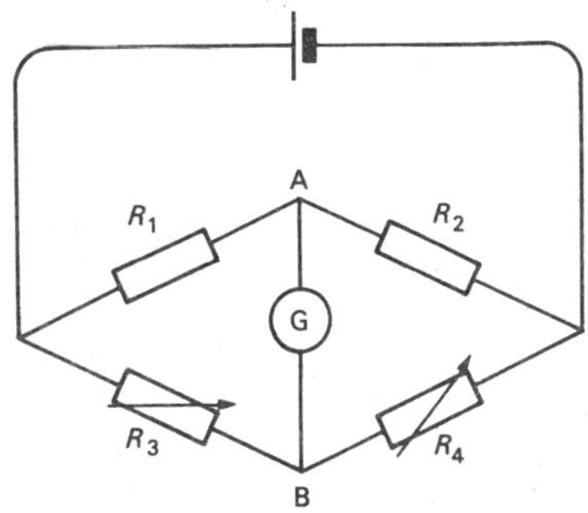

Wheatstone bridge

Wheatstone bridge An electrical circuit for measuring the value of a resistance. In the illustration, R_1 is a resistance of unknown value, R_2 is a fixed resistance of known value, R_3 and R_4 are variable resistances with known values. When no current flows between A and B the bridge is said to be balanced, the galvanometer registers no deflection, and $R_1/R_2 = R_3/R_4$. R_1 can therefore be calculated. The Wheatstone bridge is used in various forms. In the *metre bridge*, a wire 1 metre long of uniform resistance is attached to the top of a board alongside a metre rule. A sliding contact is run along the wire, which corresponds to R_3 and R_4, until the galvanometer registers zero. Most practical forms use one or more rotary rheostats to provide the variation. The device was popularized though not invented by Sir Charles Wheatstone (1802–75).

white dwarf A star of low mass in its final phase of *stellar evolution, resulting from its gravitational collapse after its thermonuclear fuel has been exhausted. *See also* Hertzsprung–Russell diagram.

Wiedemann–Franz law The ratio of the thermal conductivity of any pure metal to its electrical conductivity is approximately constant at a given temperature. The law is fairly well obeyed, except at low temperatures.

Wien's displacement law For a *black body, $\lambda_m T$ = constant, where λ_m is the wavelength corresponding to the maximum radiation of energy and T is the thermodynamic temperature of the body. Thus as the temperature rises the maximum of the spectral energy distribution curve is displaced towards the short-wavelength end of the spectrum. The law was stated by Wilhelm Wien (1864–1928).

Wigner energy Energy stored in a crystalline substance as a result of irradiation. This phenomenon is known as the *Wigner effect*. For example, some of the energy lost by neutrons in a *nuclear reactor is stored by the graphite moderator. As a result, the crystal lattice is changed and there is a consequent change in the physical dimensions of the moderator. It is named after E. P. Wigner (1902–).

Wigner nuclides Pairs of isobars with odd nucleon numbers in which the atomic number and the neutron number differ by one. ^{3}H and ^{3}He are examples.

Wilson cloud chamber *See* cloud chamber.

Wimshurst machine A laboratory electrostatic generator. It consists of two insulating discs to which radial strips of metal foil are attached. After a few strips have been charged individually, the discs are rotated in opposite directions and the charge produced on the strips by induction is collected by metal combs or brushes. It was invented by J. Wimshurst (1836–1903).

window **1.** A band of electromagnetic wavelengths that is able to pass through a particular medium with little reflection or absorption. For example, there is a *radio window* in the atmosphere allowing radio waves of wavelengths 5 mm to 30 m to pass through. This radio window enables *radio telescopes to be used on the surface of the earth. **2.** A period of time during which an event may occur in order to achieve a desired result. For example a *launch window* is the period during which a space vehicle must be launched to achieve a planned encounter.

wind power The use of winds in the earth's atmosphere to drive machinery, especially to drive an electrical generator. Practical land-based *wind generators* (*aerogenerators*) are probably capable of providing some 10^{20} J (10^{14} kW h) of energy per year throughout the world and interest in this form of renewable energy is increasing. The power, P, available to drive a wind generator is given by $P = kd^2v^3$, where k is the air density, d is the diameter of the blades, and v is the average wind speed.

Wollaston prism A type of quartz prism for producing plane-polarized light. It deviates the ordinary and extraordinary rays in opposite directions by approximately the same amount. The Wollaston prism, like the *Rochon prism, can be used with ultraviolet radiation. It is named after the inventor W. H. Wollaston (1766–1828).

Wood's metal A low-melting (71°C) alloy of bismuth (50%), lead (25%), tin (12.5%), and cadmium (12.5%). It is used for fusible links in automatic sprinkler systems. The melting point can be changed by varying the composition.

woofer A large loudspeaker designed to reproduce sounds of relatively low frequency, in conjunction with a *tweeter and often a mid-range speaker, in a high-fidelity sound reproducing system.

word A number of *bits, often 32, 48, or 64, processed by a computer as a single unit.

work The work done by a force acting on a body is the product of the force and the distance moved by its point of application in the direction of the force. If a force F acts in such a way that the displacement s is in a direction that makes an angle θ with the direction of the force, the work done is given by: $W = Fs\cos\theta$. Work is the scalar product of the force and displacement vectors. It is measured in joules.

work function A quantity that determines the extent to which *thermionic or *photoelectric emission will occur according to the Richardson equation or Einstein's photoelectric equation. It is sometimes expressed as a potential difference (symbol ϕ) in volts and sometimes as the work to be done by the exciting electron (symbol W) in electronvolts or joules. The former has been called the *work function potential* and the latter the *work function energy*.

work hardening An increase in the hardness of metals as a result of working them cold. It causes a permanent distortion of the crystal structure and is particularly apparent with iron, copper, aluminium, etc., whereas with lead and zinc it does not occur as these metals are capable of recrystallizing at room temperature.

W-particles A pair of elementary particles (W^+ and W^-), classified as *intermediate vector bosons* that are believed to transmit the weak interaction (*see* fundamental interactions) in much the same way as photons transmit the electromagnetic interaction. They are not, however, massless like photons, and are believed to have a rest mass of the order of 10^{-25} kg (70 GeV/c^2). The existence of these particles has not been confirmed experimentally.

wrought iron A highly refined form of iron containing 1–3% of slag (mostly iron silicate), which is evenly distributed throughout the material in threads and fibres so that the product has a fibrous structure quite dissimilar to that of crystalline cast iron. Wrought iron rusts less readily than other forms of metallic iron and it welds and works more easily. It is used for chains, hooks, tubes, etc.

X

X-ray astronomy The study of *X-ray sources by rockets and balloons in the earth's atmosphere and by satellites beyond it. The first nonsolar X-ray source was detected during a rocket flight in 1962, and this observation heralded an entirely new branch of astronomy which developed rapidly with the availability of satellites in the 1970s.

X-ray crystallography The use of *X-ray diffraction to determine the structure of crystals or molecules. The technique involves directing a beam of X-rays at a crystalline sample and recording the diffracted X-rays on a photographic plate. The diffraction pattern consists of a pattern of spots on the plate, and the crystal structure can be worked out from the positions and intensities of the diffraction spots. X-rays are diffracted by the electrons in the molecules and if molecular crystals of a compound are used, the electron density distribution in the molecule can be determined. *See also* neutron diffraction.

X-ray diffraction The diffraction of X-rays by a crystal. The wavelengths of X-rays are comparable in size to the distances between atoms in most crystals, and the repeated pattern of the crystal lattice acts like a diffraction grating for X-rays. Thus, a crystal of suitable type can be used to disperse X-rays in a spectrometer. X-ray diffraction is also the basis of X-ray crystallography. *See also* Bragg's law.

X-ray fluorescence The emission of *X-rays from excited atoms produced by the impact of high-energy electrons, other particles, or a primary beam of other X-rays. The wavelengths of the fluorescent X-rays can be measured by an X-ray spectrometer as a means of chemical analysis. X-ray fluorescence is used in such techniques as *electron-probe microanalysis.

X-rays Electromagnetic radiation of shorter wavelength than ultraviolet radiation and longer wavelength than gamma radiation. The range of wavelengths is 10^{-11} m to 10^{-9} m. Atoms of all the elements emit a characteristic *X-ray spectrum* when they are bombarded by electrons. The X-ray photons are emitted when the incident electrons knock an inner orbital electron out of an atom. When this happens an outer electron falls into the inner shell to replace it, losing potential energy (ΔE) in doing so. The wavelength λ of the emitted photon will then be given by $\lambda = ch/\Delta E$, where c is the speed of light and h is the Planck constant. *See also* Bremsstrahlung.

X-rays can pass through many forms of matter and they are therefore used medically and industrially to examine internal structures. X-rays are produced for these purposes by an *X-ray tube.

X-ray sources Sources of X-radiation from outside the solar system. Some 100 sources within the Galaxy have been observed as objects that emit most of their energy in the X-ray region of the electromagnetic spectrum and only a relatively small proportion of their energy in the visible spectrum. Many of these X-ray sources appear to be members of a binary system, consisting of one predominantly visual star and one X-ray star gravitationally bound together. Owing to the absorption of X-rays by the earth's atmosphere these X-ray sources are only visible by *X-ray telescopes* carried by space probes and satellites, although some high-energy X-rays can penetrate the upper atmosphere and are detectable by X-ray telescopes mounted on balloons.

X-ray spectrum *See* X-rays.

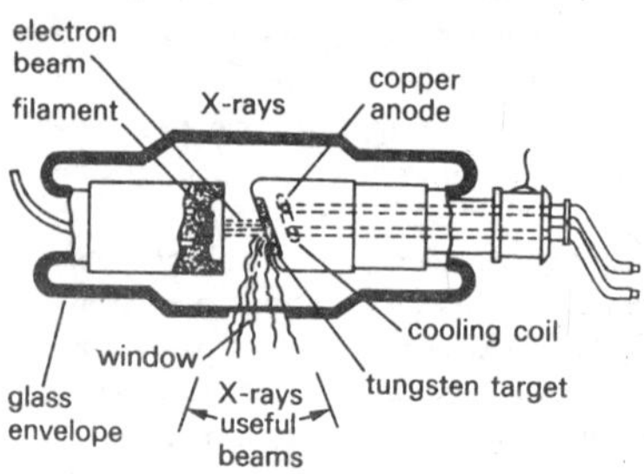

X-ray tube

X-ray tube A device for generating *X-rays by accelerating electrons to a high energy by an electrostatic field and making them strike a metal target either in a tube containing a low-pressure gas or, as in modern tubes, in a high vacuum. The target is made from a heavy metal, usually tungsten, and is backed by a massive metal anode to conduct the heat away (see illustration showing a liquid-cooled copper anode). The electron beam is produced by heating the cathode by means of a white-hot tungsten filament. A transformer supplies the high voltage, often 100 kV, the tube acting as its own rectifier. On the half-cycles when the target is negative nothing happens. When the target becomes positive, the electrons bombarding it generate X-rays.

Y

Yagi aerial A directional aerial array widely used for television and *radio telescopes. It consists of one or two dipoles, a parallel reflector, and a series of closely spaced directors (0.15–0.25 wavelength apart) in front of the dipole. When used for reception this arrangement focuses the incoming signal on the dipole. For transmission, the output of the dipole is reinforced by the directors. It is named after Hidetsuga Yagi (1886–1976).

yard The former Imperial standard unit of length. In 1963 the yard was redefined as 0.9144 metre exactly.

year The measure of time on which the calendar is based. It is the time taken for the earth to complete one orbit of the sun. The *calendar year* consists of an average of 365.25 mean solar days – three successive years of 365 days followed by one (leap) year of 366 days. The *solar year* (or *astronomical year*) is the average interval between two successive returns of the sun to the first point of Aries; it is 365.242 mean solar days. The *sidereal year* is the average period of revolution of the earth with respect to the fixed stars; it is 365.256 mean solar days. The *anomalistic year* is the average interval between successive perihelions; it is 365.259 mean solar days. *See also* ephemeris time.

yield point *See* elasticity.

Young modulus of elasticity *See* elastic modulus.

Young's slits *See* interference.

Z

Zeeman effect The splitting of the lines in a spectrum when the source of the spectrum is exposed to a magnetic field. It was discovered in 1896 by P. Zeeman (1865–1943). In the *normal Zeeman effect* a single line is split into three if the field is perpendicular to the light path or two lines if the field is parallel to the light path. This effect can be explained by classical electromagnetic principles in terms of the speeding up and slowing down of orbital electrons in the source as a result of the applied field. The *anomolous Zeeman effect* is a complicated splitting of the lines into several closely spaced lines, so called because it does not agree with classical predictions. This effect is explained by quantum mechanics in terms of electron spin.

Zener diode A type of semiconductor diode, consisting of a *p-n* junction with high doping concentrations on either side of the junction. It acts as a rectifier until the applied reverse voltage reaches a certain value, the *Zener breakdown voltage*, when the device becomes conducting. This effect occurs as a result of electrons being excited directly from the valence band into the conduction band (*see* energy bands). Zener diodes are used in voltage-limiting circuits; they are named after C. M. Zener (1905–   ).

zenith The point on the *celestial sphere that lies directly above an observer. *Compare* nadir.

zero-point energy The energy remaining in a substance at the *absolute zero of temperature (0 K). This is in accordance with quantum theory, in which a particle oscillating with simple harmonic motion does not have a stationary state of zero kinetic energy. Moreover, the *uncertainty principle does not allow such a particle to be at rest at exactly the centrepoint of its oscillations.

zeroth law of thermodynamics *See* thermodynamics.

zodiac A band that passes round the *celestial sphere, extending 9° on either side of the *ecliptic. It includes the apparent paths of the sun, moon, and planets (except Pluto). The band is divided into the twelve *signs of the zodiac*, each 30° wide. These signs indicate the sun's position each month in the year and were named by the ancient Greeks after the *zodiacal constellations* that occupied the signs some 2000 years ago. However, as a result of the *precession of the equinoxes the constellations have since moved eastwards by over 30° and no longer coincide with the signs.

zodiacal light A faint luminous glow in the sky that can be observed on a moonless night on the western horizon after sunset or on the eastern horizon before sunrise. It is caused by the scattering of sunlight by dust particles in interplanetary space.

zone refining A technique used to reduce the level of impurities in certain metals, alloys, semiconductors, and other materials. It is based on the observation that the solubility of an impurity may be different in the liquid and solid phases of a material. To take advantage of this observation, a narrow molten zone is moved along the length of a specimen of the material, with the result that the impurities are segregated at one end of the bar and the pure material at the other. In general, if the impurities lower the melting point of the material they are moved in the same direction as the molten zone moves, and vice versa.

zwitterion (ampholyte ion) An ion that has a positive and negative charge on the same group of atoms. Zwitterions can be formed from compounds that contain both acid groups and basic groups in their molecules. For example, aminoethanoic acid (the amino acid glycine) has the formula $H_2N.CH_2.COOH$. However, under neutral conditions, it exists in the different form of the zwitterion $^{+}H_3N.CH_2.COO^{-}$, which can be regarded as having been produced by an internal neutralization reaction (transfer of a proton from the carboxyl group to the amino group). Aminoethanoic acid, as a consequence, has some properties characteristic of ionic compounds; e.g. a high melting point and solubility in water. In acid solutions, the positive ion $^{+}H_3NCH_2COOH$ is formed. In basic solutions, the negative ion $H_2NCH_2COO^{-}$ predominates. The name comes from the German *zwei*, two.

Appendix 1 SI units

Table 1.1 Base and supplementary SI units

Physical quantity	*Name*	*Symbol*
length	metre	m
mass	kilogram	kg
time	second	s
electric current	ampere	A
thermodynamic temperature	kelvin	K
luminous intensity	candela	cd
amount of substance	mole	mol
*plane angle	radian	rad
*solid angle	steradian	sr

*supplementary units

Table 1.2 Derived SI units with special names

Physical quantity	*Name of SI unit*	*Symbol of SI unit*
frequency	hertz	Hz
energy	joule	J
force	newton	N
power	watt	W
pressure	pascal	Pa
electric charge	coulomb	C
electric potential difference	volt	V
electric resistance	ohm	Ω
electric conductance	siemens	S
electric capacitance	farad	F
magnetic flux	weber	Wb
inductance	henry	H
magnetic flux density (magnetic induction)	tesla	T
luminous flux	lumen	lm
illuminance	lux	lx
absorbed dose	gray	Gy
activity	becquerel	Bq
dose equivalent	sievert	Sv

Table 1.3 Decimal multiples and submultiples to be used with SI units

Submultiple	*Prefix*	*Symbol*	*Multiple*	*Prefix*	*Symbol*
10^{-1}	deci	d	10	deca	da
10^{-2}	centi	c	10^{2}	hecto	h
10^{-3}	milli	m	10^{3}	kilo	k
10^{-6}	micro	μ	10^{6}	mega	M
10^{-9}	nano	n	10^{9}	giga	G
10^{-12}	pico	p	10^{12}	tera	T
10^{-15}	femto	f	10^{15}	peta	P
10^{-18}	atto	a	10^{18}	exa	E

Table 1.4 Conversion of units to SI units

From	*To*	*Multiply by*
in	m	2.54×10^{-2}
ft	m	0.3048
sq. in	m^2	6.4516×10^{-4}
sq. ft	m^2	9.2903×10^{-2}
cu. in	m^3	1.63871×10^{-5}
cu. ft	m^3	2.83168×10^{-2}
l(itre)	m^3	10^{-3}
gal(lon)	m^3	$4.546\ 09 \times 10^{-3}$
gal(lon)	l(itre)	4.546 09
miles/hr	$m\ s^{-1}$	0.477 04
km/hr	$m\ s^{-1}$	0.277 78
lb	kg	0.453 592
$g\ cm^{-3}$	$kg\ m^{-3}$	10^{3}
lb/in^3	$kg\ m^{-3}$	$2.767\ 99 \times 10^{4}$
dyne	N	10^{-5}
kgf	N	9.806 65
poundal	N	0.138 255
lbf	N	4.448 22
mmHg	Pa	133.322
atmosphere	Pa	$1.013\ 25 \times 10^{5}$
hp	W	745.7
erg	J	10^{-7}
eV	J	$1.602\ 10 \times 10^{-19}$
kW h	J	3.6×10^{6}
cal	J	4.1868

Appendix 2 Fundamental constants

Constant	Symbol	Value in SI units
acceleration of free fall	g	$9.806\ 65$ m s^{-2}
Avogadro constant	L, N_A	$6.022\ 52 \times 10^{23}$ mol^{-1}
Boltzmann constant	$k = R/N_A$	$1.380\ 622 \times 10^{-23}$ J K^{-1}
electric constant	ϵ_0	8.854×10^{-12} F m^{-1}
electronic charge	e	$1.602\ 192 \times 10^{-19}$ C
electronic rest mass	m_e	$9.109\ 558 \times 10^{-31}$ kg
Faraday constant	F	$9.648\ 670 \times 10^{4}$ C mol^{-1}
gas constant	R	$8.314\ 34$ J K^{-1} mol^{-1}
gravitational constant	G	6.664×10^{-11} N m^{2} kg^{-2}
Loschmidt's constant	N_L	$2.687\ 19 \times 10^{25}$ m^{-3}
magnetic constant	μ_0	$4\pi \times 10^{-7}$ H m^{-1}
neutron rest mass	m_n	$1.674\ 92 \times 10^{-27}$ kg
Planck constant	h	$6.626\ 196 \times 10^{-34}$ J s
proton rest mass	m_p	$1.672\ 614 \times 10^{-27}$ kg
speed of light	c	$2.997\ 924\ 58 \times 10^{8}$ m s^{-1}
Stefan–Boltzmann constant	σ	5.6697×10^{-8} W m^{-2} K^{-4}

Appendix 3 The solar system

Planet	Equatorial diameter (km)	Mean distance from sun (10^6 km)	Sidereal period
Mercury	4878	57.91	87.969 days
Venus	12 100	108	224.7 days
Earth	12 756	149.60	365.256 days
Mars	6762	227.94	686.980 days
Jupiter	142 700	778	11.86 years
Saturn	120 800	1430	29.46 years
Uranus	51 800	2869.6	84.01 years
Neptune	49 400	4496	164.8 years
Pluto	3500	5900	248.4 years